GENETIC
TOXICOLOGY OF
COMPLEX MIXTURES

ENVIRONMENTAL SCIENCE RESEARCH

Series Editor:

Herbert S. Rosenkranz
Department of Environmental and Occupational Health
Graduate School of Public Health
University of Pittsburgh
130 DeSoto Street
Pittsburgh, Pennsylvania

Founding Editor:
Alexander Hollaender

GENETIC TOXICOLOGY OF COMPLEX MIXTURES

Edited by
Michael D. Waters
US Environmental Protection Agency
Research Triangle Park, North Carolina

F. Bernard Daniel
US Environmental Protection Agency
Cincinnati, Ohio

Joellen Lewtas
Martha M. Moore
and
Stephen Nesnow
US Environmental Protection Agency
Research Triangle Park, North Carolina

Technical Editor
Claire Wilson & Associates
Washington, D.C.

PLENUM PRESS • NEW YORK AND LONDON

Library of Congress Cataloging-in-Publication Data

International Conference on Genetic Toxicology of Complex Mixtures
 (1989 : Washington, D.C.)
  Genetic toxicology of complex mixtures / edited by Michael D.
Waters ... [et al.].
     p.   cm. -- (Environmental science research ; v. 39)
  "Proceedings of the International Conference on Genetic Toxicology
of Complex Mixtures, held July 4-7, 1989, in Washington, D.C., a
satellite symposium of the Fifth International Conference on
Environmental Mutagens"--T.p. verso.
  Includes bibliographical references.
  Includes index.
  ISBN 0-306-43683-3
  1. Genetic toxicology--Congresses. 2. Mixtures--Toxicology-
-Congresses.   I. Waters, Michael D. II. International Conference
on Environmental Mutagens (5th : 1989 : Cleveland, Ohio)
III. Title. IV. Title: Complex mixtures. V. Series.
  [DNLM: 1. Carcinogens, Environmental--adverse effects--congresses.
2. Genes--drug effects--congresses. 3. Environmental Pollutants-
-adverse effects--congresses. 4. Mutagens--adverse effects-
-congresses.   W1 EN986F v. 39 / WA 671 I6025G 1989]
RA1224.3.I574 1989
615.9'02--dc20
DNLM/DLC
for Library of Congress   90-7812
                                                              CIP

The research described in this volume has been reviewed by the Health Effects
Research Laboratory, U.S. Environmental Protection Agency, and approved for
publication. Approval does not signify that the contents necessarily reflect the views
and policies of the U.S. Environmental Protection Agency, nor does mention of trade
names or commerical products constitute endorsement or recommendation for use

Proceedings of the International Conference on Genetic Toxicology of Complex
Mixtures, held July 4-7, 1989, in Washington, D.C., a Satellite Symposium of the
Fifth International Conference on Environmental Mutagens

© 1990 Plenum Press, New York
A Division of Plenum Publishing Corporation
233 Spring Street, New York, N.Y. 10013

FOREWORD

Contained in this volume are the proceedings of the international conference on the "Genetic Toxicology of Complex Mixtures," held from July 4-7, 1989, in Washington, DC. This meeting was a satellite of the "Fifth International Conference on Environmental Mutagens" and the seventh in a biennial series of conferences on "Short-term Bioassays in the Analysis of Complex Environmental Mixtures."

Our central objective in calling together key researchers from around the world was to extend our knowledge of the application of the methods of genetic toxicology and analytical chemistry in the evaluation of chemical mixtures as they exist in the environment. This conference emphasized the study of genotoxicants in air and water, and the assessment of human exposure and cancer risk. The latest strategies and methodologies for biomonitoring of genotoxicants (including transformation products) were described in the context of the ambient environment. Source characterization and source apportionment were discussed as an aid to understanding the origin and relative contribution of various kinds of complex mixtures to the ambient environment. Similarly, investigations of genotoxicants found in the indoor environment (sidestream cigarette smoke) and in drinking water (chlorohydroxyfuranones) were given special attention in terms of their potential health impacts. New molecular techniques were described to enable more precise quantitation of internal dose and dose-to-target tissues. The emphasis of presentations on exposures/effects assessment was on integrated quantitative evaluation of human exposure and potential health effects. It is clear that the sophistication of complex mixture research technologies has increased dramatically since the first conference in 1978 with the application of state-of-the-art genetic and molecular methods. It is now apparent that interdisciplinary approaches are essential in order to assess the contribution of mixtures of genotoxic agents in the environment to total human exposure and potential cancer risk.

We are indebted to the speakers and chairpersons who presented their data at the meeting and in the excellent chapters that follow.

A conference of this type requires the cooperative efforts of many individuals. The organizing committee is grateful for the generous support of Dr. Don Hughes of Procter & Gamble, Cincinnati, Ohio, and Dr. Steve Haworth of Hazelton Laboratories, Kensington, Maryland, who contributed to these proceedings. I would like to acknowledge Dr. Bruce Casto of the Environmental Health Research and Testing, Inc., who helped to organize the meeting, and his associates, Dale Churchill and Kathy Rous, who helped on-site in Washington, DC. Special thanks to Claire Wilson & Associates, Washington, DC, for conference management and technical editing of these proceedings.

Michael D. Waters, Ph.D.
Senior Editor

CONTENTS

Complex Mixtures of Genotoxicants in Waters

Exposure/Effects Assessment

DEVELOPMENT AND APPLICATION OF NEW METHODOLOGIES APPLICABLE

TO RESEARCH ON COMPLEX ENVIRONMENTAL MIXTURES

P.H.M. Lohman, E.W. Vogel, B. Morolli, A.A. v. Zeeland, and H. Vrieling

MGC-Department of Radiation Genetics and
Chemical Mutagenesis
University of Leiden
Wassenaarseweg 72, 2333 AL Leiden, The Netherlands

Although the induction of cancer in man and in experimental animals by exposure to ionizing radiation or chemicals has been known for a long time, major insights into the mechanisms that underlie naturally occurring and induced cancers began to emerge only since the early 1970s. There is now persuasive evidence which documents that (i) many carcinogens are mutagens; (ii) most forms of cancer are due, at least in part, to changes (mutations) in the DNA (genetic material) contained in cells; and (iii) such genetic changes play a pivotal role in the initiation of cancer at the cellular level. A wide variety of test systems developed during the last 20 years--ranging from bacteria and mammalian cells including human cells in culture to whole mammals--is now available to examine the "mutagenic potential" of different chemicals, but they are only suitable for a qualitative determination of the level of cancer risk resulting from exposure of man to such agents. Agents that are capable of damaging the DNA are called "genotoxic" and a general characteristic of these is their electrophilic reactivity towards DNA and other cellular macromolecules. Interaction of chemicals with DNA has been considered as the initial step in the formation of cancer and hereditary effects in mammals, in spite of the (often spectacularly efficient) DNA repair processes in the individual cells of the organism (Fig. 1). The assumption is made that DNA lesions may escape correction by DNA repair processes, others may be erroneously repaired or not repaired at all. Moreover, the spectrum of lesions in the DNA of exposed cells is often complex and strongly dependent on the agent involved. Furthermore, DNA repair processes are found to be dependent on organ and cell type, chromosomal structure, and whether genes are active or inactive.

Mutation induction is only considered a first step (initiation) in the long chain of events leading to malignant transformation of cells. Still many other, mostly unknown, steps are involved in the complicated pathway leading from an initiated cell to an established tumor. Especially noticeable among unknown factors are those that influence the progression

Genetic Toxicology of Complex Mixtures
Edited by M. D. Waters *et al.*
Plenum Press, New York, 1990

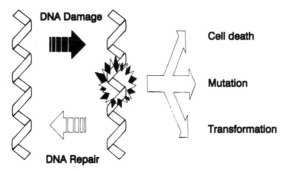

Fig. 1. DNA repair processes.

of transformed cells and those that determine the often strong strain, organ, tissue, or cell dependence of tumor formation as a result of exposure of a mammal to a particular genotoxic agent and the role of immunological defense mechanisms. Therefore, it is more surprising than logical to find that primary genotoxic damage to DNA, especially after low acute or low chronic exposures, can stochiometrically lead to mutation and transformation of cells, ultimately leading to cancer.

The complexity of the effects of environmental mixtures containing genotoxic components is not only due to the fact that exposure may occur due to a variety--both in amount and nature--of chemicals. The lack of knowledge of the number, kind, and nature of steps involved in tumor formation does not allow any prediction of the possible additivity or synergism of the biological effects of identified components of such mixtures. The situation is so uncertain that one even has to consider seriously whether the addition of any piece of knowledge on the genotoxicity of mixtures will necessarily lead to the resolution of scientific conflicts and problems in the interpretation of what is safe or unsafe; paradoxically increasing knowledge in this respect may make the world seem to be a more hazardous place than it actually may be.

One possible way to overcome part of the paradox may be a better appreciation of general toxicological principles in the evaluation of the results of genotoxic tests; the field has been too dominated by the theory that in carcinogenesis no safe dose exists. As an example, in Fig. 2 an experiment is shown where gene mutation [at the hypoxanthine-guanine phosphoribosyltransferase (HPRT) locus] is measured in cultured Chinese hamster cells exposed to ethyl methane sulphonate (ENU), 254 nm ultraviolet (UV) light, X-rays, and methyl methane sulphonate (MMS), respectively. It will be obvious from the results of this experiment that X-rays, for instance, are much less mutagenic per surviving fraction than ultraviolet light and also that relatively small changes in the structure of otherwise similar alkylating agents can lead to a dramatic difference in biological response. Simply, one may conclude from this experiment that sometimes genotoxic agents are much more lethal than mutagenic (e.g., X-rays versus ENU or UV). Obviously, there seem to be "key" lesions causing lethality and "key" lesions being less lethal but more mutagenic. The actual number of such kinds of "key" lesions can differ dramatically from agent to agent (10). After irradiation with X-rays, for instance, mammalian cells are killed when more than 3,000 DNA lesions per cell are

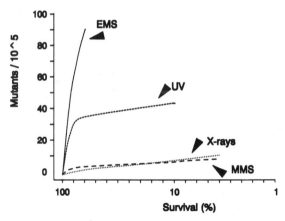

Fig. 2.   Gene mutation in Chinese hamster cells.

introduced.   The same kind of cells will still survive after the initiation
of more than 300,000 DNA lesions are induced in their genome by UV.

    From a toxicological standpoint experiments like the one shown in
Fig. 2 suggest that relative hazard assessment (quantitative ranking of
test results from various chemicals or mixtures of chemicals) may be much
more relevant for practical situations than absolute risk assessment of a
single chemical or situation.   Several attempts have been undertaken to
design approaches for relative hazard estimation of genotoxic agents such
as the comparison of genetic activity profiles (11,23) and the relation
between chemical structure and test responses (13,14).   Such analysis
can be strengthened by adding information on the actual mechanism by
which chemicals react with cellular macromolecules.   An example of such
an approach is given in Fig. 3.   In this experiment, Vogel and Zijlstra
(21,24) compared the response of a large number of bifunctional agents
(chemicals that can cause intra- and inter-crosslinks in DNA) in two tests
with the fruitfly Drosophila melanogaster.   Each test measures a different
independent biological endpoint, i.e., chromosomal aberrations (ring-X
loss) and gene mutation (recessive lethals), and the tests are done at
doses which are not toxic for the animal.   Although all chemicals depicted
in Fig. 3 have a similar reaction mechanism towards interaction with DNA,
it will be obvious that there are large differences in the test results.
Moreover, a significant ranking of the chemicals is observed as a function
of the response in the two independent Drosophila tests.   Interestingly,
the ranking of these cross-linking agents corresponds with the carcinogenic potential in rodents and/or humans of a significant number of the
indicated chemicals.

    It would be interesting to test whether the genotoxic properties of
chemicals with known reaction mechanisms with DNA as a function of the
response in two or more independent biological responses in experimental
animals would, just as in Drosophila, also quantitatively mimic carcinogenic potential.   There are still only a limited number of techniques
available that allow testing in vivo, especially at low nontoxic dose
ranges, although recently a number of promising tests have been developed.   For studying gene mutation, the molecular analysis of point mutations has become available after the introduction of the polymerase chain

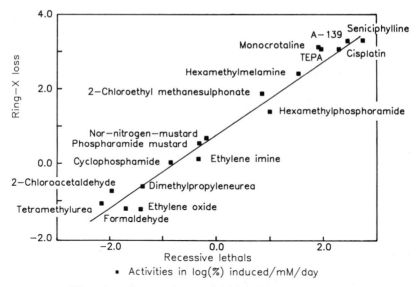

Fig. 3.   Comparison of bifunctional agents.

reaction (PCR) by Saiki et al. (16).   Vrieling et al. (22) have adapted
the PCR technique for the use of cDNA derived from mRNA of an expressed gene instead of genomic DNA; the gene used in these studies was
HPRT which is located on the X-chromosome of rodent and human cells.
By comparison of the known DNA sequence of the undamaged HPRT gene
with the DNA sequence in the mutants, it is possible to determine exactly
which mutations (deletions, point mutations = single base changes) have
been formed in the isolated mutants.   An example of the results obtained
with rodent cells treated with two different agents is given in Fig. 4.   It
can be seen that the kind of DNA sequence alterations are totally different for different agents: X-rays induce many more deletions than single
base changes; ultraviolet light, more point mutations than deletions.
Further evidence obtained by using alkylating lesions has recently revealed that the spectrum of mutations as scored in the HPRT gene of
mammalian cells is unique for the genotoxic compound used (data not
shown here), suggesting that the PCR technique for measuring mutations
in endogenous genes may be a promising new biological endpoint for in
vivo studies on the effect of low concentrations of genotoxic agents and
mixtures of agents.

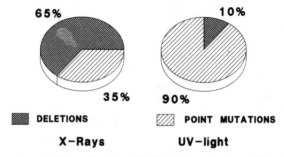

Fig. 4.   Induced mutations in the HPRT gene in irradiated rodent cells.

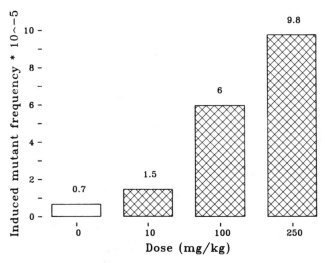

Fig. 5. ENU mutations in the brain of transgenic mice.

A limiting factor in using the PCR technique for the in vivo detection of the spectrum of DNA alterations induced by genotoxic agents is caused by the fact that the technique can so far only be applied to the HPRT gene as present in special cells of the body, primarily the T-cells isolated from the blood. Gossen et al. (5), however, developed a new method which, at least for experimental animals, can be applied to measure mutation spectra in all cells of the body. This method uses so-called "marker genes" which are being incorporated in fertilized eggs of mice. Subsequently, these eggs are reimplanted in pseudopregnant foster mice. The offspring of these mice, called transgenic animals, contain in all cells of the body the foreign genes that were transferred to the fertilized eggs. The authors used in their experiment the bacterial marker gene LacZ located on a bacteriophage lambda shuttle vector. This vector can be rescued from total genomic DNA from all cells of the body with high efficiency. If the marker gene is mutated in transgenic animals treated with a genotoxic agent, also the mutated genes can be analyzed in Eschericia coli and the spectrum of DNA alterations can be determined. In Fig. 5, the frequency of mutations induced in marker genes isolated from brain cells of transgenic mice one week after treatment with the genotoxic agent ENU are depicted; a dose-dependent increase of LacZ mutations is found with the peculiar spectrum of DNA lesions (predominantly G.C $\rightarrow$ A.T transitions). In the same animals it was found that the induction of mutations in organs other than the brain was much less. This emphasizes the applicability of this system to detect organ/tissue-specific mutation induction.

Methods for measuring mutation induction in cells of humans exposed to genotoxic agents are so far restricted to the above-mentioned PCR technique of measuring DNA alterations in a single gene (the HPRT gene of T-cells). Uitterlinden et al. (20) have developed a new method that allows the detection of "hotspots" of mutations in a large part of the genome of mammalian cells. The technique these investigators used is called "DNA fingerprinting" and is based on the fractionation of genomic

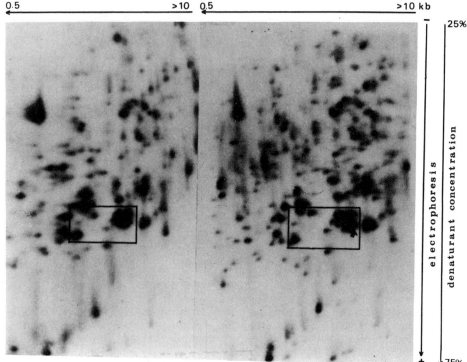

Fig. 6.   (Left panel) "DNA fingerprint" of nonirradiated T-cells; (right
            panel) "DNA fingerpring" of a T-cell clone of the same
            individual irradiated with 0.4 Gy X-rays.

DNA, cut by restriction enzymes, by two-dimensional gel electrophoresis:
in the first-dimension, the DNA fragments are fractionated on size and,
subsequently, in the second-dimension, the DNA fragments are fractionated on a denaturing gradient gel on base-composition.   The DNA fragments are visualized after blotting with a probe consisting of minisatellite
core sequences (6).   The minisatellite core sequences are spread out over
a large part of the human genome and, therefore, if restricted genomic
DNA of human cells is analyzed on the two-dimensional gels, information
is obtained on a large number of different DNA fragments in which the
minisatellite core sequences are present.   We have applied this technique
to human T-cell clones from various persons before and after irradiation
with X-rays.   In Fig. 6 (left panel), the "DNA fingerprint" is shown of a
T-cell clone of the same individual irradiated with 0.4 Gy X-rays.   In one
out of six T-cell clones irradiated with 0.4 Gy X-rays, we found an additional spot to be present in the "DNA fingerprint" (arrow in the box indicated in the righthand panel of Fig. 6).   This finding may be an indication that the two-dimensional gel electrophoresis system is sensitive
enough to measure mutations in a large fraction (about 10-15%) of the
genomic DNA of human T-cells after treatment with genotoxic agents.
However, at this moment there is not yet sufficient evidence available that
demonstrates that indeed such mutations can reproducibly be observed
through "DNA fingerprinting."

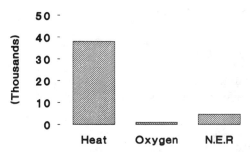

Fig. 7. Endogenous sources of DNA damage.

The new developments described in this chapter suggest that it will be possible in the near future to quantitatively measure the induction of mutations in experimental animals and humans exposed to low levels of genotoxic agents. Together with already existing sensitive techniques to measure chromosomal damage, these techniques may then allow, as in Drosophila (see Fig. 3), a comparison of the response of independent biological endpoints in mammals exposed to genotoxic agents with a common mechanism of interaction with DNA. However, one has to realize that induction of DNA damage by exogenous sources is on a daily basis often minor in comparison to DNA lesions induced in the human body due to endogenous sources. Perhaps the most ubiquitous natural cause of DNA damage is heat. Due to the thermodynamic instability of DNA, certain alterations in its structure can occur at the normal body temperature of 37°C (4,7,8,9,17). In Fig. 7 the estimated extent of possible DNA damage/cell/day in humans due to body heat is depicted. Another important class of endogenous DNA damage is free radicals, in particular the active oxygen species (2,12,18,19), although the average number of DNA lesions formed in cells of the human body per day is estimated to be lower than what would be expected due to body heat (Fig. 7). There are other important endogenous sources that cause spontaneous DNA damage, for instance due to the nonenzymatic reaction [(NER) Fig. 7] of glucose and other reducing sugars with the amino groups of DNA bases (3) or spontaneous alkylation of DNA by S-adenosyl-L-methionine, the normal methyl-group donor in cells (15).

In summary it can be stated that, surprisingly, during the lifetime of humans, the induction of DNA damage and DNA alterations may be rather more abundant than rare. A major task still remains to identify the "key" DNA lesions and/or DNA alterations caused by exogenously applied genotoxic agents or mixtures of genotoxic agents that lead to induction of cancer and/or heritable disease in addition to those induced by endogenous causes or unavoidable exposure to "natural" agents. Prioritizing chemicals or situations to be studied seems at this moment more relevant than rigorous attempts at ad hoc and absolute risk assessment. Ames (1) has suggested that exposure levels, the toxicological profile, and the use of chemicals are a more important criteria than a simplified interpretation of hazards on the basis of available short-term and long-term tests. However, as indicated in this paper, new technologies are under development that may allow a more definitive ranking of the genotoxic properties of chemicals and chemical mixtures and, therefore, will provide a significant improvement in setting priorities for the study of specific chemicals or groups of chemicals.

REFERENCES

1. Ames, B.N. (1989) Chemicals, cancers, causalities, and cautions. ChemiTech, pp. 590-598.
2. Cathcart, R., E. Schwiers, R.L. Saul, and B.N. Ames (1984) Thymine glycol and thymidine glycol in human and rat urine: A possible assay for oxidative DNA damage. Proc. Natl. Acad. Sci., USA 81:5633-5637.
3. Cerami, A. (1986) Aging of proteins and nucleic acids: What is the role of glucose? Trends in Biochem. Sci. 11:311-314.
4. Crine, P., and W.G. Verly (1976) A study of DNA spontaneous degradation. Biochim. Biophys. Acta 442:50-57.
5. Gossen, J.A., W.J.F. de Leeuw, C.H.T. Tan, E.C. Zwarthoff, F. Berends, P.H.M. Lohman, F. Berends, D.L. Knook, and J. Vijg (1989) Efficient rescue of integrated shuttle vectors from transgenic mice: A model for studying mutations in vivo. Proc. Natl. Acad. Sci., USA 86:7971-7975.
6. Jeffreys, A.J., V. Wilson, and S.L. Thein (1985) Individual-specific "fingerprints" of human DNA. Nature 314:67-70.
7. Lindahl, T., and B. Nyberg (1972) Rate of depurination of native deoxyribonucleic acid. Biochemistry 11:3610-3618.
8. Lindahl, T., and O. Karlstrom (1973) Heat-induced depyrimidination of deoxyribonucleic acid in neutral solution. Biochemistry 12:5151-5154.
9. Lindahl, T., and B. Nyberg (1974) Heat-induced deamination of cytosine residues in deoxyribonucleic acid. Biochemistry 13:3405-3410.
10. Lohman, P.H.M., R.A. Baan, A.M.J. Fichtinger-Schepman, M.A. Muysken-Schoen, M.J. Lansbergen, and F. Berends (1985) Molecular dosimetry of genotoxic damage: Biochemical and immunochemical methods to detect DNA-damage in vitro and in vivo. TIPS-FEST Supplement, pp. 1-7, Elsevier.
11. Richard, A.M., J.R. Rabinowitz, and M.D. Waters (1990) Strategies for the use of computational SAR methods in assessing genotoxicity. Mutat. Res. 221:181-196.
12. Richter, C., J.-W. Park, and B.N. Ames (1988) Normal oxidative damage to mitochondrial and nuclear DNA is extensive. Proc. Natl. Acad. Sci., USA 85:6465-6467.
13. Rosenkranz, H.S., C.S. Mitchell, and G. Klopman (1985) Artificial intelligence and Bayesian decision theory in the prediction of chemical carcinogens. Mutat. Res. 150:1-11.
14. Rosenkranz, H.S., M.R. Frierson, and G. Klopman (1986) Use of structure-activity relationships in predicting carcinogenesis. In Long-term and Short-term Assays for Carcinogens: A Critical Appraisal, R. Montesano et al., eds. IARC Scientific Publication No. 83, International Agency for Research on Cancer, Lyon, France, pp. 497-577.
15. Rydberg, B., and T. Lindahl (1982) Non-enzymatic methylation of DNA by intracellular methyl group donor S-adenosyl-L-methionine is a potentially mutagenic reaction. EMBO J. 1:211-216.
16. Saiki, R.K., S. Scharf, F. Faloona, K.B. Mullis, G.T. Horn, H.A. Ehrlich, and N. Arnheim (1985) Enzymatic amplification of X-globin genomic sequences and restriction site analysis for diagnosis of sickle cell anaemia. Science 230:1350-1354.

17. Saul, R.L., and B.N. Ames (1985) Background levels of DNA damage in the population. In Mechanisms of DNA Damage and Repair, M. Simic, L. Grossman, and A. Upton, eds. Plenum Press, New York, pp. 529-536.

18. Saul, R.L., P. Gee, and B.N. Ames (1987) Free radicals, DNA damage, and aging. In Modern Biological Theories of Aging, H.R. Warner, R.N. Butler, R.L. Sprott, and E.L. Schneider, eds. Raven Press, New York, pp. 113-130.

19. Shigenaga, M.K., C.J. Gimeno, and B.N. Ames (1989) Urinary 8-hydroxy-2'-deoxyguanosine as a biological marker of in vivo oxidative DNA damage. Proc. Natl. Acad. Sci., USA 86:9697-9701.

20. Uitterlinden, A.G., E. Slagboom, D.L. Knook, and J. Vijg (1989) Two-dimensional DNA fingerprinting of human individuals. Proc. Natl. Acad. Sci., USA 86:2742-2746.

21. Vogel, E.W. (1989) Nucleophilic selectivity of carcinogens as a determinant of enhanced mutation response in excision repair-defective strains in Drosophila: Effects of 30 carcinogens. Carcinogenesis 10:2093-2106.

22. Vrieling, H., M.L. van Rooijen, N.A. Groen, M.Z. Zdzienicka, J.W.I.M. Simons, P.H.M. Lohman, and A.A. v. Zeeland (1989) DNA strand specificity for UV-induced mutations in mammalian cells. Molec. and Cell. Biol. 9:1277-1283.

23. Waters, M.D., H.F. Stack, J.R. Rabinowitz, and N.E. Garrett (1988) Genetic activity profiles and pattern recognition in test battery selection. Mutat. Res. 205:119-138.

24. Zijlstra, J.A., and E.W. Vogel (1988) The ratio of induced recessive lethals to ring-X loss has prognostic value in terms of functionality of chemical mutagens in Drosophila melanogaster. Mutat. Res. 201:27-38.

IDENTIFICATION OF GENOTOXIC AGENTS IN COMPLEX MIXTURES

OF AIR POLLUTANTS

Dennis Schuetzle[1] and Joan M. Daisey[2]

[1]Ford Motor Company
Dearborn, Michigan 48033

[2]Indoor Environment Program
Lawrence Berkeley Laboratory
Berkeley, California 94720

INTRODUCTION

Recent surveys have shown that global environmental effects are ranked as a major issue of public concern (109). Numerous newspaper and magazine articles on the greenhouse effect, ozone depletion, acid rain, hazardous waste, and air toxics are some of the recent issues which have helped to raise this level of concern.

Air pollutants are a complex mixture of many thousands of compounds distributed over many organic classes (14,16,21,87,88). In addition, these pollutants may be in the gas-phase, the particle-phase, or in both of these phases. Only a small number of these many compounds are responsible for human health effects.

The Clean Air Act of 1970 gave the Environmental Protection Agency (EPA) the responsibility to protect public health from the growing assortment of toxic substances released into the atmosphere. The potential human health effects caused by air toxics range from short-term respiratory ailments to genotoxicity such as cancer and reproductive effects.

Human exposure to airborne particles has been of concern for many years because of their carcinogenic activity in animal bioassays (30,43, 54,55,59). Since the mid 1970s, in vitro bioassays in bacterial (4) and mammalian cells have been used widely to investigate the mutagenic and cell-transforming properties of extracts from ambient air and emission source samples (42,48,112,113). In 1976, Tokiwa et al. (103) reported that organic extracts of airborne particles collected in Japan exhibited indirect activity (with S-9 mix) in the Ames test. Strong direct-acting mutagenic activity (without S-9 mix) was found for extracts of ambient air particulates (77,101,104). Similar results were reported for diesel emissions (44,90,111), wood smoke (64), and coal fly ash (33).

It became apparent in the late 1970s that these short-term bioassays could be used in combination with chemical fractionation to greatly simplify the process of identifying mutagenic compounds in complex environmental samples. The use of these bioassays in combination with analytical measurements (referred to as bioassay-directed chemical analysis) constitutes a powerful tool for identifying potentially genotoxic compounds in ambient air samples (17,63,94).

Specific compounds are regularly identified as potential human genotoxic environmental agents as a result of animal toxicology, industrial hygiene, and epidemiological, microbiological, and biochemical studies. Such studies have indicated that genotoxic health effects may occur when humans are exposed to dioxins, benzo(a)pyrene (BaP), N-nitrosamines, 2-aminobiphenyl, butadiene, and other compounds. Once these agents have been identified as having the potential for causing significant health effects, sampling and analysis programs are usually undertaken to determine the distribution of these compounds in source samples and ambient air.

Although this method can signal a potential environmental problem, it is obviously a hit or miss approach. Therefore, many research programs have established an objective of identifying the major genotoxic compounds which are present in ambient air environments by using bioassays in combination with chemical analysis procedures.

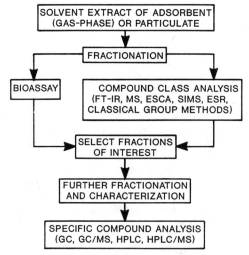

Fig. 1.   Bioassay-directed chemical analysis
          scheme for the identification of
          genotoxic agents in complex mixtures.

The purpose of this paper is to describe current progress on the development and use of bioassay-directed chemical analysis techniques to identify specific genotoxic compounds in ambient air. A new methodology using compound class analysis has been added to this procedure to determine the presence or absence of certain groups (e.g., chlorinated organics) of genotoxic compounds.

IDENTIFICATION OF GENOTOXIC AGENTS IN AMBIENT AIR

Bioassay-directed chemical analysis has allowed investigators to focus attention and resources on the identification of fractions and compounds which may be of significance to human health. This protocol has also proven to be a powerful technique for identifying the presence of strongly mutagenic compounds which could not be readily detected by conventional analytical methods.

A simplified schematic of this approach is illustrated in Fig. 1. This protocol can be used to help identify genotoxic compounds in extracts of particulate matter collected on filters or gas-phase pollutants collected on adsorbent samples. Solvent extracts of the adsorbent (gas-phase) or particulate matter are separated into fractions of increasing polarity using open-column liquid chromatography (LC) or high performance liquid chromatography (HPLC). Short-term bioassays are used to determine which fractions are the most mutagenic. Compound class analyses are used to determine if certain types of compounds are present or absent.

HPLC or open column LC, using normal phase silica-type materials, has been the method of choice for separating extracts into fractions of varying polarity. 1-Nitronaphthalene and 1,6-pyrenequinone have been established as markers to distinguish the separations between nonpolar, moderately polar, and polar fractions (93).

Using this criterion, it has been found that polar fractions account for most of the direct-acting mutagenicity of air particulate matter samples (2,5,17,20,21,62,65,105). These polar fractions have also been shown to transform mammalian cells, rat embryo cells, and mouse embryo cells, in vitro (36). Zelikoff et al. (118) have reported that polar, acetone-fractions transform Balb$_c$/3T3 cells in vitro.

Another approach has been to separate extracts into acidic, basic, and neutral fractions before fractionation. Table 1 presents some data for the distribution of mass and mutagenic activity for a Washington, D.C., air particulate sample (National Institute of Standards and Technology, SRM 1649). These studies have shown that very little direct-acting mutagenicity is present in the basic fractions (94).

The bioassay-directed chemical analysis technique has been most successful for the identification of genotoxic compounds in source samples (56,85,90,94). Much of this work has been done on diesel exhaust. The nonpolar fraction has been shown to contain polycyclic aromatic hydrocarbons (PAHs), many of which are carcinogenic in animals and mutagenic in in vitro bioassays. The PAH and other nonpolar organics are particularly amenable to analysis by capillary gas chromatography/mass spectrometry (GC/MS).

Tab. 1. Distribution of mass and mutagenic activity for an ambient air
         particulate matter extract (National Institute of Standards and
         Technology SRM 1649) *

Fraction	% Mass	% Mutagenicity −S−9	% Mutagenicity +S−9
Organic acids	7	38	21
Organic bases	1	1	1
Hexane	21	0	0
Hexane:Benzene	14	8	12
Methylene chloride	9	23	45
Methanol	34	29	21
Acidic methanol	14	1	0

*Schuetzle and Lewtas (94).

Classes of moderately polar organics identified in dichloromethane
(DCM) extracts of airborne particles or combustion emissions include aldehydes, ketones, quinones, nitro-PAH, hydroxynitro-PAH, phenols, anhydrides, and carboxylic acids (14,15,90). The nitro-PAH, dinitro-PAH,
and hydroxynitro-PAH compounds which have been identified are mutagenic and account for up to 50% of the direct-acting mutagenic activity of
these fractions (78,85,91). The rest of the activity appears to be due to
the presence of many compounds with low to moderate mutagenic activity.

Recent results from our laboratory have shown that most of the
direct-acting mutagenicity of diesel particle extracts is due to dinitro-PAH
and possibly hydroxynitro-PAH and acetoxy nitro-PAH (7). This finding
is in agreement with the work of Kittleson et al. (51) who demonstrated
that engines which run without nitrogen (argon/oxygen mixtures) emit
particles which have very low direct-acting mutagenicity.

Considerable progress has been made in the analysis and identification of moderately polar and polar organics in airborne particles. However, most of the genotoxic activity (using the Ames bioassay as a criterion) has not been accounted for. In addition, these fractions contain
hydrocarbons and PAH substituted with more than one functional group.
Thus, the number of possible isomers is greatly increased which makes
identification of specific mutagens very difficult.

Some of the major studies undertaken to determine the contribution
of specific compounds to total ambient air genotoxicty are summarized in
Tab. 2. Most of the significant mutagenic species identified to date are
nitro-PAH. These results are due in part to the sensitivity of the Ames
assay to this group of compounds.

The bioassay used to guide fractionation can limit the classes of
compounds detected in complex mixtures. The TA98 strain of Salmonella
typhimurium, most commonly used in the Ames bioassay, is not sensitive

Tab. 2.   Genotoxic agents identified in ambient air using bioassay-directed chemical analysis procedures

Reference	Sample Location	Results
Ramdahl et al. (81)	Los Angeles	2-nitrofluoranthene accounts for up to 5% of the total TA98 (-S-9) activity.
Siak et al. (96)	Detroit	Dinitropyrenes account for up to 3% of total sample TA98 (-S-9) activity.
Nishioka et al. (73)	Washington, DC (NBS SRM 1649)	Nitrohydroxyfluoranthenes account for up to 20% of the total sample TA98 (-S-9) activity.
Sicherer-Roetman et al. (97)	Netherlands	XAD-2 fractionation gave good mass and mutagenicity recovery. Ames assay chromatogram showed the presence of many mutagens.
Tuominen et al. (105)	Finland	The genotoxic activity of gas-phase components accounted for 0 to 72% of the Ames mutagenicity and 72-86% of the SCE activity. Nitrohydrobenzenes may be responsible for some of this activity.
Atkinson et al. (6)	Los Angeles	2-Nitronaphthalene and 4-nitro- were identified--the amino analogs of which are known carcinogens.

to genotoxic agents such as intercalating agents, bifunctional alkylating agents, and N-nitrosamines. Ideally, several strains of bacteria should be used, but there is rarely sufficient material in a subfraction for multiple strain screening (94).

Rossman et al. (83,84) have recently described a bioassay which requires only 5-7% of the material needed with a single Ames strain and detects many more classes of genotoxic agents than does a single Ames strain, including cross-linking and intercalating agents. V. Houk and D. DeMarini (pers. comm.) have recently found that several pesticides which are not readily detected in the Ames assay, such as Mirex, DDT, and chlordane, are positive in the Microscreen. The Microscreen assay should be particularly useful for screening subfractions from HPLC separations.

Only limited characterization of the polar fractions has been undertaken. Such polar compounds are generally nonvolatile and consequently cannot be analyzed by GC/MS without derivatization. Wauters et al. (114) extracted particulate samples first with dichloromethane (DCM) and then with methanol. The methanol-soluble fraction was then methylated and analyzed by GC/MS. The methylated derivatives detected indicated the presence of difunctional compounds such as hydroxyvaleric acid, succinic acid, and polyols.

Difunctional polar compounds were also reported (19,87) for particulate samples collected in southern California. The particulate organic compounds were volatilized directly into a high resolution mass spectrometer using a temperature programmed probe. Di-substituted alkanes with hydroxy, aldehyde, carboxyl, nitrate, and nitrite esters were identified (Tab. 3).

Eatough and Hansen (26) have identified a number of highly polar organosulfur compounds in airborne particulates using a combination of fractionation, GC/MS, NMR, and wet chemical analyses. Calorimetric titrations of airborne particulate samples collected from New York City provided the first indication of the presence of organosulfur compounds.

COMPOUND CLASS ANALYSIS

Compound class analysis is an approach which uses instrumental techniques and classical chemical class tests to determine the presence (or absence) of specific classes (e.g., aldehydes, acids) of compounds in complex mixtures. Such a screening method might be used to select samples for more complex and complete analysis of individual compounds in classes shown to be present at a significant level. A chemical class analysis system might also be useful for semiquantitative comparative assessments of the potential health hazards of aerosols from different sources or cities, for relating the relative biological activity of a sample to its composition, and for identifying sources of particulate organic matter.

Tab. 3.   Chemical tests for the analysis of compound classes in air pollutant samples

Functional Group	Reagents	Reference
Aldehydes	2,4-Dinitrophenylhydrazine	Grosjean and Fung (35)
Alkylating/Acylating agents	4-(P-Nitrobenzyl)pyridine Nicotinamide and Acetophenone	Hammock et al. (39) Sano et al. (86)
Azaarenes	Fluorescence	Dong et al. (24)
Peroxides, Hydro-peroxides	Potassium iodide Dichlorofluorescein and Hematin	Deelder et al. (23) Cathcart et al. (13)
Sulfones	Oxidize with Ag_2O; determine sulfate	Eatough and Hansen (26)
Dinitroarenes	Reduction with sodium borohydride	Ball et al. (9)

Chemical Class Analysis Techniques

A wide variety of classical tests are available to selectively determine the presence of organic compound classes in mixtures. Many of these tests may not be selective enough for ambient air studies because of sample complexity. However, a few of these tests have been applied to the characterization of genotoxic compounds in ambient air samples, as summarized in Tab. 3.

Aldehydes. Dinitrophenylhydryazone (DNPH) adducts of aldehydes can readily be formed using 2,4-dinitrophenylhydrazine (35). Spectrophotometric measurements can be made for a quantitative determination of total aldehyde concentration, or HPLC analysis can be used to separate adducts for a determination of individual species.

Alkylating/acylating agents. The term alkylating/acylating agents describes those classes of compound which can alkylate or acylate DNA. Alkylating/acylating agents include compounds such as epoxides, lactones, sulfuric acid esters, alkyl sulfates, aldehydes, and aziridines (31). Many of the compounds in this broad class are carcinogenic and/or mutagenic (31,110). These classes of compounds have generally been detected with the reagent 4-(p-nitrobenzyl)pyridine (1) but can be detected with other reagents.

Alkylating/acylating agents have been detected in airborne particles by a number of investigators (22,52,101). Earlier (52) and more recent work from our laboratories have shown that the highest concentrations are in the moderately polar (DCM) and polar (ACE) fractions. Despite the known presence of these potential carcinogens in airborne particulate matter, only a few specific compounds in this class have been identified, e.g., monomethyl sulfate, dimethyl sulfate (58).

This procedure is currently being used to help determine the presence of genotoxic compounds in chemical fractions of organic particulate extracts (12). Duplicate high performance thin-layer chromatography (HPTLC) plates were used to fractionate the extracts. One plate was used for chemical class tests while the second plate was used for in situ Ames bioassays. Earlier studies (11) had shown that the distribution of mutagenic activity among three organic fractions differed significantly for samples from different cities. More recent work has revealed some differences in the classes of organic compounds present in the fractions as well as in the mutagenic activity of subfractions separated on the HPTLC plates. For example, polar organic fractions from Elizabeth, NJ, and New York City gave two spots with similar R_f values when a reagent for detecting alkylating agents was used; no alkylating agents were detected in a composite sample from Philadelphia, one spot (different R_f) was detected for a Mexico City composite sample, and three spots were detected in a composite of samples collected in a tunnel. At present, this is a qualitative technique; relatively large amounts of organic extract (200 to 400 µg of extract per class) are required, and only a single strain of bacteria (-S-9) has been used.

Instrumental Class Analysis Techniques

Spectroscopic techniques can be used to determine the presence of compound groups in HPLC fractions. The techniques under development for this purpose include high performance mass spectrometry (HPMS), electron spectroscopy for chemical analysis (ESCA), Fourier transform infrared spectroscopy (FTIR), and GC with selective detectors.

Electron spectroscopy for chemical analysis. ESCA is a highly sensitive surface analysis technique which has been used in our laboratory for the analysis of airborne particulates (89) and by Novakov and his group (18,29,74). Since those earlier studies, the capabilities of ESCA have been greatly improved in terms of sensitivity and resolution. This technique is readily applicable to the determination of the presence of compound functional groups in HPLC fractions of airborne particulate matter extracts (Tab. 4). Since different functional groups affect the polarity of the molecule, the combination of HPLC fractionation and analysis of fractions by ESCA greatly simplify spectral interpretation. For instance, hydroxyarenes and peroxides give similar O(1s) spectra, but elute in different HPLC fractions. This technique can provide semiquantitative results down to 0.1-0.2 wgt% of a 100 Angstrom sample film deposited on an inert sample holder. Samples must be cooled to reduce volatilization under the high vacuum conditions of the instrument.

Fluorescence spectroscopy. The use of a fluorescent label to react with a specific functional group can greatly increase the sensitivity of the class test (by one or two orders of magnitude over UV or visible light absorbing derivatives). In addition, by measuring the fluorescent label, a measurement independent of the individual compounds in a class is, in principle, possible; i.e., the response for any compound in the class mixture will be that of the label. This, however, requires that the extent of reaction between the tagging reagent and the compounds be identical for all compounds within the class. This condition is almost met for some reagents and classes of compounds, such as carboxylic acids with 4-methyl-7-methoxycoumarin (28), but will not be met for all classes of interest. Consequently, such a method can only be considered semiquantitative.

Tab. 4. Electron spectroscopy for chemical analysis (ESCA) X-ray lines used for detecting compound classes in bioassay-directed chemical analysis

Compound Class	HPLC Fraction	X-Ray Lines Element	Energy (eV)
Organic sulfates	polar (8,9)	S(2p)	166.0
Chlorinated organics	nonpolar (2,3)	Cl(2s)	201.0
Azarenes	polar (7-9)	N(1s)	398.0
Aminoarenes	polar (8,9)	N(1s)	399.0
Nitroarenes	moderately-polar(4-6)	N(1s)	406.0
Nitrohydroxyarenes	polar (7,8)	N(1s)	406.0
Alkylnitrates	polar (7-8)	N(1s)	408.5
Aldehydes	moderately-polar(4-6)	O(1s)	533.0
Hydroxyarenes	polar (7,8)	O(1s)	533.5
Peroxides	moderately-polar(4-6)	O(1s)	534.0

Fourier transform infrared spectroscopy. HPLC coupled with a Fourier transform infrared (FTIR) spectrometer appears to be a promising approach for the analysis of polar organics. The HPLC/FTIR can provide a very sensitive, highly specific and almost universal detector for organic compounds and can provide information on functional groups of unknown compounds. Solvent interference in the infrared, however, remains a major problem.

Gas chromatography coupled with a matrix isolation FTIR has produced some excellent results for moderately polar compounds in particulate matter. There is no solvent interference and sensitivity in the low ng range. By monitoring certain IR frequencies, it is possible to determine the presence of many group substituents. We have recently identified a series of alkyl PAH anhydrides in polar fractions using this technique. However, more polar compounds are difficult to elute, due to the limitations of current GC columns.

Group-specific GC detectors. A number of GC detectors have been developed which are very specific for certain classes of compounds. Chlorinated organics can be determined in complex environmental mixtures using a coulometric detector with a Ag/AgCl electrode. Nitrogen selective detectors have been used for the selective detection of nitroarenes (76).

Mass spectrometry. High resolution mass spectrometry (HRMS), coupled with either direct temperature programmed insertion or gas chromatography, has been used to determine if certain compound groups are present in particulate matter, particulate extracts, or fractions of extracts (87). Certain ions can be monitored with a high degree of specificity and sensitivity (approx. 100 ng) to determine the presence of certain compound groups (see Tab. 5). For example, all chlorinated compounds lose a HCl^+ ion which can be monitored using low or high resolution mass analysis.

GENOTOXIC COMPOUNDS IDENTIFIED IN AMBIENT AIR

The classes of compounds of potential genotoxic interest in air pollution samples are given in Tab. 6. Most of the research dedicated to the identification of specific genotoxic compounds in ambient air has focused on 4- to 6- ring polycyclic aromatic hydrocarbons (37,45,46,71) and nitroarenes (47,115). Only limited studies have been undertaken to identify aminoarenes, nitroalkanes, epoxides, nitrobenzenes, nitrosamines, peroxides, and free radicals in ambient air samples. The knowledge base gathered to date on these compounds is summarized in this section.

Aldehydes. The gas-phase aldehydes, formaldehyde, acetaldehye, and acrolein are the major aldehydes of interest with respect to health effects (68). These aldehydes are reactive electrophiles that form adducts with DNA and proteins.

Aldehyde derivatives of PAH and C_4-C_{10} aliphatics have been identified in combustion emissions (90) and ambient air. However, the toxicological importance of these compounds has not been fully established.

Alkenes, Alkadienes. These compounds are produced in incomplete combustion processes and naturally by microorganisms, plants, and animals. They are also constituents of petroleum and petroleum refining.

Tab. 5.   High-resolution mass spectrometry (HRMS) fragment ions used for detecting the presence of compound classes in bioassay-directed chemical analysis

Compound Class	Ions Monitored
Chlorinated organics	HCl^+ (35.975;37.973)
Brominated organics	HBr^+ (79.926; 81.924)
Organic sulfates, Sulfones	SO_2 (63.962); SO^+ (47.967)
Organic nitrates	NO_2^+ (45.993); NO^+ (29.998)
PAH derivatives	$(PAH - H)^+$; BaP (251.086); pyrene (202.070)

The only alkenes and alkadienes of current genotoxic interest are the gas-phase C_2-C_4 compounds.

Ethene has been the subject of a recent review (72). It has been shown that this compound is metabolized to ethylene oxide by mammals. Ethene is a mutagen (116) and carcinogen (25,66,99). Ethene concentrations in urban and industrial areas are in the 5-100 ppb range.

1,3-Butadiene has also been shown to form an epoxide with microsomal preparations and to be mutagenic in the Salmonella assay. 1,3-Butadiene is an animal carcinogen as observed in mice.

Alkylnitrates. Nitroalkanes are formed in the atmosphere (32,87) via atmospheric reactions. These compounds are quite stable in the atmosphere and thus they are considered as one set of end products of NOx and hydrocarbon reactions in the atmosphere. We have identified a number of difunctionally substituted alkane derivatives in photochemical smog including nitroalkanes (19,87) using a direct-probe HRMS technique. O'Brien (75) used IR spectroscopy to determine the presence of nitroalkanes in Los Angeles. The genotoxic significance of these compounds is not known.

Aminoarenes. Several aminoarenes have been identified as human and animal carcinogens (95). Aminoarenes have not been identified in engine emissions or in other high temperature combustion processes. These compounds are easily oxidized during combustion and have a short lifetime in the atmosphere. It is probably for these reasons that aminoarenes have not been detected in the ambient atmosphere.

Azaheterocyclics. Azaheterocyclic hydrocarbons (AHHs) are known for their carcinogenic and mutagenic properties (95). AHHs, with the exception of neutral carbazole and indole homologues, are found in the basic fractions of urban atmospheric particulate matter (117).

Epoxides. Epoxides are powerful direct-acting frameshift mutagens and animal carcinogens. Pitts (77) found that BaP exposed to sub-ppm

Tab. 6.   Genotoxic compound classes of potential interest in air pollution samples

Compound Class	Example	Gas Phase	Particle Phase
Organics			
Aldehydes	Formaldehyde	x	–
Alkenes, Alkadienes	Ethylene	x	–
Alkylnitrates	Methylnitrate	x	x
Aminoarenes	4-Aminobiphenyl	x	x
Azoheterocyclics	Dibenz(a,j)acridine	x	x
Chlorinated and brominated alkanes and arenes	Dichloroethylene	x	x
Epoxides	Ethylene oxide	x	x
Nitroarenes	1,8-Dinitropyrene	x	x
Nitrobenzenes	2-Nitrophenol	x	–
Nitrosamines	Dimethylnitrosamine	x	–
Organic sulfates	Dimethylsulfate	x	–
Peroxides	Peroxyacetylnitrate	x	x
Polynuclear aromatic hydrocarbons (PAHs)	Benzo(a)pyrene	–	x
Inorganics			
Metals	Lead and lead salts	–	–
Oxidants	ozone	x	–
Other			
Free radicals	Hydroxy radicals	x	x

levels of ozone formed BaP-4,5-oxide. However, such epoxides have yet to be identified in the atmosphere. This is not surprising since epoxides are highly reactive and probably have very short lifetimes in the atmosphere.

Halogenated organics. Many halogenated organic compounds are toxic or genotoxic (57). Most of the compounds identified in the atmosphere to date resulted from stationary source emissions. Dichlorethane, chloroform, and carbon tetrachloride are the most common chlorinated compounds found in the gas-phase. Dioxins, pesticides, and PCBs have been identified in urban particulate matter (34). The dioxins are formed primarily from combustion of organic material containing chlorinated material.

Recent work from our laboratory (8) has shown that chloromethyl isomers of benzo(a)pyrene are highly mutagenic in the Ames assay. The direct-acting mutagenicity of the 11-chloromethyl-BaP is 69,000 revertants/mole in TA98, which is the highest direct-acting mutagenicity of any compound measured to date. In addition, this compound is a potent direct-acting mutagen in mammalian cells. Lawley et al. (57) have found that 7-bromo-methylbenz(a)anthracene and 7-chloromethyl-12-methylbenzo-(a)anthracene and other isomers are animal carcinogens.

Tab. 7. Some difunctionally substituted alkane derivatives found in submicron ambient air particles*

Compound	n
$HOOC-(CH_2)_n-COOH$	1-5
$HOOC-(CH_2)_n-CHO$	3-5
$HOOC-(CH_2)_n-CH_2OH$	3-5
$HOOC-(CH_2)_n-CH_2ONO$ or $CHO-(CH_2)_n-CH_2ONO_2$	3-5
$CHO-(CH_2)_n-CH_2OH$	3-5
$CHO-(CH)_n-CHO$	3-5
$HOOC-(CH_2)_n-COONO$ or $CHO-(CH_2)_n-COONO_2$	3-5
$CHO-(CH_2)_n-COONO$	3,4
$HOOC-(CH)_n-COONO_2$	4,5
$HOOC-(CH_2)_n-CH_2ONO_2$	3,4

*Source: Schuetzle et al. (75) and Cronn et al. (19).

It is possible that chloromethyl isomers of BaP and other PAHs can be formed as a result of atmospheric reactions. HCl is present in the atmosphere, the source of which is the particle-phase reaction of H_2SO_4 and NaCl (32). HCl can react with OH radicals to form Cl. However, the probability of Cl reactions with PAH-methyl radicals is very low because of competing reactions (T. Wallington, pers. comm.). It is more likely that the source of these compounds in the atmosphere, if they are present, would be from the combustion of chlorine containing organic materials.

Because of the high genotoxic activity exhibited by the chloromethyl-PAH, we believe that it is important to determine its presence in the atmosphere. GC coupled with a chlorine specific coulometric detector, and GC/MS (via monitoring of the HCl^+ ion at m/z 36) can be used to readily determine the presence of these compounds in particulate extract fractions.

The presence of numerous halogenated organics has been detected in particulates collected from gasoline and diesel engines and urban ambient air (3,100) by monitoring bromine and chlorine ions during GC/MS analysis of extract fractions. Trace concentrations of dioxins have also been detected in diesel particles (W. Crummett, pers. comm.).

COMPOUND TYPE	RINGS	EXAMPLE
NO_2 - PAH	2 - 5	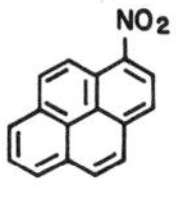 1- NITROPYRENE
NO_2 -ALKYL- PAH	2 - 4	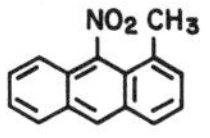 1- METHYL-9- NITROANTHRACENE
NO_2 -OXY- PAH	3 - 4	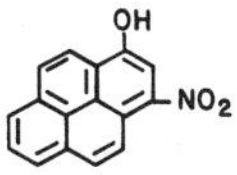 1-HYDROXY-3-NITROPYRENE
NO_2-METHOXY- PAH	3 - 4	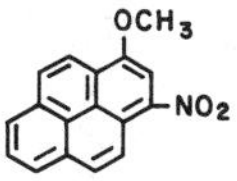 1- METHOXY-3- NITROPYRENE
$2NO_2$ - PAH	2 - 4	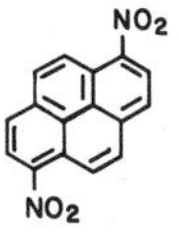 1,6 DINITROPYRENE
NO_2 - S - PAH	2 - 3	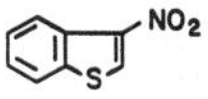 3- NITROBENZOTHIOPHENE
NO_2 - N - PAH	2 - 3	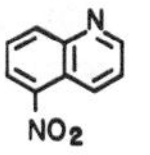

Fig. 2. Types of nitroarenes identified to date in combustion emissions and ambient air.

Nitroarenes. Nitroarenes are emitted from engines and other sources which combust organic material (115). Several of these compounds (e.g., dinitropyrenes) are highly mutagenic and carcinogenic (47,67). The types of nitroarenes identified to date in combustion emissions and ambient air are summarized in Fig. 2. Nitroarene isomers emitted from combustion processes are different than those which are formed as a result of atmospheric chemistry. Ambient air studies undertaken to date have shown that atmospheric formation is the predominant source of nitroarenes in the atmosphere (6,32,79).

Two predominant nitroarenes formed in the gas-phase are 2-nitronaphthalene and 4-nitrobiphenyl. As discussed by Tannenbaum in this volume, these nitroarenes could be reduced in vivo to the 2-aminonaphthalene and 4-aminobiphenyl in the body. These compounds in turn readily form DNA and protein adducts.

Nitrobenzenes. Derivatives of nitrobenzenes have been identified in ambient air (60). Some of these compounds could be produced by atmospheric transformation of alkylbenzenes and alkylphenols. Nitrophenols act as uncoupling agents in oxidative phosphorylation and they are known to affect cell metabolism at low concentrations (82).

Nitrosamines. Nitrosamines, including dimethylnitrosamine (DMN) and diethylnitrosamine (DEN) are considered potent carcinogens in experimental animals (107). DMN and DEN have been identified in ambient air (88). Little work has been undertaken to identify higher molecular weight nitrosamines, although there is some evidence of their presence in airborne particles (53).

Peroxides. A number of peroxides and hydroperoxides exhibit genotoxic activity (55). Benzoyl peroxide has been shown to be a promoter of tumors in mice (98).

The predominant source of peroxides in the atmosphere is most likely by atmospheric formation. Peroxides and hydroperoxides have been identified in the vapor phase in smog chambers (40). Although these compounds are probably emitted from combustion sources, no measurements have been made to date. Gas-phase peroxyacetyl nitrate (PAN), peroxypropionyl nitrate (PPN), and hydrogen peroxide are the only peroxides measured to date in the atmosphere.

Kneip et al. (52) have found evidence for the possible presence of these classes in nonpolar and polar fractions using the chemical class tests described in Tab. 3, but no additional work has been undertaken to confirm those results.

Organic sulfates. Dimethyl sulfate and monomethyl sulfuric acid have been detected in ambient air samples (26,27). Whether these are primary pollutants or secondary pollutants is unknown. Dimethyl sulfate and diethyl sulfate are carcinogenic in animals (106).

Polynuclear aromatic hydrocarbons (PAHs). As discussed earlier, comprehensive studies have been reported on the abundance of PAH in ambient air samples. Much less work has been undertaken on the identification of specific alkyl-PAH derivatives. Studies of alkyl-PAH derivatives in emission samples show that the methyl and dimethyl derivatives are in higher abundance than the parent compounds (90). Such studies are warranted since the methyl and dimethyl derivatives of certain PAH are often more mutagenic and carcinogenic than the parent PAH (41,49, 95).

Metals. A number of metal and metal salts are recognized as human carcinogens. Inorganic nickel has been recognized as carcinogenic in humans. Chromite, chromium, and several chromium compounds are suspected or confirmed animal carcinogens (45). Other metals of interest include platinum and palladium. Methods to measure low concentrations of metals in air particulate matter are well established, and extensive data sets exist for many locations throughout the world.

Oxidants. The term photochemical oxidants refers to ozone and other secondary oxidants formed in photochemical smog. A topic of current speculation is that exposure to low levels of ozone accelerates the aging of lung tissue by the oxidation of certain compounds in proteins and DNA (108). Generally, it is not considered that oxidants are present in the particle phase.   However, organic free radicals, hydroperoxides, peroxynitrates, and peroxides can all react with lung tissue.

Organic free radicals.   Particles generated from combustion sources contain organic free radicals and it is suspected that these radicals are involved in pulmonary diseases (80).   In fact, the evidence for radical involvement for promotion of cancer (50,69,70) and emphysema (102) has become stronger in recent years.   Several recent studies have indicated that the free radicals which occur in combustion particles react with DNA. The DNA damage appears to be due to the hydroxyl radical (10).   Preliminary studies have shown that long-lived free radicals are present in ambient air samples (W. Pryor, pers. comm., 1989).

CONCLUSIONS

Although much progress has been made on the identification of genotoxic compounds in ambient air, most of the genotoxic activity of these samples has not been accounted for.   In addition to PAH, azaheterocyclics, and nitroarenes, there are other classes of potentially genotoxic compounds which warrant further attention.   Chloromethyl substituted PAHs are potent mutagens and carcinogens and may be present in the atmosphere.   Peroxides, epoxides, and hydroperoxides are strong alkylating agents, and more extensive efforts are needed to determine their presence in the atmosphere.   Some of the group analysis techniques described in this paper, combined with bioassay-directed chemical analysis, can be used to help determine the presence and importance of these compounds in the atmosphere.

ACKNOWLEDGEMENTS

This work was supported in part by Grant R-815755-01-0 from the U.S. EPA and by the Assistant Secretary for Conservation and Renewable Energy of the U.S. Department of Energy under contract No. DE-AC03-76F00098.   The information in this paper has not been reviewed by the U.S. EPA and does not necessarily reflect the views of that organization.

REFERENCES

1.   Agarwal, S.C., B.L. VanDuuren, and T.J. Kneip (1979) Detection of epoxides with 4-(p-nitrobenzyl)pyridine.   Bull. Env. Contam. Toxicol. 23:825-829.
2.   Alfheim, I., G. Löfroth, and M. Moller (1983) Bioassay of extracts of ambient particulate matter.   Env. Health Perspec. 47:227-238.
3.   Alsberg, T., R. Westerholm, U. Stenberg, M. Strandell, and B. Jansson (1983) Particle associated organic halides in exhausts from gasoline and diesel fueled vehicles.   In Polynuclear Aromatic Hydrocarbons: Mechanisms, Methods, and Metabolism, M. Cooke and A.J. Dennis, eds.   Battelle Columbus Laboratories, Columbus, OH, pp. 87-97.
4.   Ames, B.N., J. McCann, and E. Yamasaki (1975) Methods for detecting carcinogens and mutagens with the Salmonella/mammalian-microsome mutagenicity test.   Mutat. Res. 31:347-364.
5.   Atherholt, T.B., G.J. McGarrity, J.B. Louis, L.J. McGeorge, P.J. Lioy, J.M. Daisey, A. Greenberg, and F. Darack (1985) Muta-

genicity studies of New Jersey ambient air particulate extracts. In Short-term Bioassays in the Analysis of Complex Environmental Mixtures IV, M.D. Waters et al., eds. Plenum Publishing Corp., New York, NY, pp. 211-231.

6. Atkinson, R., S.M. Aschmann, J. Arey, B. Zielinska, and Dennis Schuetzle (1989) Gas-phase atmospheric chemistry of 1- and 2-nitronapthalene and 1,4-naphthoquinone. Atmos. Env. (in press).

7. Ball, J.C., W.C. Young, and I.T. Salmeen (1987) Catalyst-free sodium borohydride reduction as a guide in the identification of direct-acting Ames assay mutagens in diesel-particle extracts. Mutat. Res. 192:283-287.

8. Ball, J.C., A.A. Leon, S. Foxall-Vanaken, P. Zacmanidis, G.H. Daug, and D.L. Vander Jagt (1988) Mutagenicity of chloromethylbenzo(a)pyrenes in the Ames assay and in Chinese hamster V79 cells. In Polynuclear Aromatic Hydrocarbons: A Decade of Progress, M. Cooke and A.J. Dennis, eds. Battelle Press, Columbus, OH, pp. 41-58.

9. Ball, J.C., W.C. Young, and I.T. Salmeen (1989) The effect of catalyst-free sodium borohydride reduction on the direct-acting Ames-assay activity of an ambient-airborne particle extract. Mutat. Res. (sub. for publ.).

10. Borish, E.T., D.F. Church, and W.A. Pryor (1989) Cigarette smoking and free radical damage to DNA. Amer. J. Med. (sub. for publ.)

11. Butler, J.P., T.J. Kneip, F. Mukai, and J.M. Daisey (1985) Interurban variations in the mutagenic activity of the ambient aerosol and their relations to fuel use patterns. In Short-term Bioassays in the Analysis of Complex Environmental Mixtures IV, M.D. Waters et al., eds., Plenum Publishing Corp., New York, NY, pp. 223-246.

12. Butler, J.P., T.J. Kneip, and J.M. Daisey (1987) An investigation of interurban variations in the chemical composition and mutagenic activity of airborne particulate organic matter using an integrated chemical class/bioassay system. Atmos. Env. 21:883-892.

13. Cathcart, R., E. Schwiers, and B.W. Ames (1983) Detection of picomole levels of hydroperoxides using a fluorescent dichlorofluorescein assay. Analyt. Biochem. 134:111-116.

14. Cautreels, W., and K. Van Cauwenberghe (1976) Determination of organic compounds in airborne particulate matter by gas chromatography-mass spectrometry. Atmos. Env. 10:447-457.

15. Cautreels, W., and K. Van Cauwenberghe (1978) Experiments on the distribution of organic pollutants between airborne particulate matter and the corresponding gas phase. Atmos. Env. 12:1133-1141.

16. Ciccioli, P., A. Brancaleoni, A. Cecinato, C. DiPalo, A. Brachetti, and A. Liberti (1986) Gas chromatographic evaluation of the organic components presnt in the atmosphere at trace levels with the aid of carbopack B for pre-concentration of the sample. J. Chomatog. 351:433-449.

17. Claxton, L. (1983) The integration of bioassay and physiochemical information for complex mixtures. In Short-term Bioassays in the Analysis of Complex Environmental Mixtures III, M.D. Waters et al., eds. Plenum Publishing Corp., New York, NY, pp. 153-162.

18. Craig, N.L., A.B. Harker, and T. Novakov (1974) Determination of the chemical states of sulfur in ambient pollution aerosols by X-ray photoelectron spectroscopy. Atmos. Env. 8:15-21.

19. Cronn, D.R., R.J. Charlson, R.L. Knights, A.L. Crittenden, and B.R. Appel (1977) A survey of the molecular nature of primary and

secondary components of particles in urban air by high resolution mass spectrometry. Atmos. Env. 11:929-937.

20. Daisey, J.M., L. Hawryluk, T.J. Kneip, and F. Mukai (1979) Mutagenic activity in organic fractions of airborne particulate matter. In Proceedings--Carbonaceous Particles in the Atmosphere, T. Novakov, ed. National Technical Information Service, Springfield, VA, pp. 187-192.

21. Daisey, J.M., T.J. Kneip, I. Hawryluk, and F. Mukai (1980) Seasonal variations in the bacterial mutagenicity of airborne particulate organic matter in New York City. Env. Sci. Tech. 14:1487-1490.

22. Daisey, J.M., T.J. Kneip, M. Wang, L. Ren, and W. Lu (1983) Organic and elemental composition of airborne particulate matter in Beijing, Spring 1981. Aerosol Sci. Tech. 2:407-415.

23. Deelder, R.S., M.G.F. Kroll, and J.H.M. Van Den Berg (1976) Determination of trace amounts of hydroperoxides by column liquid chromatography and colorimetric detection. J. Chromatog. 125:307-314.

24. Dong, M., D.C. Locke, and D. Hoffmann (1977) Characterization of aza-arenes in basic organic portion of suspended particulate matter. Env. Sci. Tech. 11:612-618.

25. Dunkelberg, H. (1979) On the oncogenic activity of ethylene oxide and propylene oxide in mice. Brit. J. Cancer 39:588-589.

26. Eatough, D.J., and L.D. Hansen (1983) Organic and inorganic S(IV) compounds in airborne particulate matter. Adv. Env. Sci. Tech. 12:221.

27. Eatough, D.J., T. Major, J. Ryder, M. Hill, N.F. Mangelson, R.G. Meisenheimer, and J.W. Fisher (1978) The formation and stability of sulfite species in aerosols. Atmos. Env. 12:263-271.

28. Elbert, W., S. Breitenbach, A. Neftel, and J. Hahn (1985) 4-Methyl-7-methoxycoumarin as a fluorescent label for high-performance liquid chromatographic analysis of dicarboxylic acids. J. Chromatog. 328:111-120.

29. Ellis, E.C., and T. Novakov (1982) Application of thermal analysis to the characterization of organic aerosol particles. Sci. Total Env. 23:227-238.

30. Epstein, S.S., S. Joshi, J. Andrea, N. Mantel, E. Sawicki, T. Stanley, and E.C. Tabor (1966) Carcinogenicity of organic particulate pollutants in urban air after administration of trace quantities to neonatal mice. Nature (London) 212:1305.

31. Fishbein, L. (1979) Potential Industrial Carcinogens and Mutagens. Elsevier Publishing Co., New York, NY.

32. Finlayson-Pitts, B., and J.N. Pitts, Jr. (1986) Atmospheric Chemistry: Fundamentals and Experimental Techniques. John Wiley and Sons, Inc., New York, pp. 1098.

33. Fisher, G.L., C.E. Chrisp, and O.G. Raabe (1979) Physical factors affecting the mutagenicity of fly ash from a coal-fired plant. Science 204:189.

34. Foreman, W.T., and T.F. Bidleman (1987) An experimental system for investigating vapor-particle partitioning of trace organic pollutants. Env. Sci. Tech. 21:869-875.

35. Fung, K., and D. Grosjean (1981) Determination of nanogram amounts of carbonyls as 2,4-dinitrophenylhydrazones by high-performance liquid chromatography. Analyt. Chem. 53:168-171.

36. Gordon, R.J., R.J. Bryan, J.S. Rhim, C. Demoise, R.G. Wolford, A.E. Freeman, and R.J. Heubner (1973) Transformation of rat and

mouse embryo cells by a new class of carcinogenic compounds isolated from particles in city air. Int. J. Cancer 12:223-232.

37. Grimmer, G., and F. Pott (1983) Occurrence of PAH. In Environmental Carcinogens: Polycyclic Aromatic Hydrocarbons, G. Grimmer, ed. CRC Press, Boca Raton, Florida, pp. 27-60.

38. Grosjean, D., K. Van Cauwenberghe, J.P. Schmid, P.E. Kelley, and J.N. Pitts, Jr. (1978) Identification of C_3-C_{10} aliphatic dicarboxylic acids in airborne particulate matter. Env. Sci. Tech. 12:313-316.

39. Hammock, L.G., B.D. Hammock, and J.E. Casida (1974) Detection and analysis of epoxides with 4-(p-nitrobenzyl)pyridine. Bull. Env. Tox. 12:759-764.

40. Hanst, P.L., and B.W. Gay, Jr. (1983) Atmospheric oxidation of hydrocarbons: Formation of hydroperoxides and peroxyacids. Atmos. Env. 17:2259-2265.

41. Hecht, S.S., S. Amin, A. Rivenson, and D. Hoffmann (1979) Tumor initiating activity of 5,11-dimethylchrysene and the structural requirements favoring carcinogenicity of methylated polynuclear aromatic hydrocarbons. Cancer Lett. 8:65.

42. Holmberg, B., and U. Ahlborg (1983) Consensus report: Mutagenicity and carcinogenicity of car exhausts and coal combustion emissions. Env. Health Persp. 47:1-30.

43. Huper, W.C., P. Kotin, and E.C. Tabor (1962) Carcinogenic bioassays on air pollutants. Arch. Pathol. and Lab. Med. 74:89-116.

44. Husingh, J., R. Bradow, R. Jungers, L. Claxton, R. Zweidinger, S. Tejada, J. Bumgarner, F. Duffield, M. Waters, V.F. Simmon, C. Hare, C. Rodriguez, and L. Snow (1979) Application of bioassay to characterization of diesel particle emissions. In Application of Short-term Bioassays in the Fractionation and Analysis of Complex Environmental Mixtures, M.D. Waters, S. Nesnow, J.L. Huisingh, S. Sandhu, and L. Claxton, eds. Plenum Publishing Corp., New York, NY, pp. 383-418.

45. International Agency for Research on Cancer (IARC) (1978) Chemicals with Sufficient Evidence of Carcinogenicity in Experimental Animals, IARC Working Group Report, Lyon, France.

46. International Agency for Research on Cancer (IARC) (1983) Evaluation of the Carcinogenic Risk of Chemicals to Humans, Vol. 32, Polynuclear Aromatic Hydrocarbons, Lyon, France.

47. International Agency for Research on Cancer (IARC) (1989) Evaluation of Carcinogenic Risks to Humans, Vol. 46, Engine Exhausts and Some Nitroarenes, Lyon, France.

48. Ishinishi, N., A. Koizumi, R.O. McClellan, and W. Stober (1986) Carcinogenic and mutagenic effects of diesel engine exhaust. In Developments in Toxicology and Environmental Science, Vol. 13. Elsevier Science Publishers, Amsterdam.

49. Iyer, R.P., J.W. Lyga, J.A. Secrist, III, G.H. Daub, and T.J. Slaga (1980) Comparative tumor-initiating activity of methylated benzo(a)pyrene derivatives in mouse skin. Cancer Res. 40:1073.

50. Kensler, T.M., and B.G. Taffe (1986) Free radicals in tumor promotion. Adv. Free Radical Biol. Med. 2:347-388.

51. Kittelson, D.B., C.J. Du, and R.B. Zweidinger (1984) Society of Automotive Engineers, Paper #840364, Warrensdale, Pennsylvania.

52. Kneip, T.J., M. Lippmann, F. Mukai, and J.M. Daisey (1979) Trace organic compounds in the New York City atmosphere. Part 1--Preliminary studies. Report EA-1121, Electric Power Research Institute, Palo Alto, California.

53. Kneip, T.J., J.N. Daisey, J.J. Solomon, and R.J. Hershman (1983) N-Nitroso compounds: Evidence for their presence in airborne particles. Science 221:1045-1047.
54. Kotin, P., H.L. Falk, P. Mader, and M. Thomas (1956) Production of skin tumors in mice with oxidation products of aliphatic hydrocarbons. Cancer 9:905.
55. Kotin, P., and H.L. Falk (1963) Organic peroxides, hydrogen peroxide, epoxides, and neoplasia. Radiat. Res. Suppl. 3:193.
56. Lamb, S.I., and N.E. Adin (1983) Polar Organic pollutants in air: A reivew of collection and analysis techniques. Env. Int. 9:225-243.
57. Lawley, P.D. (1976) Carcinogenesis by alkylating agents. In Chemical Carcinogens, C.E. Searle, ed. ACS Monographs, 173, American Chemical Society, Washington, DC, pp. 156-162.
58. Lee, M.L., D.W. Later, D.K. Rollins, D.J. Eatough, and L.D. Hansen (1980) Dimethyl and monomethyl sulfate: Presence in coal fly ash and airborne particulate. Science 207:186-188.
59. Leiter, J., M.B. Shimkin, and M.J. Shear (1942) Production of subcutaneous sarcomas in mice with tars extracted from atmospheric dusts. JNCI 3:155-175.
60. Leuenberger, C., J. Czucwa, J. Tremp, and W. Giger (1989) Chemosphere (in press).
61. Lewtas, J., and K. Williams (1986) A retrospective view of the value of short-term genetic bioassays in predicting and chronic effects of diesel soot. In Carcinogenic and Mutagenic Effects of Diesel Engine Exhaust, N. Ishinishi, A. Koizumi, and R.O. McClellan, eds. Elsevier Science Publishers, New York, pp. 119-141.
62. Lewtas, J., A. Austin, L. Claxton, R. Burton, and R. Jungers (1982) The relative contribution of PNA's to the microbial mutagenicity of respirable particles from urban air. In Proceedings of the VI International Symposium, M. Cooke, A. Dennis, and G. Fisher, eds. Battelle Press, Columbus, Ohio.
63. Lewtas, J. (1988) Genotoxicity of complex mixtures: Strategies for the identification and comparative assessment of airborne mutagens and carcinogens from combustion sources. Fund. and Appl. Tox. 10:571-589.
64. Löfroth, G. (1978) Mutagenicity assay of combustion emissions. Chemosphere 7:791.
65. Lioy, P.J., and J.M. Daisey (1983) The New Jersey project on airborne toxic elements and organic substances (ATEOS): A summary of the 1981 summer and 1982 winter studies. Air Pollut. Control Assoc. J. 33:649-657.
66. Lynch, D.W., TR. Lewis, W.J. Morrman, J.R. Burg, D.H. Groth, A. Khan, L.J. Ackerman, and B.Y. Cockrell (1984) Carcinogenic and toxicologic effects of inhaled ethylene oxide and propylene oxide in F344 rats. Appl. Pharm. 76:69-84.
67. Manabe, Y., T. Kinouchi, and Y. Ohnishi (1985) Identification and quantification of highly mutagenic nitroacetoxypyrenes and nitrohydroxypyrenes in diesel-exhaust particles. Mutat. Res. 158:3-18.
68. Marnett, L.J. (1988) Health effects of aldehydes and alcohols in mobile source emissions. In Air Pollution. The Automobile and Public Health, National Academy Press, Washington, DC, pp. 579-605.
69. McBrien, D.C.H., and T.F. Slater (1982) Free Radicals, Lipid Peroxidation and Cancer. Academic Press, New York.
70. Nagata, C., M. Kodama, Y. Loki, and T. Kumura (1982) Free radicals produced from chemical carcinogens and their significance in

carcinogenesis. In Free Radicals and Cancer, R.A. Floyd, ed. Marcel Dekker, New York, pp. 1-58.

71. National Academy of Sciences (1983) Polycyclic Aromatic Hydrocarbons: Evaluation of Sources and Effects, National Acdemy Press, Washington, DC.

72. National Research Council Canada (1985) Ethylene in the environment: Scientific criteria for assessing its effects on environmental quality. NRCC No. 22496:1-183.

73. Nishioka, M.G., C.C. Howard, D.A. Contos, L.M. Ball, and J. Lewtas (1988) Detection of hydroxylated nitro aromatic and hydroxylated nitro polycyclic aromatic compounds in an ambient air particulate extract using bioassay-directed fractionation. Env. Sci. Tech. 22:908-915.

74. Novakov, T.S., S.G. Chang, and R.L. Dod (1977) Application of ESCA to the analysis of atmospheric particulates. In Contemporary Topics in Analytical and Clinical Chemistry, Vol. 1, D.M. Hercules, G.M. Hiefje, L.R. Snyder, and M.A. Evenson, eds. Plenum Press, New York, pp. 249-286.

75. O'Brien, R.J., J.H. Crabtree, J.R. Holmes, M.C. Hoggan, and A.H. Bockian (1975) Formation of photochemical aerosol from hydrocarbons. Atmospheric analysis. Env. Sci. Tech. 9:577.

76. Paputa-Peck, M.C., R.M. Marano, D. Schuetzle, T. Riley, C.M. Hampton, T.J. Prater, L.M. Skewes, T.E. Jensen, T.H. Ruehle, L.C. Bosch, and W.P. Duncan (1983) Determination of nitrated-polynuclear aromatic hydrocarbons in particulate extracts using capillary column gas chromatography with nitrogen selective detection. Analyt. Chem. 55:1946-1954.

77. Pitts, Jr., J.N., D. Grosjean, T.M. Mischke, V.F. Simmon, and D. Poole (1977) Mutagenic activity of airborne particulate organic pollutants. Tox. Lett. 1:65-70.

78. Pitts, Jr., J.N., D.M. Lokensgard, W. Harger, T.S. Fisher, V. Mejia, J.J. Shuler, G.M. Scorziell, and Y.A. Katzenstein (1982) Mutagens in diesel exhaust: Identification and direct activities of 6-nitro-benzo(a)pyrene, 9-nitroanthracene, 1-nitropyrene, and 5-H-phenanthro(4,5-bcd)pyran-5-one. Mutat. Res. 103:241-249.

79. Pitts, J.r, J.N. (1983) Formation and fate of gaseous and particulate mutagens and carcinogens in real and simulated atmospheres. Env. Health Persp. 47:115-140.

80. Pryor, W.A. (1986) Cancer and free radicals. In Antimutagenesis and Anticarcinogenesis Mechanisms, D. Shankel, P. Hartman, T. Kada, and A. Hollaender, eds. Plenum Publishing Corp., New York, pp. 45-59.

81. Ramdahl, T., J.A. Sweetman, B. Zielinska, W.P. Harger, A.M. Winer, and R. Atkinson (1985) Determination of nitrofluoranthenes and nitropyrenes in ambient air and their contribution to direct mutagenicity. Presented at the tenth anniversary of the International Symposium on Polynuclear Aromatic Hydrocarbons, Columbus, Ohio.

82. Rippen, G., E. Zietz, R. Frank, T. Knacker, and W. Klopffer (1987) Do airborne nitrophenols contribute to forest decline? Env. Tech. Lett. 8:475-482.

83. Rossman, T.G., L.W. Meyer, J.P. Butler, and J.M. Daisey (1985) In Short-term Bioassays in the Analysis of Complex Environmental Mixtures IV, M. Waters et al., eds. Plenum Publishing Corp., New York, pp. 211-231.

84. Rossman, T.G., L.W. Meyer, and M. Molina (1989) Ann. N.Y. Acad. Sci. (in press).
85. Salmeen, I.T., A.M. Pero, R. Zator, D. Schuetzle, and T.L. Riley (1984) Ames assay chromatograms and the identification of mutagens in diesel particle extracts. Env. Sci. Tech. 18:375-382.
86. Sano, A., and S. Takitani (1985) Fluorometric determination of sulfide ion with methyl p-toluenesulfonate, taurine and ophthaldehyde. Analyt. Chem. 57:1687-1690.
87. Schuetzle, D. (1975) Analysis of complex mixtures by computer controlled high resolution mass spectrometry--Application to atmospheric aerosol composition. Biomed. Mass Spectr. 2:288-298.
88. Schuetzle, D. (1980) Air pollutants. In Biochemical Applications of Mass Spectrometry, First Supplementary Volume, G.R. Waller and O.C. Dermer, eds. John Wiley and Sons, Inc., New York, pp. 969-1005.
89. Schuetzle, D., L.M. Skewes, G.E. Fisher, S.P. Levine, and R.L. Gorse (1981) The determination of sulfates in diesel particulates. Analyt. Chem. 9:93-144.
90. Schuetzle, D., F.S.-C. Lee, T.J. Prater, and S.B. Tejada (1981) The identification of polynuclear aromatic hydrocarbon derivatives in mutagenic fractions of diesel particulate extracts. Int. J. Env. Analyt. Chem. 9:93-144.
91. Schuetzle, D. (1983) Sampling of vehicle emissions for chemical analysis and biological testing. Env. Health Persp. 47:65-80.
92. Schuetzle, D., T.E. Jensen, and J.C. Ball (1985) Polar polynuclear aromatic hydrocarbon (PAH) derivatives in extracts of particulates: Biological characterization and techniques for chemical analysis. Env. Int. J. 11:169-181.
93. Schuetzle, D., and T. Jensen (1985) Analysis of nitrated polycyclic aromatic hydrocarbons by mass spectrometry. In Nitrated Polycyclic Aromatic Hydrocarbons, C. White, ed. Hutig Verlag Publishers, Heidelberg, Germany, pp. 122-167.
94. Schuetzle, D., and J. Lewtas (1986) Bioassay-directed chemical analysis in environmental research. Analyt. Chem. 58:1060A.
95. Searle, C.E. (1976) Chemical Carcinogens, ACS Monograph 173. American Chemical Society, Washington, DC.
96. Siak, J., T.L. Chan, T.L. Gibson, and G.T. Wolff (1985) Contribution to bacterial mutagenicity from nitro-PAH compounds in ambient aerosols. Atmos. Env. 19:369.
97. Sicherer-Roetman, A., M. Ramlal, C.E. Voogd, and H.J. Bloemen (1988) The fractionation of extracts of ambient particulate matter for mutagenicity testing. Atmos. Env. 12:2803-2808.
98. Slaga, T.J., A.J. Klein-Szanto, L.L. Triplett, L.P. Yotti, and J.E. Trosko (1981) Skin tumor-promoting activity of benzoyl peroxide, a widely used free-radical generating compound. Science 213:1023-1025.
99. Snelling, W.M., C.S. Weil, and R.R. Maronpot (1984) A two-year inhalation study of the carcinogenic potential of ethylene oxide in Fischer 344 rats. Tox. and Appl. Pharm. 75:105-117.
100. Stenberg, U., T. Alsberg, and R. Westerholm (1983) Applicability of a cryogradient technique for the enrichment of PAH from automobile exhausts: Demonstration of methodology and evaluation experiments. Env. Health Persp. 47:43-51.
101. Talcott, R., and E. Wei (1977) Airborne mutagens bioassayed in Salmonella typhimurium. JNCI 58:449-451.

102. Taylor, J.C., and C. Mittman, eds. (1986) Pulmonary Emphysema and Proteolysis. Academic Press, New York.
103. Tokiwa, H., H. Takeyoshi, K. Morita, K. Takahanshi, N. Saruta, and Y. Ohnishi (1976) Detection of mutagenic activity in urban air particulates. Mutat. Res. 38:351.
104. Tokiwa, H., S. Kitamori, K. Takahashi, and Y. Ohinishi (1980) Mutagenic and chemical assay of extracts of airborne particulates. Mutat. Res. 77:99-108.
105. Tuominen, J., S. Salomaa, H. Pyysalo, E. Skytta, L. Tikkanen, T. Nurmela, M. Sorsa, V. Pohjola, M. Sauri, and K. Himberg (1988) Polynuclear aromatic compounds and genotoxicity in particulate and vapor phases of ambient air: Effect of traffic season and metero-logical conditions. Env. Sci. Tech. 22:1228-1234.
106. Fourth Annual Report on Carcinogens (1985) U.S. Department of Health and Human Services, pp. 86 and 93.
107. U.S. Environmental Protection Agency (1977) Scientific and Technical Assessment Report on Nitrosamines, Office of Research and Develop-ment, EPA-600/6-77-001, Washington, DC.
108. U.S. Environmental Protection Agency (1978) Air Quality Criteria for Ozone and Other Photochemical Oxidants, Report No. EPA-600/8-78-004, Washington, DC.
109. U.S. News and World Report, June 12, 1989, Washington, DC.
110. VanDuuren, B.L., S. Melchionne, R. Blair, B.M. Goldschmidt, and C. Katz (1971) Co-carcinogenesis studies on mouse skin and inhibi-tion of tumor initiation. JNCI 46:143-149.
111. Wang, C.Y., Mei-Sie Lee, Charlie M. King, and Peter O. Warner (1980) Evidence for nitroaromatics as direct-acting mutagens of air-borne particles. Chemosphere 9:83-87.
112. Waters, M.D., J.L. Huisingh, L. Claxton, and S. Nesnow, eds. (1979) Short-term Bioassays in the Fractionation and Analysis of Complex Environmental Mixtures, Plenum Publishing Corp., New York.
113. Waters, M.D., S.S. Sandhu, J. Lewtas, L. Claxton, N. Chernoff, and S. Nesnow, eds. (1983) Short-term Bioassays in the Analysis of Complex Environmental Mixtures, III, Plenum Publishing Corp., New York.
114. Wauters, E., F. Vangaever, P. Sandra, and M. Verzele (1979) Polar organic fraction of air particulate matter. J. Chromatogr. 170:133-138.
115. White, C., ed. (1985) Nitrated Polycyclic Aromatic Hydrocarbons, Hutig Verlag Publishers, Heidelberg, Germany.
116. Wolman, S.R. (1979) Mutational consequences of exposure to ethylene oxide. J. Env. Pathol. Tox. 2:1289-1303.
117. Yamauchi, T., and T. Handa (1987) Characterization of Aza hetero-cyclic hydrocarbons in urban atmospheric particulate matter. Env. Sci. Tech. 21:1177-1181.
118. Zelikoff, J.T., J.M. Daisey, K.A. Traul, and T.J. Kneip (1985) BalB$_c$/3T3 cell transformation response to extracts of organic air samples as seen by their survival in aggregate form. Mutat. Res. 144:107-116.

HUMAN EXPOSURE TO AIRBORNE MUTAGENS INDOORS AND OUTDOORS

USING MUTAGENESIS AND CHEMICAL ANALYSIS METHODS

Hidetsuru Matsushita, Sumio Goto, Yukihiko Takagi,
Ōsamu Endo, and Kiyoshi Tanabe

Department of Community Environmental Sciences
National Institute of Public Health
Minato-ku, Tokyo, Japan

INTRODUCTION

The lung cancer mortality rate has been steadily increasing in indus-trialized countries of the world. There are many risk factors for the lung cancer induction in our environments. Airborne carcinogens and mutagens are considered to be one of the major risk factors. However, quantitative contribution of airborne carcinogens and mutagens to the lung cancer induction has not yet been accurately evaluated, because of the lack of data on long-term human exposure to these chemicals in many areas which are in different pollution levels. For this purpose, there is an urgent need to develop methodologies suitable for monitoring long-term exposure to these carcinogens and mutagens.

Airborne carcinogens and mutagens can be measured and evaluated by chemical analysis and biological methods such as mutation assay. We have been investigating over a decade to develop chemical and biological monitoring techniques for airborne carcinogens and mutagens indoors and outdoors, to establish the optimum monitoring protocols, and to evaluate human exposure to these chemicals. In this paper, after reviewing brief-ly the monitoring status for airborne carcinogens and mutagens in Japan, we will discuss (i) optimum sampling frequency to estimate accurately the annual averages of carcinogens and mutagens in environmental air, (ii) trials for making standard procedures for monitoring airborne particulate mutagens by mutation assay, (iii) sensitive and automatic analytical meth-ods for polynuclear aromatic hydrocarbons (PAHs) in airborne particulates collected with indoor and personal samples, and (iv) highly sensitive mutation assays for indoor and personal particulate samples.

MONITORING OF CARCINOGENS AND MUTAGENS IN AMBIENT AIR

Ambient PAH Concentrations Over Two Decades in Japan

Many kinds of PAHs have been found in ambient air and exhausts from air pollution emission sources (5,6,15,27,28,40,42,43,56).

Genetic Toxicology of Complex Mixtures
Edited by M. D. Waters *et al.*
Plenum Press, New York, 1990

Benzo(a)pyrene (BaP) is a representative carcinogen and mutagen among PAHs and its air concentration has been widely monitored in many cities of the world (16,37,39,56). In Japan, ambient BaP concentrations were first measured in January 1957 by Suzuki (47), who found high concentrations of 32-108 ng/m³ in Sapporo, capital of Hokkaido, a northern island of Japan. Tsunoda (54) monitored BaP concentrations in Sapporo in the winter and summer of 1961. The BaP concentrations in January and August were 175-202 ng/m³ and 21 ng/m³, respectively. He suggested that the high wintertime concentrations may be due to coal combustion for space heating. An annual average of 36.9 ng/m³ in 1966 was reported for the ambient BaP concentrations in Kitakyushu, a large city in a southern island of Japan, where large iron and steel industries have been operated (19).

Ambient BaP monitoring by a local government was first started in Osaka, the second largest city in Japan, in 1967. Figure 1 shows the trend of BaP concentrations in industrial and commercial areas in Osaka from 1967 to 1983. Showing seasonal variation, high in wintertime and low in summertime, the BaP concentrations have steadily decreased from 1970, and in 1983 no remarkable difference of the BaP concentrations was observed between the industrial and commercial areas. In particular, the BaP concentrations in both areas decreased sharply in the first half of

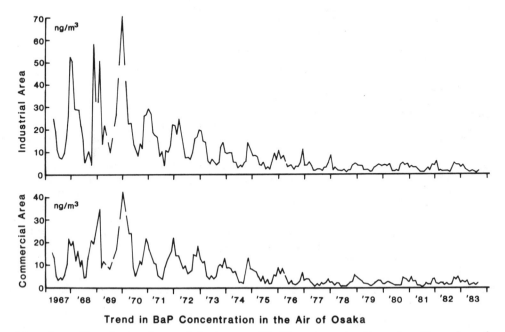

Trend in BaP Concentration in the Air of Osaka

Fig. 1.   Trend in benzo(a)pyrene (BaP) concentration in the air of Osaka in Japan from 1967 to 1983.

1970s, due to great regulatory and technological efforts. The Air Pollution Control Law was amended in 1970 and enforced strict emission controls to the stationary and automobile exhaust gases. To meet the strict
regulation, counter measures to air pollution have been taken, such as
fuel conversion from coal to petroleum oil, and installation of many kinds
of equipment for removing particulates, sulfur dioxide, nitrogen oxides,
etc. in a stack gas. Catalytic converter reduced the PAHs emissions of a
gasoline engine significantly.

The BaP monitoring in the National Air Surveillance Network (NASN)
was started in 1974. The annual averages and their maximum and minimum concentration during 1983 through 1987 were shown in Fig. 2 (16).

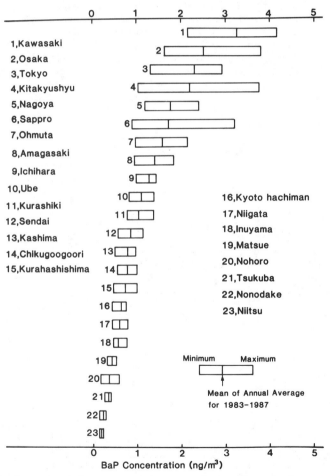

Fig. 2.   Maximum, minimum, and mean values of annual averages of BaP
concentrations in the National Air Surveillance Network (NASN)
in Japan for 1983 to 1987.

The 23 NASN monitoring sites can be classified into three categories: (i) large cities of more than one million population, (ii) industrial cities of fewer than one million population, and (iii) nonpolluted rural areas. The large cities, including the sites numbered 1 through 6 and 12 in Fig. 2, tend to have higher BaP concentrations than the industrial cities (Numbers 7, 11, 13, 17, and 19) some of which gave a severe air pollution in the past. This suggests strongly that automobile exhaust is becoming the major factor for BaP air pollution. Figure 2 shows also that annual averages of BaP concentrations in Japan are in a low and narrow range of 0.16-4.17 ng/m³ (0.18-3.25 ng/m³ in the five-year averages), showing a great reduction of BaP concentration during the past two decades.

Optimum Sampling Frequency to Estimate Accurately the
Annual Averages of Ambient PAH Concentrations

The ambient PAH concentrations have been measured every two or four weeks at monitoring stations of local governments and at NASN. However, as shown in Fig. 3, the ambient PAH concentrations fluctuate widely due to human activities and meteorological conditions (11). These large fluctuations give rise to the following questions: (i) How do the monitoring data represent the real annual averages of BaP concentrations? and (ii) What is the optimum interval for monitoring the PAHs in order to estimate accurately the annual averages?

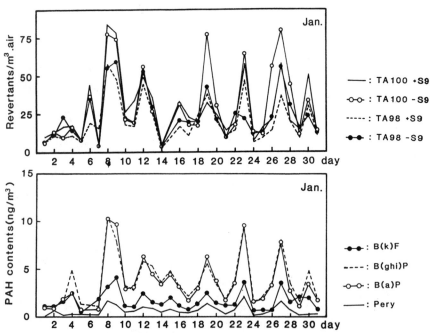

Fig. 3.   Daily variation of mutagenic activity and PAH concentrations in airborne particulates in Tokyo.

We conducted a year-round BaP monitoring in 1983 and looked for the optimum and practical sampling frequencies to represent the annual average concentration. Suspended particulates were collected every day from January 1 to December 31, 1983, with a high volume sampler for each 24 hr at the National Institute of Public Health in central Tokyo. BaP in the particulates was extracted by sonication into benzene-ethanol (3:1, v/v) solution, and analyzed by HPLC/spectrofluorometry (44). By the comparison of the annual average based on the year-round daily measurements with those on the intermittent measurements, we found the following results (Tab. 1): (i) If BaP concentrations are measured once a month (every third Thursday) like a NASN protocol, the geometric mean of the once-a-month measurements has about 30% bias to the real annual average; (ii) If BaP concentrations are measured biweekly, the bias becomes about 20%; and (iii) If BaP concentrations are measured for six consecutive days from Monday to Saturday in all four seasons, the bias becomes about 40%. We also found that BaP concentrations differed between weekdays and weekends.

According to the results of the year-round survey, we propose the sampling on every 15 days from viewpoints of accuracy and practice. Sampling frequency of the proposed sampling is nearly the same with that of biweekly sampling, and the bias is expected to be less than 10% from Tab. 1 (32). This sampling protocol can be applied to the monitoring of PAHs having five or more condensed rings because of high correlation coefficients between these PAHs and BaP (32,56). Furthermore, this sampling protocol may be useful to the monitoring of mutagenicity of particulates in ambient air, because the mutagenic activities vary in a similar manner with the BaP concentrations, as shown in Fig. 3, and correlation coefficients between them are generally high (11,32,35).

Tab. 1. Effect of sampling schedule on the accuracy of the BaP annual average

Sampling	Deviation from true value	No. of samplings per year
Ordinary Sampling		
Once a month (the third Thursday)	29%	12
Twice a month		
(first and third Thursdays)	18%	24
(second and fourth Thursdays)	22%	24
Consecutive six days for each season (e.g., Jan., April, July, and Oct.)	39%	24
New Sampling		
15-day intervals	7%	25

True value: Geometric average calculated from whole BaP data obtained from particulates by every day sampling throughout one year. (1.18 ng/cubic meter; G.S.D. 2.53).

TRIAL OF MAKING STANDARD PROCEDURES FOR MUTATION ASSAYS
FOR AIR MONITORING

The latent period of occupational cancer is generally considered to
be 20 yr or more (6,21). The concentrations of ambient carcinogens and
mutagens are fairly low as compared with those in working environments,
like coke-oven topside, where occupational lung cancer has been observed
(6). Then it may be reasonable that the latent period of lung cancer
caused by the exposure to airborne carcinogens and mutagens is far longer than 20 yr. Therefore, it is important to monitor quantitatively airborne carcinogens and mutagens over several decades in order to elucidate the relationship between the lung cancer induction and dosage or
exposure to these environmental chemicals. The monitoring methods for
this purpose should be simple, sensitive, accurate, and precise. The
Ames assay (1,60) may have a potentiality to be one of the appropriate
methods to meet these requirements. It is simple and sensitive enough to
measure mutagenicity of airborne particulate extracts, so that many
researchers have applied this method to the survey of airborne carcinogens and mutagens (2-4,7-9,11,20,23,33,35,36,38,46,51-53,57,58). However, the repeatability of the Ames assay seems not to be satisfactory
when comparing it with various chemical monitoring methods (4,10). This
poor repeatability may be partly due to the lack of standard operating
procedures for the mutagenicity assay of airborne particulate matter;
therefore, researchers have been testing the mutagenicity with somewhat
different procedures.

We organized a working group with experts from eight laboratories in
order to make standard operating procedures for the mutagenicity monitoring of airborne particulate matter under the contract with Environment
Agency, Japan. The standard operating procedures have been determined based on the following steps: (i) selection of suitable techniques
for each unit process involved in the mutagenicity monitoring, such as
airborne particulate sampling, storage of particulate samples, extraction of
mutagens from particulate samples, preparation of test solutions for the
Ames assay, preculture of the tester strains, and each step of mutagenicity testing, etc.; (ii) making up tentative standard operating procedure
by combining systematically the selected techniques; (iii) evaluation of the
tentative procedures by intra- and interlaboratory cross-checking; and
(iv) modification of the procedures to reach the optimum combination.

Table 2 shows the selected techniques for the unit processes involved in the standard operating procedures. We selected ultrasonic
techniques using dichloromethane as a solvent for the extraction process,
because this technique is simple and can treat many samples in a short
time (12). This step should be done using a screw-cupped glassware in
order to prevent loss of the solvent during sonication. Solvent-exchange
method was used for the preparation of a test sample in order to minimize
loss of volatile mutagnes in the condensation procedure involved (24).
For a testing method, we selected the Ames preincubation method, using
Salmonella typhimurium strains TA98 and TA100 with and without S-9 mix,
because this method is widely used in Japan. The test guideline for the
Ames assay by the Ministry of Labor (22) was slightly modified, that is,
the preculture condition of the tester strains was controlled so as to get
approximately 3×10^9 cells/ml in the concentration of tester strains.
Furthermore, airborne particulate extract was used as an internal standard for normalization of test results.

Tab. 2. Recommended mutagenicity monitoring procedures for airborne particulates

Sampling	High-volume sampler, 24 hr, quart fiber filter
Storage	-80°C, shield from light
Extraction	Ultrasonic extraction (dichloromethane)
Preparation of Test Solution	Solvent-exchange method from CH_2Cl_2 to DMSO
Preculture	To control to be a ca. 3×10^9 cell/ml. (In many cases, the preculture time for TA98 and TA100 was 11 hr when L-shaped tube was used.)
Test Method	Preincubation method according to the test guideline by the Ministry of Labor, Japan
Internal Standard	Airborne particulates or airborne particulates extract stored at -80°C.
Mutagenicity	Rev./m³, air and rev./mg, particulates. Normalization should be done by using internal standard.

Tab. 3. Repeatability of mutagenic activity obtained by intra- and inter-laboratory checking

Sample	Coefficient of Variation of Mutagenic Activity (%)			
	TA100 (+S9)	TA100 (-S9)	TA98 (+S9)	TA98 (-S9)
Intra-laboratory Checking				
P3	11.9 ± 7.5	15.9 ± 8.1	9.9 ± 4.0	15.5 ± 4.2
A1	8.9 ± 4.8	9.0 ± 4.3	8.1 ± 3.6	11.1 ± 6.4
BaP	14.0 ± 3.9	-------	9.9 ± 5.4	-------
AF-2	-------	12.0 ± 4.5	-------	5.4 ± 5.1
Inter-laboratory Checking				
P3	36.3	45.7	17.5	28.6
A1	31.4	34.8	14.4	21.4

P3: Airborne particulate, A1: Airborne particulate extract, Number of laboratories: 7.

The intra- and interlaboratory cross-checking for the mutagenicity method was carried out using four test samples at the seven laboratories participating in the working group. The test samples were airborne particulate, airborne particulate extract, BaP and AF-2. As shown in Tab. 3, the intralaboratory test resulted in a good repeatability. Coefficient of variations for this test was in the range of 5.4% and 15.9%. On the other hand, the interlaboratory test resulted in somewhat worse repeatability, giving coefficient of variations of 31.4% to 45.7% for the strain TA100, and 14.4% to 28.6% for the strain TA98. We examined the reason why the interlaboratory test resulted in a poor repeatability and found that a laboratory that reported high mutagenicity for the particulate sample, too, as shown in Fig. 4. This fact suggests that better agreement among the laboratories is expected if the sensitivity of the mutation assay in each laboratory is normalized with the internal standard.

The suitable internal standard was sought from the Soxhlet extracts of various airborne particulates and dusts, such as airborne particulates collected by a Hi-volume sampler, dusts of an office building collected with a bag filter, and dusts collected in an automobile tunnel with a bag filter. Synthesized compounds such as BaP, AF-2 and a mixture of PAHs, and 1-nitropyrene were also tested. The best result was obtained when the Soxhlet extract of airborne particulates was used as the internal standard.

The interlaboratory test using the internal standard was conducted among the seven laboratories participating in the working group. Seven kinds of test samples, which were prepared from airborne particulates

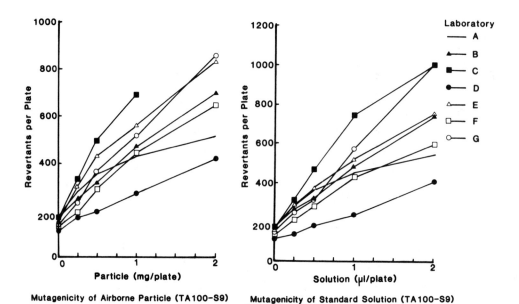

Mutagenicity of Airborne Particle (TA100-S9)   Mutagenicity of Standard Solution (TA100-S9)

Fig. 4.   Comparison of dose. Mutagenic responses for airborne particulates and internal standard solution among seven laboratories.

$$(M.A.)_{Pa,i}^{Nor} = \frac{(M.A.)_{Pa,i}^{obs}}{(M.A.)_{St,i}^{obs}} \times \frac{\sum\limits_{1}^{n}(M.A.)_{St,i}^{obs}}{n}$$

M.A.: Mutagenic Activity
Pa : Particulate Sample
St   : Internal Standard
i   : i th Laboratory

Fig. 5.   Normalization of mutagenic activity of airborne particulates.

collected by the Hi-volume sampler at seven places located in high and low air pollution areas, and the internal standard (Soxhlet extract of airborne particulates) were sent to these laboratories, where their mutagenicities were measured according to the protocol in each laboratory. Mutagenicity data were sent to my laboratory and analyzed statistically. Normalization was carried out by the use of the equation described in Fig. 5. That is, the mutagenic activity for the test sample measured by each laboratory was normalized by multiplying the ratio of the mutagenic activity for the internal standard of the laboratory with the average of mutagenic activities for the internal standard among all laboratories. Figure 6 shows a part of the results obtained in this test. Before the normalization, the ratio of maximum mutagenic activity to minimum one ranged from 1.7 to 5.0 (mean, 3.0). This value reduced to be in the range of 1.4 to 1.9 (mean, 1.65) by the normalization. That is, great improvement of repeatability was achieved by the normalization.

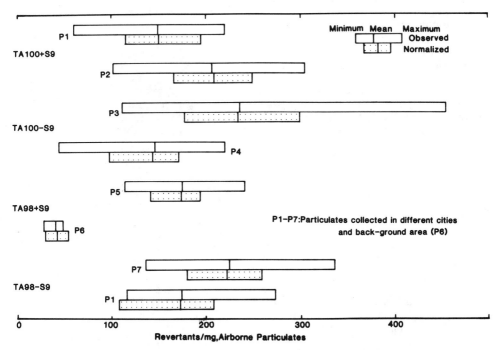

Fig. 6.   Comparison of mutagenic activities of airborne particulates measured at seven laboratories before and after normalization.

After conduction feasibility tests for the proposed method by various institutes, our working group will reevaluate the method to finalize the standard operating procedures for the mutagenicity monitoring of airborne particulates.

SENSITIVE AND AUTOMATIC POLYCYCLIC AROMATIC HYDROCARBON ANALYSIS OF INDOOR AND PERSONAL SAMPLES

Development of sensitive and simple analytical methods for carcinogens and mutagens is needed in order to evaluate human exposure to indoor carcinogens and mutagens. These methods should be sensitive enough to analyze small amounts of the target carcinogens and mutagens, and simple enough to easily analyze a large number of samples in a relatively short time.

People spend most of their time in indoor environments (34,55) where large temporal and spatial variations of carcinogen and mutagen concentrations have been reported due to varieties of smoking intensities, space heating methods, ventilation rates, and cooking habits (24,41,55). It is, therefore, important to measure indoor carcinogen and mutagen concentrations at various locations, and to directly measure personal exposure to them at various occasions in order to obtain accurate information of the exposure levels to the carcinogens and mutagens.

As to the sensitivity of the analytical methods, the amount of sampled air in the indoor monitoring is very small as compared with ambient air monitoring. The air sampling rate for particulates suspended in indoor environments is limited to less than 30 1/min in order not to disturb the indoor environment and to maintain noise levels as low as possible. Furthermore, for personal monitoring, the sampling rate is around 1 1/min, which is one-thousandth of the flow rate of the Hi-volume sampler used for ambient air monitoring.

PAHs are regarded as one of the most important carcinogenic and mutagenic groups in airborne particulates indoors and outdoors. Their concentrations in airborne particulates are generally higher than the other carcinogens and mutagens such as nitroarenes and azaheterocyclic hydro-

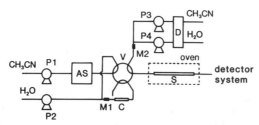

Fig. 7.   Multicolumn HPLC system for automatic multicomponent analysis of ultra micro PAHs. P_1-P_4: HPLC pump. D: degasser. AS: automatic sample injector. V: high pressure six-way valve. M_{10} and M_2: mixer. C: concentration column (4.6 mmφ x 30 mm; PO-60-5). S: separation column (4.6 mmφ x 250 mm; PO-60-5).

carbons. We developed an automatic HPLC/fluorometry/computer system for PAH analysis to meet the requirements for indoor and personal exposure monitoring. Figure 7 illustrates the flow diagram of the HPLC system (29,50).

PAHs in particulate samples are extracted by sonication with 1 to 4 ml of acetonitrile. The extracts are kept in each vial and set at an injection part of an auto-sampler as shown in Fig. 7. A 300 to 500 μl of the extract is sequentially taken and injected into HPLC at 50-min intervals. The injected sample is conveyed with acetonitrile from Pump P_1 immediately before the concentration column (C in Fig. 7), mixed with equal amounts of water from Pump P_2, concentrated into the column, and cleaned up by washing the column with acetonitrile-water (1:1, v/v) for 5 min. After that, a high pressure six-way valve (V) is turned to the other position and the PAHs in the concentration column are transferred to the separation column (S) and separated into each component with the mobile phase from Pumps 3 and 4. The composition of the mobile phase was usually acetonitrile-water (80:20, v/v), although a linear gradient composition was also used for the separation analysis. After 15 min from that time, when the PAHs are transferred to the separation column, the concentration column is back-flushed for cleaning up with acetonitrile-water (9:1, v/v) for 15 min by switching the high pressure valve (V) to the original position and then conditioned to the initial mode by washing with acetonitrile-water (1:1, v/v) for 15 min. Each PAH eluted from the separation column is detected by a spectrofluorometer whose excitation and emission wavelengths are set automatically so as to detect the target PAH with a high sensitivity and selectivity. These sequences are automatically controlled by a computer program.

Figure 8 shows a typical HPLC chromatogram for seven PAHs in an indoor particulate sample collected by a low noise indoor sampler. Of these PAHs, benz(a)anthracene (BaA), benzo(k)fluoranthene (BkF), BaP, and dibenzo(a,e)pyrene (DBaeP) are carcinogenic, and pyrene and benzo-(ghi)perylene (BghiP) are cocarcinogenic to experimental animals (15). All PAHs in Fig. 8 are mutagenic to Salmonella typhimurium TA100 in the presence of S-9 mix (30). The optimum spectrofluorometrical conditions are summarized in Tab. 4 along with detection limits for the PAHs shown in Fig. 8. Lower limits of PAH concentrations in air for quantitative analysis are in the range of 0.01-0.07 ng/m^3 for the particulate sample collected by an indoor air sampler, and 0.03-0.19 ng/m^3 for the sample collected by a personal sampler. Recovery of these PAHs from airborne particulates is 90% or more and its coefficient of variation is less than 5%. These results demonstrate clearly that this method is useful for the indoor and personal exposure monitoring of PAHs.

Using the developed method, we measured hourly variation of PAHs in an office, in which a smoker worked, for a week. The particulates in the office air were hourly sampled at a flow rate of 15 1/min for 55 min each with a β-ray dust monitor during seven consecutive days, and PAHs in the 168 samples collected were analyzed by the developed method. Figure 9 shows hourly variation of BaP concentration in the room. The BaP concentration increased remarkably with smoking and decreased with opening an entrance door and operating an exhaust fan and air cleaner. Figure 9 also suggests that outdoor pollution affects the PAH concentration indoors. Profiles of BkF and BghiP concentrations were similar to that of BaP concentrations. Furthermore, we found that 90% or more of

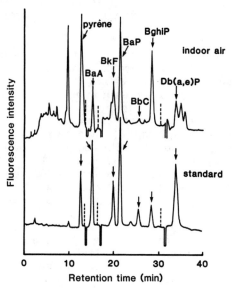

Fig. 8.   HPLC chromatogram for seven PAHs in airborne particulates indoors.

Tab. 4.   Detection wavelengths and limits for detection and quantitative analysis

Compound	Detection Wavelengths (nm)		Detection Limit* (ng)	Lower Limit of PAH Concentration in Air for Quantitative Analysis** (ng/m³)	
	Excitation	Emissions		Indoor Sample[a]	Personal Sample[b]
Pyrene	339	373	0.02	0.07	0.19
Benz(a)anthracene	292	412	0.01	0.03	0.09
Benzo(k)fluoranthene	370	406	0.007	0.02	0.06
Benzo(a)pyrene	370	406	0.003	0.01	0.03
Benzo(c)chrysene	293	401	0.02	0.07	0.19
Benzo(ghi)perylene	385	408	0.015	0.05	0.14
Dibenzo(a,e)pyrene	305	398	0.02	0.07	0.19

* S/N = 2, ** S/N = 20

[a] Indoor sample: air volume 28.8 m³ (20 l/min x 24 hr), volume of extraction solvent 3.0 ml, injection volume 300 µl.

[b] Personal sample: air volume 2.16 m³ (1.5 l/min x 24 hr), volume of extraction solvent 1.0 ml, injection volume 500 µl.

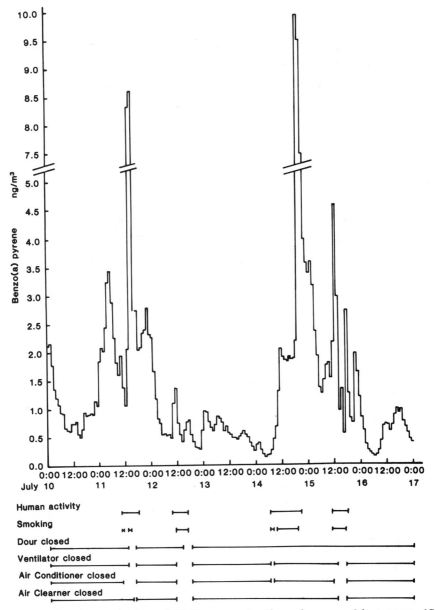

Fig. 9.   Hourly variation of BaP concentrations in a smoking room (July, 1988).

the 3 PAHs (BaP, BkF, and BghiP) indoors are present in fine particles less than 1.25 μm in diameter which have a high deposition rate into the lung (32).

Daily average of personal exposures to PAHs were monitored for a week in summertime (July 18 through 24) to demonstrate performance of

the developed method, because PAH concentrations in summertime are lower than in wintertime. Six people participated in this study and carried
a personal sampler which collects airborne particulates at the flow rate of
1.5 1/min for 24 hr. The personal exposure to PAHs was calculated from
the daily average concentrations obtained from each subject and 20 m^3 of
inhaled air volume per day. They were in the range of 6-38 ng/day
(mean, 15 ng/day) for pyrene, 4-47 ng/day (12 ng/day) for BaA, 3-22
ng/day (9 ng/day) for BkF, 6-72 ng/day (22 ng/day) for BaP, and 8-47
ng/day (26 ng/day) for BghiP. The personal exposure concentrations of
PAHs were similar to the outdoor PAH concentrations. This may be due
to high ventilation rates by keeping windows open in summertime (29).

DEVELOPMENT OF HIGHLY SENSITIVE MUTATION ASSAYS FOR INDOOR AND PERSONAL PARTICULATE SAMPLES

The Ames assay has been widely used to survey the mutagenicity of
particulates in ambient air (2-4,11,20,35,38,53,58). However this assay
requires a relatively large amount of particulates collected from more than
400 m^3 of urban air. It is particularly difficult to collect such amounts of
particulates with the indoor sampler. For this reason, the Ames assay is
hardly applicable to the survey of mutagenicity of airborne particulates
indoors.

Lewtas et al. (24) developed a sensitive micro forward mutation
assay using Salmonella typhimurium strain TM677 by the modification of
the original method by Skopec et al. (45), and showed its usefulness for
the mutagenicity survey of the indoor environment. We applied this assay
to indoor samples, and confirmed that this assay is applicable to the
measurement of mutagenicity of indoor particulates collected with the low
noise sampler at a flow rate of 20 1/min for 24 hr (48). This assay was
also applied to the survey of distribution of mutagenicity in different
diameters of indoor particulates collected by a 13-step Andersen-Impactor
low noise sampler at a flow rate of 25 1/min for 72 hr, and revealed that
a large part of mutagenicity of the indoor particulates were present in the
fine particles less than 1.25 μm in diameter (48).

Concentrations of carcinogens and mutagens indoors fluctuate largely
with many factors, as shown in Fig. 8. Therefore, it is necessary to
develop a more sensitive mutation assay in order to survey in more detail
the mutagenicity of particulates indoors and also to measure personal
exposure to airborne mutagens. We have developed the following two
assays for this purpose: the micro-suspension assay using Salmonella
typhimurium strains TA98/pYG219 and TA100/pYG219 (13) and the ultramicro forward mutation assay using Salmonella typhimurium strain TM677
(49).

Microsuspension Assay Using A New Tester Strain

Salmonella microsuspension assay has ten times or higher sensitivity
than the Ames assay using the same tester strains (17,18), and has been
used to measure exposure to mutagens from environmental tobacco smoke
(25,26). Recently, Watanabe et al. (59) established new tester strains,
Salmonella typhimurium TA98/pYG219 and TA100/pYG219, by transforming
a plasmid, which contains an acetyltransferase gene, into Salmonella

typhimurium strains TA98 and TA100. They named the strains TA98/-pYG219 and TA100/pYG219 as YG1024 and YG1029, respectively. These new strains have high sensitivity to nitroarenes and aromatic amines (31,59). For example, mutagenic activities of 21 nitro-derivatives of benzene, biphenyl, naphthalene, anthracene, fluorene, and pyrene in TA100/pYG219 (YG1029) were ten to 600 times higher than TA100 in the absence of S-9 mix (31). We applied these new tester strains to measure mutagenicity of airborne particulates and found that the mutagenic activities in the new strains were several times higher than those in the parent strains, TA98 and TA100. Furthermore, the mutagenic activities in the new strains were well correlated with those in the parent strains. The correlation coefficients were in the range of 0.705-0.961 (n = 33, p < 0.001) (32). These results suggest the usefulness of the new strains in the mutagenicity survey of airborne particulates.

We applied the new tester strains to microsuspension assay in order to obtain a highly sensitive mutation assay for airborne particulates (13). Figure 10 shows dose-response curves for an ambient particulate extract in the microsuspension assay and the Ames preincubation assay using Salmonella typhimurium TA98 and TA98/pYG219 in the absence of S-9 mix. The microsuspension assay using the strain TA98/pYG219 gave far higher sensitivity in the mutagenic response than the Ames assay using strain TA98. The sensitivity suggests that this assay may be applicable to measure mutagenicity of indoor particulates collected from about 10 m³ of air which corresponds to the air volume of 8-hr indoor sampling.

In order to confirm this suggestion, we collected airborne particulates indoors by a low noise sampler at a flow rate of 20 m³ during working time (8 hr) for two weeks, and measured mutagenic activities of these particulate samples by the microsuspension assay using TA98/pYG219. All particulate samples collected in smoking and nonsmoking rooms gave positive mutagenic responses. As shown in Fig. 11, the mutagenic activities in the smoking room were significantly high and distributed in a wide range as compared with those in the nonsmoking room when an exhaust fan did not work. The operation of the exhaust fan resulted in the large reduction of mutagenic activity in the smoking room, but did not affect the activity in the nonsmoking room, suggesting that mutagens in the nonsmoking room came mainly from outdoor air. These results demonstrate clearly that this microsuspension assay is useful for the mutagenicity monitoring indoors during working times.

Ultramicro Forward Mutation Assay

Figure 12 illustrates the mutagenicity test procedures by the ultramicro forward mutation assay using Salmonella typhimurium strain TM677. This assay was developed by modification of the preincubation step of the micro forward mutation assay (49). That is, volume of solution in the preincubation step was reduced to 1/10th of the micro forward mutation assay. Difficulty due to this volume reduction was overcome by the use of (i) a solvent-exchange method for test sample preparation and (ii) a 100 µl test vial with teflon stopper in the preincubation step. Repeatability of this method was nearly the same with that of the micro forward mutation assay; that is, the coefficient of variation of the mutation frequency of an airborne particulate extract was 13.7% and 12.3% with and without S-9 mix, respectively (49).

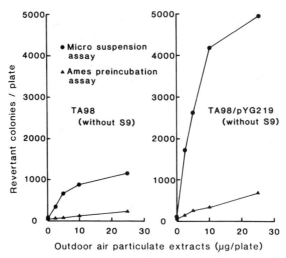

Fig. 10. Dose-mutagenic responses for airborne particulate extract obtained by microsuspension assay and the Ames assay using S.
typhimurium TA98 and TA98/pYG219 (YG1024).

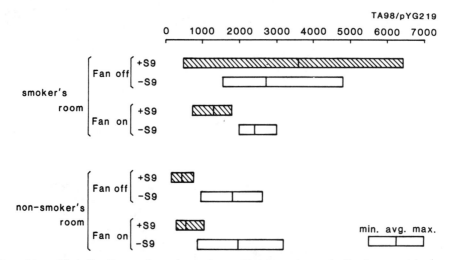

Fig. 11. Distribution of mutagenic activities (rev./m³) in smoking and
nonsmoking rooms during working hours (8 hr).

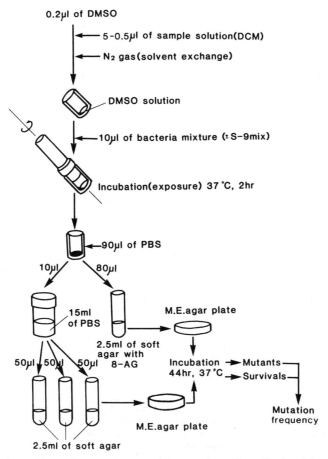

Fig. 12. Ultramicro forward mutation assay using S. typhimurium TM677.

The right figure of Fig. 13 shows relative mutagenic responses for an airborne particulate extract by the ultramicro forward mutation assay (49), the micro forward mutation assay (24), and the Ames preincubation assay (22,60), respectively. The ultramicro forward mutation assay gave about ten times higher sensitivity than the micro forward mutation assay, and about 100 times higher than the Ames preincubation assay. This high sensitivity of the developed method brought the potentiality to measure mutagenicity of airborne particulates collected from about 3 m³ of air which corresponds to the air volume of 2-hr indoor air sampling at a flow rate of 20-25 1/min and personal sampling at a flow rate of 2 1/min for 24 hr.

We conducted every 2-hr sampling of airborne particulates in a smoking room in December, 1987, and measured the mutagenic activity of these samples. At the same time, particulates outdoors were also collected and their mutagenic activity was measured. The left figure of

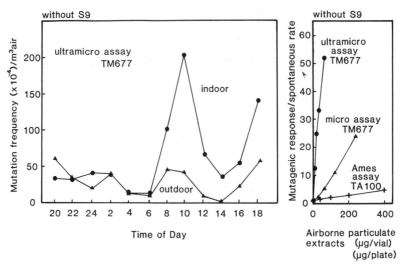

Fig. 13.   Comparison of sensitivity of three mutation assay methods
(right) and hourly variation of mutagenic activities in indoor
and outdoor air by the ultramicro forward mutation assay (left).

Fig. 13 shows the diurnal variation of mutagenic activities indoors and
outdoors measured by the ultramicro forward mutation assay without S-9
mix.   Mutagenic activities indoors were nearly the same as those outdoors
in the nighttime, but were enhanced largely in the daytime.   This suggests that indoor factors, such as smoking, largely affect the mutagenicity level indoors.   As shown in Fig. 12, the mutagenic profile of the
indoor environment resembles somewhat that of the outdoors, suggesting
that outdoor particulate mutagens are also an important factor affecting
the mutagenicity level indoors.

A preliminary survey on personal exposure to airborne mutagens was
performed in order to confirm the utility of the developed assay.   Four
subjects, two nonsmokers and two smokers, carried each sampler on a
winter day and collected airborne particulates at a flow rate of 2 1/min
for 24 hr.   Mutagens in the particulates collected were extracted by
sonication with benzene-ethanol (3:1, v/v), dried up under a reduced
pressure, dissolved in a small amount of dichloromethane, and offered to
the mutation assay shown in Fig. 12.

Table 5 shows the results.   Personal exposure levels to airborne
mutagens were highest first in the ten-cigarette smoker, the six-cigarette
smoker, the nonsmoker who received passive smoke, and the nonsmoker
who lived in a nonsmoking environment during the sampling (14).   Air
ventilation of a livingroom is generally poor in wintertime.   Therefore,
these results demonstrate that, in addition to the usefulness of the
developed method, smoking and poor ventilation remarkably enhance the
personal exposure level to airborne mutagens.

Tab. 5. Personal exposure to particulate mutagens obtained by the ultramicro forward mutation assay

Subject	Smoking	Mutation Frequency (x 10^{-5}/m³) + S-9 mix	- S-9 mix
Nonsmoker (I)	non	13.5	67.3
Nonsmoker (II)	passive smoking	72.1	207.7
Smoker (I)	6 cigarettes	132.7	605.6
Smoker (II)	10 cigarettes	338.5	1,056.7

CONCLUSION

Assessment of human exposure to airborne carcinogens and mutagens can be satisfactorily performed through the evaluation of long-term monitoring data obtained by chemical and biological methods. The former method provides accurate information on the target carcinogens and mutagens, and the latter method integrates information on biological responses to complex mixtures present in indoor and outdoor air. We spend most of our time in indoor environments where concentrations of carcinogens and mutagens fluctuate widely by many factors such as smoking, heating, cooking, ventilation rate, pollution levels outdoors, etc. Furthermore, amounts of particulate samples collected by indoor and personal samplers are generally very small. Therefore, there is an urgent need to develop chemical and biological method to meet the requirements for the long-term monitoring of airborne carcinogens and mutagens indoors and outdoors.

We described the following technological developments on the exposure monitoring of carcinogens and mutagens in airborne particulates.

(i) A practical sampling of airborne particulates for getting accurate annual averages of PAH concentrations in environmental air is a uniform periodic sampling with 15-day intervals. This sampling may also be useful to estimate the annual average of mutagenic activities of airborne particulates.

(ii) The use of internal standards is useful for quantitatively monitoring the mutagenicity of airborne particulates by the Ames assay. The best internal standard was airborne particulate extracts.

(iii) A sensitive automatic PAH analysis was developed. This method was proved to be useful for the routine analysis of PAHs in small amounts of airborne particulates collected by low noise indoor and personal samplers.

(iv) The following two highly sensitive mutation assays were developed: the microsuspension assay using Salmonella typhimurium strains TA98/pYG219 and TA100/pYG219, and the ultramicro forward mutation assay using <u>Salmonella typhimurium</u> strain TM677. The former assay is useful for the survey of mutagenicity indoors during working hours, and the latter assay for the survey of diurnal variation of mutagenic activities indoors, and also for the survey of personal exposure to airborne particulate mutagens.

These technological developments may offer useful tools for human exposure assessment of airborne carcinogens and mutagens.

ACKNOWLEDGEMENTS

The authors gratefully acknowledge the valuable discussions and encouragement of Dr. M. Murata (School of Veterinary Medicine, Azabu University), Dr. S. Imamiya (School of Health Science, Kitasato University), and Dr. J. Lewtas (Health Effects Research Laboratory, U.S. EPA). The authors also gratefully appreciate the generous gift of <u>S. typhimurium</u> strains YG1024 and YG1029 from Dr. M. Ishidate, Jr. (National Institute of Hygienic Sciences). This research was supported partly by research grants from the Ministry of Education and the Ministry of Health and Welfare.

REFERENCES

1. Ames, B.N., W.E. Durston, E. Yamasaki, and F.D. Lee (1973) Carcinogens are mutagens: A simple test system combining liver homogenates for activation and bacteria for detection. <u>Proc. Natl. Acad. Sci., USA</u> 70:2281-2285.
2. Alfheim, I., G. Löfroth, and M. Moller (1983) Bioassay of extracts of ambient particulate matter. <u>Env. Health Persp.</u> 47:227-238.
3. Alink, G.M., H.H. Smit, J.J. van Houdt, J.R. Kolkman, and J.S.M. Boleij (1983) Mutagenic activity of airborne particulates at nonindustrial locations. <u>Mutat. Res.</u> 116:21-34.
4. Chrisp, C.E., and G.L. Fisher (1980) Mutagenicity of airborne particles. <u>Mutat. Res.</u> 76:143-164.
5. Claxton, L.D. (1983) Characterization of automotive emissions by bacterial mutagenesis bioassay: A review. <u>Env. Mutagen.</u> 5:609-631.
6. Committee on Biological Effects of Atmospheric Pollutants (1972) <u>Particulate Polycyclic Organic Matter</u>, National Academy of Sciences, Washington, DC, pp. 191-236.
7. Crebelli, R., S. Fuselli, A. Menequz, G. Aquilina, L. Conti, P. Leopardi, A. Zijno, F. Baris, and A. Carere (1988) In vitro and in vivo mutagenicity studies with airborne particulate extracts. <u>Mutat. Res.</u> 204:565-575.
8. de Wiest, F., D. Rondia, R. Gol-Winkler, and J. Gielen (1982) Mutagenic activity of non-volatile organic matter associated with suspended matter in urban air. <u>Mutat. Res.</u> 104:201-207.
9. Fukino, H., S. Mimura, K. Inoue, and Y. Yamane (1982) Mutagenicity of airborne particles. <u>Mutat. Res.</u> 102:237-247.
10. Gogglemann, W., A. Grafe, J. Vollman, M. Baumeister, P.J. Kramer, and B.L. Pool (1983) Criteria for the standardization of <u>Salmonella</u> mutagenicity tests: Results of a collaborative study. <u>Terato., Carcinog., Mutagen.</u> 3:205-213.

11. Goto, S., U. Kato, A. Orii, K. Tanaka, Y. Hisamatsu, and H. Matsushita (1982) Daily variation of mutagenicities of airborne particulates. J. Jpana Soc. Air Pollut. 17:295-303.

12. Gogo, S., A. Kawai, T. Yonekawa, and H. Matsushita (1982) Ultrasonic extraction method: A technique for mutagenicity monitoring of airborne particulates. J. Japan Soc. Air Pollut. 17:53-57.

13. Goto, S., O. Endo, J. Lewtas, and H. Matsushita (1989) Micro suspension assay using a new Salmonella typhimurium strain for the evaluation of mutagenicity of airborne particulates indoors. Abstracts of the International Conference on Genetic Toxicology of Complex Mixtures, Washington, DC, p. 29.

14. Goto, S., Y. Takagi, M. Murata, J. Lewtas, and H. Matsushita (1989) Mutation assay for personal airborne particulate samples by a highly sensitive ultramicro forward mutation method. Env. Molec. Muta. 14:74.

15. International Agency for Research on Cancer (1983) IARC Monographs on the Evaluation of the Carcinogenic Risk of Chemicals in Humans: Polynuclear Aromatic Hydrocarbons, Part I, IARC, Lyon, France, Vol. 32, pp. 33-91.

16. Japan Environmental Sanitation Center (1988) Data for Suspended Particulates and its Components in National Air Surveillance Network (NASN) in Japan, Report to Environment Agency, Kawasaki, pp. 83-101.

17. Kado, N.Y., D. Langley, and E. Eisenstadt (1983) A simple modification of the Salmonella liquid-incubation assay: Increased sensitivity for detecting mutagens in human urine. Mutat. Res. 121:25-32.

18. Kado, N.Y., G.N. Guirguis, C.P. Flessel, R.C. Chan, K. Chang, and J.J. Wesolowski (1986) Mutagenicity of fine (0.5 um) air particles: Diurnal variation in community air determined by a Salmonella micro pre-incubation (micro suspension) procedure. Env. Muta. 8:53-66.

19. Kodama, Y., and N. Ishinishi (1976) Analysis of benzo(a)pyrene in the air. J. Japan Soc. Air Pollut. 10:732-741.

20. Kolber, A., T. Wolf, T. Hughes, E. Pellizzari, C. Sparacino, M.D. Waters, J.L. Lewtas, and L. Claxton (1980) Collection, chemical fractionation and mutagenicity bioassay of ambient air particulate. In Short-term Bioassays in the Analysis of Complex Environmental Mixtures, II, M.D. Waters, S.S. Sandhu, J.L. Huisingh, L. Claxton, and S. Nesnow, eds. Plenum Publishing Corp., New York, pp. 21-43.

21. Kuratsune, M., S. Tokudome, T. Shirakusa, and M. Yoshida (1974) Occupational lung cancer among copper smelters. Int. J. Cancer 13:552-558.

22. Labor Standard Bureau, Ministry of Labour, ed. (1987) Guidebook for Mutagenicity Test Using Microorganisms, 2nd ed., Japan Industrial Safety and Health Assoc., Tokyo, Japan, pp. 33-81.

23. Lewis, C.W., R.E. Baumgardner, R.K. Stevens, L.D. Claxton, and J. Lewtas (1988) Contribution of woodsmoke and motor vehicle emissions to ambient aerosol mutagenicity. Env. Sci. Tech. 22:986-971.

24. Lewtas, J., S. Goto, K. Williams, J.C. Chauang, B.A. Peterson, and N.K. Wilson (1987) The mutagenicity of indoor air particulates in a residential pilot field study: Application and evaluation of new methodologies. Atmos. Env. 21:443-449.

25. Ling, P.I., G. Löfroth, and J. Lewtas (1987) Mutagenic determination of passive smoking. Toxicol. Lett. 35:147-151.

26. Löfroth, G., R.M. Burton, L. Forehand, S.K. Hammond, R.L. Seila,
     R.B. Zweidinger, and J. Lewtas (1989) Characterization of environ-
     mental tobacco smoke. Env. Sci. Tech. 23:610-614.
27. Matsushita, H., Y. Esumi, and K. Yamada (1970) Identification of
     polynuclear hydrocarbons in air pollutants. Japan Analyst 19:951-
     966.
28. Matsushita, H. (1978) Analytical methods for moitoring polycyclic
     aromatic hydrocarbons in the environment. In Polycyclic Hydrocar-
     bons and Cancer, H.V. Gelboin and P.O.P. Ts'o, eds. Academic
     Press, New York, Vol. 1, pp. 71-81.
29. Matsushita, H., K. Tanabe, and S. Goto (1989) Development of
     sensitive automatic analytical methods for toxic chemicals in airborne
     particulates. In Environmental Research in Japan, Environment
     Agency, Tokyo, Japan, Vol. 1, Chap. 6, pp. 1-19.
30. Matsushita, H., S. Goto, O. Endo, J.H. Lee, and A. Kawai (1986)
     Mutagenicity of diesel exhaust and related chemicals. In Carcino-
     genic and Mutagenic Effects of Diesel Engine Exhaust, N. Ishinishi,
     A. Koizumi, R.O. McClellan, W. Strober, eds. Elsevier Sci. Pub-
     lishers, B.V., Amsterdam, pp. 103-118.
31. Matsushita, H., O. Endo, H. Matsushita, Jr., M. Mochizuki, M.
     Watanabe, and M. Ishidate, Jr. (1989) Mutagenicity of nitroarenes by
     new tester strains, TA100/pYG216, TA100/pYG219, TA98/pYG216 and
     TA98/pYG219. Env. Molec. Muta. 14:123-124.
32. Matsushita, H., Y. Shiozaki, S. Goto, O. Endo, and K. Tanabe (un-
     published data).
33. Moller, M. and I. Alfheim (1980) Mutagenicity and PAH analysis of
     airborne particulate matter. Atmos. Env. 14:83-88.
34. Mori, T., M. Kikkawa, and H. Matsushita (1986) Effects of various
     living environments on personal exposure levels of nitrogen dioxide.
     J. Japan Soc. Air Pollut. 21:446-453.
35. Ohe, T. (1982) Studies on mutagenicity of tar derived from airborne
     dust collected throughout one year. Nihon Koshu Eisei Zasshi
     29:261-272.
36. Ohtani, Y., Y. Shimada, A. Ujiiye, T. Nishimura, and H. Matsushita
     (1985) Comparison between mutagenic activities of airborne particu-
     lates in Maebashi and in Minato-ku, Tokyo. J. Japan Soc. Air
     Pollut. 20:463-469.
37. Oka, M., and T. Kawaraya (1988) Trend in air pollution by benzo-
     (a)pyrene and related chemicals in Osaka and Kyoto district. In Air
     Monitoring--Collection and Evaluation of Data on Carcinogens in the
     Environmental Air in Japan, Report to the Environmental Agency,
     Japan, Japan Environmental Sanitation Center, Kawasaki, pp. 153-
     192.
38. Pitts, Jr., J.N., W. Harger, D.M. Lokensgard, D.R. Fitz, G.M.
     Scorziell, and V. Mejia (1982) Diurnal variations in the mutagenicity
     of airborne particulate organic matter in California's south coast
     basin. Mutat. Res. 104:35-41.
39. Pott, F. (1983) Environmental contamination by PAH: Air. In Envi-
     ronmental Carcinogens: Polycyclic Aromatic Hydrocarbons, G.
     Grimmer, ed. CRC Press, Florida, pp. 84-105.
40. Sawicki, E. (1977) Chemical composition and potential genotoxic
     aspects of polluted atmospheres. In Air Pollution and Cancer in
     Man, U. Mohr, D. Schmähl, L. Tomatis, and W. Davis, eds. IARC
     Sci. Publications No. 16, International Agency for Research on
     cancer, Lyon, France, pp. 127-157.

41. Schmeltz, I., D. Hoffmann, and E.L. Wynder (1975) The influence of tobacco smoke on indoor atmospheres: An overview. Prev. Med. 4:66-82.

42. Schuetzle, D., F.S.C. Lee, and T.J. Prater (1981) The identification of polynuclear aromatic hydrocarbon (PAH) derivatives in mutagenic fractions of diesel particulate extract. Int. J. Env. Analyt. Chem. 9:93-144.

43. Schuetzle, D. (1983) Sampling of vehicle emissions for chemical analysis and biological testing. Env. Health Persp. 47:65-80.

44. Shiozaki, T., K. Tanabe, and H. Matsushita (1984) Analytical method for polynuclear aromatic hydrocarbons in airborne particulates by high performance liquid chromatography. J. Japan Soc. Air. Pollut. 19:300-307.

45. Skopec, T.R., H.L. Liber, D.A. Kaden, and W.G. Thilly (1978) Relative sensitivities of forward and reverse mutation assays in Salmonella typhimurium. Proc. Natl. Acad. Sci., USA 75:4465-4469.

46. Suter, W. (1988) Ames Test: Mutagenic activity of airborne particles collected on air conditioner filters during the fire at a Sandoz storehouse in Schweizerhalle on November 1, 1986. Mutat. Res. 260:411-427.

47. Suzuki, A. (1958) Study on urban air pollution by carcinogenic hydrocarbons, espeically 3,4-benzopyrene. Hoppo Sangyo Eisei 18:49-72.

48. Takagi, Y., S. Goto, M. Murata, H. Matsushita, and J. Lewtas (1988) Application of the micro-forward mutation assay to assess mutagenicity of airborne particulate in indoor air. J. Japan Soc. Air Pollut. 23:24-31.

49. Takagi, Y., S. Goto, C.T. Kuo, S. Sugita, M. Murata, J. Lewtas, and H. Matsushita (1989) Ultramicro forward mutation assay and its application to the survey of indoor air pollution. J. Japan Soc. Air Pollut. 24:244-251.

50. Tanabe, K., C.T. Kuo, S. Imamiya, and H. Matsushita (1987) Micro analysis of PAHs in airborne particulates by column concentration-- High performance liquid chromatography with spectrofluorometric detection. J. Japan Soc. Air Pollut. 22:334-339.

51. Teranishi, K., K. Hamada, and H. Watanabe (1978) Mutagenicity in Salmonella typhimurium mutants of the benzene soluble organic matter derived from airborne particulate matter and its five fractions. Mutat. Res. 56:273-280.

52. Tokiwa, H., K. Morita, H. Takeyoshi, K. Takahashi, and Y. Ohnishi (1977) Detection of mutagenic activity in particulate air pollutants. Mutat. Res. 48:237-248.

53. Tokiwa, H., S. Kitamori, K. Horikawa, and R. Nakagawa (1983) Some findings on mutagenicity in airborne particulate pollutants. Env. Muta. 5:87-100.

54. Tsunoda, F. (1961) Study on air pollution by carcinogenic hydrocarbons. Hoppo Sangyo Eisei 29:21-34.

55. Turiel, I. (1985) Indoor Air Quality and Human Health, Stanford University Press, Stanford, California, pp. 3-82.

56. U.S. Environmental Protection Agency (1975) Scientific and Technical Assessment Report on Particulate Polycyclic Organic Matter (PPOM), Chap. 5, pp. 1-24, EPA-600/6-75-001.

57. van Houdt, J.J., L.H.J. de Haan, and G.M. Alink (1989) The release of mutagens from airborne particles in the presence of physiological fluids. Mutat. Res. 222:155-160.

58. Watts, R.R., R.J. Drago, R.G. Merrill, R.W. Williams, E. Perry, and J. Lewtas (1988) Wood smoke impact air: Mutagenicity and chemical analysis of ambient air in a reidential area of Juneau, Alaska. J. Air Pollut. Contr. Assoc. 38:652-660.

59. Watanabe, M., P. Einiströ, M. Ishidate, Jr., and T. Nohmi (1989) Establishment and an application of new substrains of S. typhimurium highly sensitive to nitroarenes and aromatic amines. Abstract of the International Conference on Genetic Toxicology of Complex Mixtures, Washington, DC, p. 41.

60. Yahagi, T., M. Nagao, Y. Seino, T. Matsushima, T. Sugimura, and M. Okada (1977) Mutagenicities of N-nitrosamine of Salmonella. Mutat. Res. 48:121-130.

GENETIC TOXICOLOGY OF AIRBORNE PARTICULATE MATTER USING

CYTOGENETIC ASSAYS AND MICROBIAL MUTAGENICITY ASSAYS

Roberto Barale,[1] Lucia Migliore,[2] Bettina Cellini,[2]
Lucia Francioni,[2] Francesco Giorgelli,[2] Italo Barrai,[1]
and Nicola Loprieno[2]

[1]Istituto di Zoologia
Università di Ferrara
Ferrara, Italy

[2]Dipartimento de Scienze dell'Ambiente e del Territorio
Università di Pisa
Pisa, Italy

ABSTRACT

The mutagenicity of airborne particulate matter was assayed using four different tests, namely, the Ames Salmonella test, sister chromatid exchanges (SCEs), chromosomal aberrations (CAs), and the micronucleus (MN) test in human lymphocytes. Two types of extracts were used for all the tests. The first was a sample obtained from urban air in winter which showed high mutagenic activity. The second was a summer sample, again from urban air, which showed very low activity. The particulate matter was assayed as a whole organic extract, and in the form of different fractions. In the first sample, we observed that SCE and CA tests were more sensitive than the Ames test. Since both tests in eukaryotic cells are time-consuming and expensive, we tried to assess the value of the MN test with a second sample showing low activity. Again, the three cytogenetic tests were more sensitive than the test on Salmonella, and in particular the MN test seems to be a useful test in conjunction with the Ames test. As a first screening, it could substitute for the other two tests.

INTRODUCTION

After the discovery that organic extracts of urban airborne particulate matter is carcinogenic in experimental animals (5,7) and the assessment that the Ames test is characterized by high predictability for carcinogenicity of many chemical classes (17), several mutagenicity tests have been performed all over the world on organic extracts from

ambient air. The Ames test was routinely employed to assess the genotoxicity of organic extracts from the particles and of different chemical
fractions in order to evaluate the amount of mutagens present and to
identify the major chemical classes possibly responsible for the observed
mutagenicity. The Ames test is able to detect only gene (point) mutations, and airborne particulate matter is a complex mixture containing
thousands of chemicals whose genotoxic specificity is unknown. 1-6-Dini-
tropyrene which has been detected in diesel and gasoline exhausts and in
urban airborne particulates is able to break chromosomes and to induce
aneuploidy (genome mutations) in mammalian cells (1,4,10). Recently
(23), an increased number of SCEs and CAs have been detected in the
lymphocytes of policemen exposed to traffic pollution suggesting that
chromosome damage can be a remarkable genetic endpoint induced by airborne complex mixtures. Up to now, only a few mutagenicity assays at
the chromosomal level on mammalian cells have been performed with
organic extracts from airborne particulate matter. This is mainly due to
their high cost because such assays are time-consuming and require
skilled cytogeneticists (2,10-15,20).

The aim of the present work, first was, to compare the sensitivity of
the standard Ames test with some cytogenetic assay (SCE; structural and
numerical CAs in human lymphocytes) by the use of two different organic
extracts with and without metabolic activation and, second, to validate the
MN test as a possible substitute of the above-mentioned cytogenetic
assays. It is known, in fact, that micronuclei can be formed from chromosome acentric fragments as well as from entire chromosomes, the former
as the consequence of chromosome damage and the latter of mitotic spindle
disturbances (3,9). Moreover, the MN test is faster and easier to perform, so it makes it possible to extend analyses on large numbers of samples. The sensitivity of this test has been recently improved by the use
of cytocalasin B (Cyt B), an inhibitor of cytokinesis, which makes it possible to distinguish cells that have divided once and so are binucleate
(8). By selecting binucleated lymphocytes during scoring, the induced
micronuclei frequency is evaluated more precisely. In fact, chromosome
damaged cells, due to the treatment toxicity, often cannot divide and then
express the potential micronuclei, yielding false negatives or weak outcomes. Moreover, the ratio of binucleated cells over mononucleated (un-
divided) cells can be used as a measure of induced cell toxicity. Recently, we have shown that by using this methodology the sensitivity of the
MN assay was comparable to that of conventional chromosomal analysis for
the assessment of the clastogenic activity of adriamycin (ADM) and much
more sensitive for the evaluation of the effects of vincristine (VCR), a
well-known poison of the mitotic spindle (18).

MATERIALS AND METHODS

Airborne particles smaller than 10 μm were collected with an HiVol
air sampler (Sierra Andersen) on precleaned glass microfiber filters. The
first sample was collected in November 1988 along an intersection in Pisa
(particulate matter: 255 μg/m^3 of air) and organic material was Soxhlet
extracted (8 hr) in sequence with n-hexane, dichloromethane (DCM), and
methanol. The three extracts were concentrated with a rotary evaporator
(40° C) and then dried under a nitrogen stream. The dried extracts
were dissolved in dimethylsulfoxide (DMSO) for mutagenicity testing. The
second sample was collected during June of the same year (particulate

matter: 99 µg/m³ of air) and Soxhlet extracted with DCM (18 hr). The organic material was divided into two parts: the first was processed for mutagenicity tests as described above and the second was chemically fractionated into acid, neutral, and basic fractions according to Crebelli et al. (6). Each fraction was dried and dissolved in DMSO for mutagenicity testing.

Mutagenicity Assays

Ames test. The organic extracts were tested as described by Maron and Ames (16) in strains TA98 and TA100 with and without S-9 mix. The S-9 mix was obtained from Aroclor-induced Sprague-Dawley rat liver homogenate and the concentration in the S-9 was 4%. Spontaneous revertants were determined in plates treated with DMSO. Positive control (2-aminofluorene, 1 µg/plate) was included in each experiment. All samples were tested over a wide range of doses in order to obtain a dose-response relationship.

Sister Chromatid Exchange and Chromosomal Analyses

Lymphocyte cultures. A 0.3 ml sample of whole blood from one healthy human male was added to 4.7 ml of Ham's F-10 medium (Flow Laboratories) supplemented with 10% fetal bovine serum containing 1% phytohemagglutinin (PHA, Wellcome Diagnostics), antibiotics (100 IU penicillin/ml and 100 µg streptomycin/ml, Sigma Chemical). 5-Bromodeoxyuridine (BrdUrd, Sigma Chemical; final concentration, 3 µg/ml) was added to whole blood for the evaluation of cell proliferation, SCE counting, and selection of first division metaphases for chromosome aberration analysis. Colchicine (Sigma Chemical; final concentration, 4 µg/ml) was added to whole blood cultures 2 hr prior to harvesting. Cyt B (Sigma Chemical; final concentration, 3 µg/ml) was added to cultures 24 hr prior to harvesting. Control cultures were grown in identical conditions and added with 1% DMSO (the same amount used for sample testing). Positive controls: ADM, VCR (both from Farmitalia-Carlo Erba, 10 ng/ml, Mitomycin C (Mit C, Kyowa), Cyclophosphamide (CPA, Asta-Werke, 25 µg/ml). S-9 mix, at the same composition for the Ames test, was added contemporaneously to the air extracts or CPA at the final concentration of 5%. Test compounds and S-9 mix were added for 3 hr, from the 44th to the 47th hour of culture. Before their addition, cultures were washed and resuspended in fresh medium without fetal calf serum. In cultures without S-9 mix, the same amount of medium was added. After treatment, cells were washed with phosphole buffered solution (PBS) and then resuspended in fresh medium until harvesting. Continuous treatments were performed by adding organic extracts throughout the culture time. Hypotonic treatment, slide preparation, and staining were performed according to standard procedures as previously described (18). For the micronucleus assay, a modification of the standard procedure was used (19).

Slide Scoring

Slides were coded and scored blind under a magnification of 1,000X. For SCE analysis, 25-second division well-spread metaphases were scored for each experimental point. For chromosome analysis, clear metaphases were scored for the presence of chromatid and chromosome aberrations. Cells with very well-preserved cytoplasms were analyzed for the presence of extra chromosomes (2n + 1 or more, aneuploid cells) or for doubling the normal chromosome set (4n, polyploid cells). Endoreduplicated cells (4n, with two pairs of homologous chromosomes, diplochromosomes) were also recorded. Two thousand binucleated (Bin) cells with intact cytoplasms were checked for the presence of micronuclei for each experimental point; the number of mononucleated (Mon) cells was recorded as well in order to evaluate possible variations of the ratio Bin/Mon cells as an index of cell toxicity.

Statistical Methods

For the Ames test, we fit a linear regression to the number of revertants over dose using our own software package (available upon request). For SCE analysis, we have compared the level of spontaneous exchnages to the average number of SCEs at different doses. Although in many cases there is a significant linear component, the curve generally flattens out at higher doses. For chromosomal aberrations and genomic mutations, we have used Fisher's exact test. The spontaneous response measured in number of micronuclei present in binucleate cells was compared to the number of micronuclei counted in the lymphocytes of the treated cells cultured by a 2 x 2 chi square.

RESULTS: WINTER SAMPLE

Ames Test

The results shown in Tab. 1 of the mutagenicity assay in Salmonella TA98 and TA100 with and without S-9 mix for the organic extracts obtained with n-hexane, DCM, and methanol. The response variable tabulated is the b coefficient or slope of the linear regression in units of revertants per cubic meter equivalent of air (m^3Eq). In the same Table, the standard errors and the t-tests are also given, the levels of significance and the doubling dose are indicated.

The response obtained with the n-hexane extract is highly significant and is doubled in the presence of S-9 mix when TA98 is used; for TA100 there is no difference of effect due to metabolic activation. The response of the DCM extract is not significant for TA100 without metabolic activation, possibly due to the toxic effect at the highest dose. In the other three tests, both with TA98 and in the case of TA100 with S-9 mix, it is highly significant. The effect of the mixture on TA98 with S-9 mix is again higher than without S-9 mix and doubles the number of revertants. The difference is significant at the 5% level, confirming the presence of a significant amount of indirect mutagens active on the TA98 strain. Finally, the mixture extracted with methanol has no significant mutagenic activity. It must be said that the nonsignificant effect is due to the range of doses we have used; when the dose is increased above 6.5 m^3Eq, there is a sharp drop, followed by an increase in the number of revertants. This phenomenon, which is not rare, destroys the significance of the increases observed at the two lower doses. However, it is less pronounced than those obtained with the two other extracts.

Sister Chromatid Exchanges and Chromosome Aberrations

The induction of SCEs in 3-hr treated cultures is reported in Tab. 2. The numbers counted at different doses were compared with the spon-

Tab. 1. Genotoxicity of three organic extracts. (Winter sample, airborne particulate)

Time (hr)	S-9 mix	Chromosomal Aberrations Percent of Cells (No./Total)	Genome Aberrations Percent of cells (No./Total)
Controls			
72	–	0.50 (2/400)	0.25 (1/400)
72	+	0.50 (2/400)	0.25 (1/400)
96	–	0.50 (2/400)	0.25 (1/400)
96	+	0.50 (2/400)	0.00 (0/400)
N-Hexane			
72	–	3.66 (11/300)^b	1.66 (5/300)$^{n.s.}$
72	+	1.66 (5/300)$^{n.s.}$	0.00 (0/400)
96	–	2.66 (6/225)^a	3.55 (8/225)^c
96	+	6.33 (19/300)^d	2.00 (6/300)^b
Dichloromethane			
72	–	5.57 (15/269)^d	1.11 (3/269)$^{n.s.}$
72	+	7.00 (21/300)^d	1.66 (5/300)^b
96	–	4.35 (12/276)^c	1.09 (3/276)$^{n.s.}$
96	+	2.67 (8/300)^b	2.00 (6/300)$^{n.s.}$
Methanol			
72	–	3.33 (10/300)^c	1.00 (3/300)$^{n.s.}$
72	+	2.25 (6/226)^a	0.37 (1/265)$^{n.s.}$
96	–	4.00 (12/300)^c	3.33 (10/300)^c
96	+	3.10 (9/300)^c	3.00 (9/300)^d
Cyclophosphamide			
(72)	+	9.33 (28/300)^d	5.33 (16/300)^d
Adriamycine			
(72)	–	3.10 (9/300)^c	2.00 (6/300)^c
Vincristine			
(72)	–	2.67 (8/300)^b	29.00 (87/300)^d

N-hexane dose: 8.5 m^{3}Eq; DCM dose: 11.3 m^{3}Eq; methanol dose: 11.3^{3}Eq; a =
p<0.05; b = p<0.025; c = p<0.01; d = p<0.001; n.s. = not significant.

Tab. 2.   Sister chromatid exchange induction by three organic extracts.
(Winter sample, airborne particulate)

Solvent	S-9 mix	m^3Eq	Mean ± S.D.	
			72 hr	96 hr
N-hexane	–	0	4.04 ± 1.64[b]	5.00 ± 1.55[b]
	–	1.22	8.76 ± 2.80[b]	7.52 ± 2.41[b]
	–	2.44	9.80 ± 3.35[b]	10.32 ± 3.68[b]
	–	4.88	8.96 ± 2.80[b]	9.20 ± 2.29[b]
	+	0	5.88 ± 2.76[b]	5.40 ± 1.68[b]
	+	1.22	8.12 ± 2.18[b]	8.40 ± 3.51[b]
	+	2.44	5.96 + 1.94[n.s.]	10.16 ± 3.33[b]
	+	4.88	9.40 ± 2.56[b]	13.32 ± 4.13[b]
DCM	–	0	5.52 ± 1.87[b]	5.68 ± 1.93[b]
	–	1.22	6.20 ± 1.65[n.s.]	8.44 ± 2.77[b]
	–	2.44	n.d.	6.80 ± 2.56[n.s.]
	–	4.88	7.96 ± 2.13[b]	7.60 ± 1.65[b]
	+	0	4.20 ± 1.87[b]	5.92 ± 1.80[b]
	+	1.22	6.92 ± 3.01[b]	9.28 ± 2.68[b]
	+	2.44	7.68 ± 2.41[b]	7.72 ± 3.15[a]
	+	4.88	10.08 ± 4.23[b]	9.32 ± 3.23[b]
Methanol	–	0	5.52 ± 1.87[b]	5.68 ± 1.93[n.s.]
	–	1.22	7.64 ± 2.78[b]	5.76 ± 1.78[n.s.]
	–	2.44	7.36 ± 3.65[a]	7.00 ± 1.89[a]
	–	4.88	7.92 ± 2.08[b]	6.72 ± 1.62[a]
	+	0	4.20 ± 1.87[b]	5.92 ± 1.80[n.s.]
	+	1.22	6.32 ± 2.59[b]	6.92 ± 2.66[n.s.]
	+	2.44	7.52 ± 2.88[b]	7.20 ± 2.36[a]
	+	4.88	5.88 ± 2.06[a]	7.32 ± 2.73[a]
Cyclophosphamide				
	+	20 µg/ml	22.33 ± 11.37[c]	n.d.

a = p<0.05; b = p<0.01; c = p<0.001; S.D. = standard deviation; n.s. = not
significant; n.d. = not determined.

taneous number of SCEs by a t-test. The n-hexane and DCM extracts
appear to be more effective than methanol extract, and the addition of
S-9 mix seems not to affect the response at both cell harvesting times.
So, the last extract is less effective in the Ames test and the SCE assay.
The results of chromosomal aberrations are given in Tab. 3. We observed
a spontaneous number of 0.5 chromosomal aberrations (gaps excluded) and
0.25 genomic aberrations per one hundred cells. The number of chromosomal aberrations, mainly of chromatid type, was significantly increased at
all doses for n-hexane and DCM extracts; the increase is also observed
for methanol extracts. All types of aberrations were observed, including
dicentrics, exchanges, and rings. Genomic aberrations are significantly
increased only at longer incubation times (96 hr), suggesting that a recovery from induced toxicity is necessary for the expression of the in-

Tab. 3.   Mutagenicity of the three organic extracts using the Ames test.
(Winter sample, airborne particulate)

Extract	Strain	S-9 mix	b Value (I)	t Value (II)	Estimated doubling dose (m^3Eq)
Hexane					
	TA98	−	6.54	5.58^b	5.4
	TA98	+	12.36	5.68^b	2.8
	TA100	−	14.00	5.34^b	10.3
	TA100	+	15.02	5.68^b	10.5
Dichloromethane					
	TA98	−	3.45	6.16^b	12.6
	TA98	+	7.75	5.45^b	6.1
	TA100	−	5.83	$3.89^{n.s.}$	(24.8)
	TA100	+	8.67	5.61^a	17.3
Methanol					
	TA98	−	2.91	$2.92^{n.s.}$	(11.8)
	TA98	+	3.30	$1.92^{n.s.}$	(16.3)
	TA100	−	1.67	$0.57^{n.s.}$	(92.3)
	TA100	+	0.88	$0.29^{n.s.}$	(180.6)
Positive controls					
	TA98	+	2AF, 5 µg/plate :	3036 ± 112 rev./plate	
	TA100	+	" "	: 2345 ± 33 " "	

(I): b represents the slope coefficient of the regression; (II): t, student value. $a = p < 0.05$; $b = p < 0.025$; n.s. = not significant.

duced damage. This is not unexpected because, in order to obtain aneuploid or polyploid cells, a cell cycle is required after the poisoning of the mitotic spindle. The increases were obtained with doses corresponding to 8.5, 11.3, and 11.3 m^3Eq for n-hexane, DCM, and methanol extracts, respectively. Methanol extract seems to be as effective as the others in inducing chromosome and genome aberrations.

RESULTS: SUMMER SAMPLE

Due to the positive results obtained for structural and numerical CAs at low doses, we were led to study the response in terms of the number of induced micronuclei after treatment with the second air sample taken during the summer.

Ames Test

The summer sample was characterized by lower activity in the Salmonella test as shown in Tab. 4. In effect, only three tests over 16 were

Tab. 4.   Mutagenicity of the summer sample of airborne particle: Total
extract and fractions using the Ames test

Sample	Strain	S-9 mix	b Value (I)	t Value (II)	Estimated doubling dose (m^3Eq)
Total					
	TA98	−	0.09	$3.23^{n.s.}$	−
	TA98	+	0.36	6.38^{b}	92
	TA100	−	0.09	$1.88^{n.s.}$	−
	TA100	+	0.53	$6.51^{n.s.}$	410
Acid					
	TA98	−	0.10	$2.05^{n.s.}$	−
	TA98	+	0.25	$3.36^{n.s.}$	−
	TA100	−	0.09	$0.42^{n.s.}$	−
	TA100	+	0.84	6.61^{b}	266
Basic					
	TA98	−	− 0.01	$- 1.32^{n.s.}$	−
	TA98	+	− 0.08	$- 0.37^{n.s.}$	−
	TA100	−	0.09	$1.57^{n.s.}$	−
	TA100	+	0.63	$2.71^{n.s.}$	−
Neutral					
	TA98	−	0.01	$0.02^{n.s.}$	−
	TA98	+	0.13	$2.76^{n.s.}$	−
	TA100	−	0.05	$0.11^{n.s.}$	−
	TA100	+	0.45	$1.28^{n.s.}$	−
Positive controls					
	TA98	+	2AF, 5 µg/plate	:	2959 ± 50 rev./plate
	TA100	+	"	"	2021 ± 51 " "

(I): b represents the slope coefficient of the regression; (II): t, student value. a = p<0.05; b = p<0.025; n.s. = not significant.

significant. The doubling doses were calculated for the significant
responses, and it was about one to two orders of magnitude lower than
that calculated for the first extract. The total extract was significantly
positive on TA98 and TA100 in the presence of S-9 mix only, whereas the
acid fraction was effective, plus S-9 mix, on TA100 only. The Ames test
also reveals the presence of indirect mutagens in this weak summer sample.

Sister Chromatid Exchanges and Chromosome Aberrations

As a first approach, we tested total extract and acid fraction because of its positive result on the Ames test for 3 hr with and without

Tab. 5.   Sister chromatid exchange induction by total extract and acid fraction.   (Summer   sample,   airborne   particle,   3-hour   treatment)

Extract	Dose m^3Eq	SCE/Cell: Mean ± S.D.	
		−S-9 mix	+S-9 mix
Total			
	0	5.64 ± 2.19	6.84 ± 2.86
	10.5	7.28 ± 2.26[a]	6.72 ± 2.37[n.s.]
	21.0	8.40 ± 2.50[b]	8.72 ± 4.56[n.s.]
	42.0	7.32 ± 3.10[a]	7.48 ± 2.77[n.s.]
	84.0	9.64 ± 4.53[a]	10.92 ± 3.39[b]
Acid			
	0	7.80 ± 3.09	8.20 ± 2.53
	10.5	8.92 ± 2.62[n.s.]	8.04 ± 3.54[n.s.]
	21.0	n.d.	7.76 ± 2.40[n.s.]
	42.0	8.44 ± 2.00[n.s.]	10.04 ± 2.80[a]
	84.0	11.76 ± 5.15[a]	11.67 ± 4.37[a]
Cyclophosphamide			
	25 µg/ml	−	18.22 ± 6.89[b]

a = p<0.05; b = p<0.01; n.d. = not determined; n.s. = not significant.

metabolic activation.   For the total extract, in absence of metabolic activation, almost all doses gave a significant increase of the revertants over the spontaneous rate; however, with S-9 mix, only the last dose was significant.   The lowest effective dose (p<0.05) is corresponding to 10 m^3Eq of air without S-9 mix.   The acid fraction appears to be less active being the lowest effective dose of 42 m^3Eq of air.   As for the total extract, the presence of S-9 mix doesn't seem to modify the induction of SCEs, while on the positive control (CPA), it is active.   The effects of continuous treatments (72 hr) on SCE induction by the total extract and the three fractions are reported in Tab. 6.   Total extract was effective since a dose corresponding to 21 m^3Eq with a dose-response relationship until the highest doses where the curve flattens and toxicity is observed. The acid fraction is effective since the lowest dose assayed (10 m^3Eq) reaches a maximum at the dose of 84 m^3Eq.   The basic fraction shows a regular induction of SCEs reaching approximately the same highest values observed for the two previous samples.   The neutral fraction appears to be the least effective in SCE induction, and the most toxic for cell division.

The influence of the S-9 mix on the induction of CAs and genome aberrations by the total extract and the acid fraction is illustrated in

Tab. 6.   Sister chromatid exchange induction by total extract and fraction.   (Summer sample, airborne particle; continuous treatment
for 72 hours)

Dose m^3Eq	SCE/cell : mean ± S.D.	
	TOTAL EXTRACT	ACID FRACTION
0.0	5.64 ± 1.75	5.64 ± 1.75
10.5	$6.96 \pm 2.44^{n.s.}$	11.20 ± 3.31^b
21.0	9.64 ± 3.20^b	10.56 ± 4.16^b
42.0	10.40 ± 2.90^b	10.48 ± 3.20^b
84.0	12.24 ± 4.54^b	14.64 ± 6.10^b
126.0	12.64 ± 4.18^b	12.08 ± 3.36^b
168.0	Toxic	11.86 ± 5.01^b
	BASIC FRACTION	NEUTRAL FRACTION
0.0	5.66 ± 1.80	5.66 ± 1.80
10.0	$7.60 \pm 2.91^{n.s.}$	$5.60 \pm 1.84^{n.s.}$
21.0	n.d.	$5.64 \pm 1.82^{n.s.}$
42.0	12.44 ± 3.93^b	9.36 ± 4.07^a
84.0	12.92 ± 5.19^b	9.00 ± 2.35^b
126.0	13.08 ± 3.82^b	Toxic
168.0	Toxic	Toxic

a = $p<0.05$; b = $p<0.01$; n.d. = not determined; n.s. = not significant.

Tab. 7.   Genotoxicity of the summer sample.   (Three-hour treatment)

Extract	Chromosomal Aberrations Percent of cells (No./Total)		Genome Aberrations Percent of cells (No./Total)	
	$-$S-9 mix	$+$S-9 mix	$-$S-9 mix	$+$S-9 mix
Control	0.27 (2/739)	0.27 (2/739)	0.27 (2/739)	0.27 (2/739)
Total	2.90^d (11/379)	2.00^c (8/400)	1.58^b (6/379)	2.50^d (10/400)
Acid	1.90^c (10/526)	2.00^c (12/600)	$0.95^{n.s.}$ (5/526)	$0.50^{n.s.}$ (3/600)

a = $p<0.05$; b = $p<0.025$; c = $p<0.01$; d = $p<0.001$; n.d. = not significant.
Assayed dose: 451 m^3Eq.

Tab. 7. Both samples induced every type of structural CAs, mainly chromatid type, including fragments, breaks, minutes, and many gaps which were recorded, but not herein reported. Dicentrics, exchanges, and rings were also observed. No significant effect was produced by the addition of exogenous metabolic activation and the acid fraction appeared to be as effective as the total extract in inducing CAs. Genomic aberrations, mainly aneuploid cells (2n + 1 or 2 and, rarely, 3) and polyploid cells (4n) were induced by the total fraction only. So, when short treatment times are involved, spindle poisons seem not to be detectable in the acid fraction. Again, the addition of S-9 mix doesn't influence the yield of anomalies observed. In Tab. 8 the induction of CAs and genome aberrations induced by the total extract and the three fractions after continuous treatments for 72 hr is reported. The relatively high proportion of the observed chromosome-type aberration is probably due to such a longer treatment time. All samples had induced substantial increases of CAs, and among fractions the acid one is the most active. Genome aberrations are induced as well, and the basic fraction seems to be the most potent. However, the significance of these differences is approaching the lower limit only ($p < 0.05$).

Micronuclei Assay

In Tab. 9, the results concerning the induction of micronuclei and the fraction (%) of Bin cells for each treatment are reported. Micronuclei are sharply increased by the total extract since the lowest dose assayed

Tab. 8. Genotoxicity of the summer sample. (Seventy-two-hour treatment)

Extract	Chromosomal Aberrations Percent of cells (No./Total)	Genome Aberrations Percent of cells (No./Total)
Control	0.27 (2/739)	0.27 (2/739)
Total	3.67[d] (22/600)	1.50[b] (9/600)
Acid	3.08[d] (18/584)	1.59[b] (9/566)
Basic	1.71[c] (9/526)	2.66[d] (14/526)
Neutral	1.82[c] (8/438)	1.82[c] (8/438)

a = p10<0.05; b =p<0.025; c = p<0.01; d =p<0.001. Assayed dose: 451 m^3Eq.

Tab. 9.   Micronuclei induction by treatment with summer sample extracts.
          (Continuous treatment for 72 hours)

Dose m^{3}Eq.	Percent Bin Cells	Mic Cells	Percent Bin Cells	Mic Cells
	TOTAL EXTRACT		ACID FRACTION	
0.0	13.8	3.0	13.0	2.0^b
10.5	10.7	15.0^c	13.6	9.0^b
21.0	6.7	13.0^c	14.2	17.0^c
42.0	8.2	14.5^c	12.0	12.0^b
84.0	9.0	19.0^c	14.4	9.5^c
126.0	8.0	10.0^c	18.0	15.0^c
168.0	6.6	7.5^b	14.4	12.5^c
	BASIC FRACTION		NEUTRAL FRACTION	
0.0	14.0	4.0	13.5	6.5^c
10.5	18.7	9.5^a	12.0	19.5^c
21.0	16.5	9.0^a	12.0	6.5$^{n.s.}$
42.0	12.9	22.0^c	8.5	11.0$^{n.s.}$
84.0	9.9	18.0^c	8.9	9.0$^{n.s.}$
126.0	3.8	16.5^b	8.0	8.5$^{n.s.}$
168.0	5.0	16.5^b	6.2	4.5$^{n.s.}$
Mitomycin C µg/ml				
0.00	20.0	3.6^c		
0.03	7.6	25.4^c		
0.17	8.3	123.6^d		

a = p<0.05; b = p<0.025; c = p<0.01; d = p<0.001; n.s. = not significant.

(10.5 m^{3}Eq of air), then the curve flattens possibly due to the toxic effects of the mixture as indicated by the reduction of the percentage of
Bin cells. The acid fraction seems to be less toxic and micronuclei frequency rises more gradually, giving a maximum yield at a dose corresponding to 20 m^{3}Eq of air.   The basic fraction is characterized by an
intermediate behavior with regard to the toxicity and by a maximum yield
of inducted micronuclei at the dose corresponding to 42 m^{3}Eq of air.   The
neutral fraction is rather toxic and highly genotoxic at the lowest does
tested.   In general, the fractionation of the total extract determines a
reduction or a different distribution of the toxicity among the single
fractions allowing a different expression of the genotoxic potential.   In
this case the toxicity seems to be mainly relegated to the neutral fraction.

DISCUSSION

The genotoxic activity of organic extracts of airborne particules at the chromosomal level has been reported, using mainly SCEs (2,13-15). In these studies significant increases of SCE induction were reported at doses of total air organic extracts corresponding to as few m³ as 0.6-1.8 m³Eq, depending on the sample potency (12). The induction of CAs has been assessed also, and doses corresponding to 3.7-5 m³Eq of air were shown to induce chromatid and chromosome breaks being toxic at higher doses (10,12). Exogenous metabolic activation systems were not included in these assays. Lockard et al. (15) had shown that the addition of S-9 mix doesn't affect the induction of SCEs and his+ revertants in the Ames test when some Lexington (Kentucky) air samples were assayed. However, in the same paper benzo(a)pyrene (BaP) was reported to induce SCEs in human lymphocytes in the absence of S-9 mix, suggesting that these cells retain some metabolizing capabilities. It is also possible that some metabolic activity was present in the red blood cells, usually included in the lymphocyte assays (21). The present study seems to confirm that the addition of S-9 mix doesn't modify the induction of SCEs when organic materials from air particles are assayed. However, it would be desirable to confirm that result with further fractionated samples. In fact, it is possible that in these large fractions indirect mutagen activities are covered by prevailing, direct ones. Analyses for CAs show that by increasing the toxicity with the dose, only a few metaphases are scorable, thus making weak the performance of the test. However, a variety of induced damages to chromosomes such as breaks, fragments, and particularly exchanges, dicentrics, and rings suggest a high clastogenic potential of the airborne particulate matter extracts. This genotoxic potential seems to elicit other adverse effects such as aneuploidy, polyploidy, and endoreduplications. These endpoints are not directly connected with possible damages of DNA base sequences or structure, but to tubulin functions or to other cellular activity involved in the control of chromosome mitotic segregation. Alteration of these mechanisms by treatment of cells with air sample organics is clearly shown in the present work. Moreover, the rareness of these events in control cultures makes very significant, from a biological point of view, the induction of undramatic increases observed in the treated cultures. It is noteworthy to observe that genomic aberrations appear more frequently at longer sampling times. This is in agreement with the hypothesis of the presence, the organic extracts, of mitotic spindle poisons that, as a consequence of a block or slowdown of the cell cycle, determine a delay of the appearance of genome anomalies. It has been shown that one of the major causes of mutagenicity of urban airborne particulate matter on Salmonella strains (-S-9 mix) are nitro-derivatives and particularly 1,6-dinitropyrene (22). Adams et al. (1) have recently shown that 1,6-dinitropyrene is also clastogenic to human lymphocytes. Baukinger et al. (4) found it caused spindle disturbances and induced micronuclei in Chinese hamster V79 cells. Nitropyrenes are expected to be present in the neutral fraction or in the DCM extract.

In our assays, however, high genotoxic activity has also been found in the basic fraction, which may contain aromatic amines and/or N-heterocyclic compounds, whereas low activity was found in the neutral fraction which is known to contain polycyclic aromatic hydrocarbons (PAHs). This strongly suggests that a variety of clastogens and spindle poisons are present in chemical classes different from PAH. The observation that S-9 mix is not required for the assessment of all the cytogenetic endpoints evaluated could mean that either airborne particulate matter, contains direct-acting clastogens or spindle poisons, or that the whole blood-lymphocyte system is capable of some metabolic activation. The combination of gene, chromosome, and genome mutagenic activity present in airborne particulate matter may make it a more hazardous carcinogen. Consequently, more detailed studies should be conducted to better define the genotoxic potential of several chemical classes present in the organic extracts of airborne particles. In our case, further fractionations and chemical characterizations of the organic extracts are therefore needed for identification of the chemical classes or specific compounds responsible for the observed genotoxicity at the chromosome and genome level.

In this paper, the induction of micronuclei by organic extracts has been demonstrated. The significance of the results obtained is indeed very high. Significant increases of micronuclei have been shown at doses comparable to those effective in inducing SCEs or metaphase chromosomal anomalies using samples moderately or weakly positive with the Ames test. Therefore, the availability of a suitable cytogenetic assay such as the MN test in human lymphocytes that is easy and time-saving, combined with the Ames test, makes possible the performance of more detailed and extended "bioassay-directed chemical analysis" in order to detect the major causes of airborne particulate matter genotoxicity at all possible levels.

ACKNOWLEDGEMENTS

This study was supported by the National Research Council (C.N.R.) and Minister of Education (M.P.I.) of Italy.

REFERENCES

1. Adams, K., A. Lafi, and James M. Parry (1988) The clastogenic activity of 1.6-dinitropyrene in peripheral human lymphocytes. Mutat. Res. 209:135-140.
2. Alink, G.M., H.A. Bmit, J.J. van Houdt, J.R. Kolkman, and J.S.M. Boleij (1983) Mutagenic activity of airborne particles at non-industrial locations. Mutat. Res. 116:21-34.
3. Bandhun, N., and G. Obe (1985) Mutagenicity of methyl 2-benzimidazolecarbamate, diethylstilbestrol and estradiol: Structural chromosomal aberrations, sister-chromatid exchanges, C-mitoses, polyploidies and micronuclei. Mutat. Res. 156:199-218.
4. Bauchinger, M., E. Schmid, F.J. Wiebel, and E. Roscher (1988) 1,6-Dinitropyrene causes spindle disturbances and chromosomal damage in V79 Chinese hamster cells. Mutat. Res. 208:213-218.
5. Clemo, G.R., E.W. Miller, and F.C. Pybus (1955) The carcinogenic action of city smoke. Brit. J. Cancer 9:137-141.
6. Crebelli, R., S. Fuselli, A. Meneguz, G. Aquilina, L. Conti, P. Leopardi, A. Zijno, F. Baris, and A. Carere (1988) In vitro and in vivo mutagenicity studies with airborne particulate extracts. Mutat. Res. 204:565-575.
7. Epstein, S.S., S. Joshi, J. Andrea, N. Mantel, E. Sawicki, T. Stanley, and E.C. Tabor (1966) Carcinogenicity of organic particu-

late pollutants in urban air after administration of trace quantities to neonatal mice. Nature 212:1305-1307.

8. Fenech, M., and A.A. Morly (1986) Cytokinesis-block micronucleus method in human lymphocytes: Effect of in vivo ageing and low dose X-irradiation. Mutat. Res. 161:193-198.

9. Haddle, J.A., M. Hite, B. Kirkhart, K. Mavournin, J.T. MacGregor, G.W. Newell, and M.F. Salamone (1983) The induction of micronuclei as a measure of genotoxicity. A report of the U.S. Environmental Protection Agency Gene-Tox Program. Mutat. Res. 123:61-118.

10. Hadnagy, W., N.H. Seemayer, and R. Tomingas (1986) Cytogenetic effects of airborne particulate matter in human lymphocytes in vitro. Mutat. Res. 175:97-101.

11. Hadnagy, W., and N.H. Seemayer (1988) Cytotoxic and genotoxic effects of extract of particulate emission from a gasoline-powered engine. Env. Molec. Mutagen. 12:385-396.

12. Hadnagy, W., N.H. Seemayer, R. Tomingas, and K. Ivanfy (1989) Comparative study of sister-chromatid exhcanges and chromosomal aberrations induced by airborne particulates from an urban and a highly industrialized location in human lymphocyte cultures. Mutat. Res. 225:27-32.

13. Krishna, G., J. Nath, and T. Ong (1984) Correlative genotoxicity studies of airborne particles in Salmonella typhimurium and cultured human lymphocytes. Env. Mutagen 6:585-592.

14. Krishna, G., J. Nath, L. Soler, and T. Ong (1986) Comparative in vivo and in vitro genotoxicity studies of airborne particle extract in mice. Mutat. Res. 171:157-163.

15. Lockard, J.M., C.J. Viau, C. Lee-Stephens, J.P. wojciechowski, H.G. Enoch, and P.S. Sabharwal (1981) Induction of sister chromatid exchanges and bacterial revertants by organic extracts of airborne particles. Env. Mutagen. 3:671-681.

16. Maron, D.M., and B.N. Ames (1983) Revised methods for the Salmonella mutagenicity test. Mutat. Res. 113:173-215.

17. McCann, J., and B.N. Ames (1976) Detection of carcinogens as mutagens in Salmonella/microsome test: Assy of 300 chemicals: Discussion. Proc. Natl. Acad. Sci., U.S.A. 73:950-954.

18. Migliore, L., R. Barale, D. Belluomini, A.G. Cognetti, and N. Loprieno (1987) Cytogenetic damage induced in human lymphocytes by adriamycin and vincristine: A comparison between micronucleus and chromosomal aberration assays. Toxicol. In Vitro 1:247-254.

19. Migliore, L., M. Nieri, S. Amodio, and N. Loprieno (1989) The human lymphocyte micronucleus assay: A comparison between whole blood and separated lymphocyte cultures. Mutat. Res. (in press).

20. Motykiewitz, G., J. Michalska, J. Szeliga, and B. Cimander (1988) Mutagenic and clastogenic activity of direct-acting component from air pollutants of the Silasian region. Mutat. Res. 204:289-296.

21. Norppa, H., H. Vainio, and M. Sorsa (1983) Metabolic activation of styrene by erythrocytes detected as increased sister chromatid exchanges in cultured human lymphocytes. Cancer. Res. 43:3579-3582.

22. Tokiwa, H., S. Kitamori, R. Nakagawa, K. Horikawa, and L. Matamala (1983) Demonstration of a powerful mutagenic dinitropyrene in airborne particulate matter. Mutat. Res. 121:107-116.

23. Wagida, A.A., and M. Kamal (1988) Cytogenetic effects in a group of traffic policemen in Cairo. Mutat. Res. 208.225-231.

INDOOR AND OUTDOOR SOURCES OF AIRBORNE MUTAGENS

IN NONURBAN AREAS

J.J. van Houdt[1,2] and G.M. Alink[1]

[1]Department of Toxicology

[2]Department of Air Pollution
Agricultural University
6703 HD Wageningen, The Netherlands

INTRODUCTION

Air pollution components are present as gases and as particulate matter. As particle deposition takes place in various parts of the respiratory system, particulate matter may have other toxicological implications than gaseous pollutants, which all may penetrate the lower part of the respiratory tract. In addition, suspended particulate matter represents a group of pollutants of variable physical as well as chemical composition. Therefore, airborne particulate matter cannot be regarded as a single, pure pollutant.

This study deals with the mutagenic potency of airborne particulate matter-bound organics and some of its toxiological implications. The assessment of health hazards is not easily possible without knowledge of the chemical character of the particles. Risk assessment through a toxicological consideration of the individual constituents has serious drawbacks because of the large number of chemicals involved and the complexity of the mixture. Since only 30-40% of the organic compounds on airborne particles have been identified, the contribution of unidentified compounds to the toxicological risk may be significant. Therefore, the assessment of the overall mutagenic or carcinogenic activity in air samples may provide a more realistic basis for the evaluation of the possible risks, than an evaluation on the basis of individual compounds.

The Salmonella/microsome assay has been a major assay used for monitoring the mutagenic potential of complex environmental mixtures, such as airborne particulate matter. Results obtained with this test system may also be useful as a general air pollution parameter, representing particle-bound extractable organics. Many studies indicate that aerosols may have mutagenic properties. Most of these investigations were conducted in ambient air of industrial or urban areas, or near specific sources. Although it is known that indoor concentrations of total suspended particles often exceed outdoor concentrations and that several

Genetic Toxicology of Complex Mixtures
Edited by M. D. Waters *et al.*
Plenum Press, New York, 1990

sources in the home produce mutagens, virtually no data concerning the mutagenicity of indoor particulate matter are available. As in most studies, the number of samples is limited, factors which cause or are related to the variations in mutgenicity of airborne particles are only poorly understood (20). Our institute has performed several studies on the mutagenic activity of airborne particles collected at nonindustrial locations in The Netherlands. Results of a total of about 250 samples of indoor and outdoor air have already been published (14-19).

The overview, which deals with the integration of the results of these studies, may achieve a more complete understanding of the processes and factors which cause or are related to the variation of indoor and outdoor mutagenicity at nonindustrial locations. The Salmonella/microsome assay involves sample collection, extraction, exposure of the test strain, and the quantitative assessment of the revertants. Extraction is carried out routinely with organic solvents, while liver homogenates are generally used to identify mutagens which require metabolic activation. In the first part of this overview, methanol and liver homogenates from Aroclor 1254-pretreated Wistar rats were used to study the occurrence of pariticle-bound mutagens collected indoors and outdoors. In the second part, physiological fluids and lung homogenates, which are more representative of the environmental particles encountered in lungs, are used as solvent and metabolizing systems, respectively, in the Salmonella/microsome assay. Finally, cytotoxicity of airborne particulate matter was determined in order to study some toxicological implications of inhalation of particle-bound organic compounds. In the second part, three samples, typical for different air pollution exposure conditions as demonstrated in the first part, were collected and tested.

MATERIALS AND METHODS

Preparation of Aerosol Extracts

Sampling of particulate matter outdoors was performed at two locations 158 km apart. The station at Terschelling, an island with no major industrial activities, is also used for background measurements of air pollution in The Netherlands. The other sampling location was located in Wageningen, a small rural town (approximately 30,000 inhabitants) without industrial sources of pollution. Both measuring sites are distant from automobile traffic. Indoor samples were collected in living rooms and kitchens from homes in Wageningen. The sampling technique was described previously (16). Extraction took place by Soxhlet extraction and sonication with methanol, newborn calf serum, and lung lavage fluid from pigs. The extracts were evaporated to dryness or filtrated and dissolved in DMSO (19).

Mutagenicity Testing

The experiments were performed using the standard plate assay (2), with minor modifications described previously (14). Enzymic systems were prepared from livers and lungs of three-month-old male Wistar rats and six-week-old male Swiss mice. Half of the animals were pretreated with Aroclor 1254 [500 mg/kg; (2)]. Two experiments were done per sample, each in triplicate. Extracts derived from different amounts of air were tested. TA98, the most sensitive tester strain for indoor and outdoor particles, was used in all experiments.

Biochemical Assays

In the liver and lung homogenates and in human lung homogenate, cytochrome P-450 content was determined by its CO-binding spectrum (32). Absorption maixma of the Soret bands were determined by applying first-derivative spectrometry. Protein concentrations were determined (27) with BSA as a standard.

Cytotoxicity Testing

Alveolar macrophages were isolated from the lungs of female Wistar rats (29). Alveolar macrophages were cultured (36) and exposed by a two-hour incubation to different amounts of aerosol extract. Phagocytosis was determined after incubation of the cells for 1.5 hr at 37°C in the presence of approximately 10^7 dead yeast cells, colored by boiling them for 30 min in congo red PBS solution. All experiments were carried out on at least five occasions in duplicate.

RESULTS, PART I: MUTAGENIC ACTIVITY OF AIRBORNE PARTICLES IN INDOOR AND OUTDOOR AIR

Outdoor Air

Much evidence exists on the mutagenicity of airborne particles at urban and industrial locations. Also, in extracts of airborne particles collected in the ambient atmosphere of nonindustrial locations, mutagenic activity was detected (1,9). Our studies show that when mutagenicity was studied over a one-year period in Wageningen (Fig. 1), the variation of mutagenicity between different days over the entire period was considerable. Our studies indicate that mutagenicity is significantly higher in the winter period than in summer (p < 0.01). This is probably due to sea-

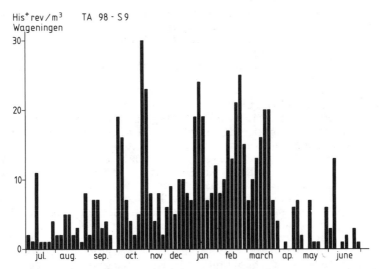

Fig. 1.   Mutagenic activity of airborne particulate matter (rev/m³) collected in Wageningen in <u>Salmonella typhimurium</u> TA98 (-S-9) (taken from Ref. 16).

Tab. 1.   Matrix of correlation coefficients between pollution levels of SO_2, NO_2, NO, CO, and O_3, temperature and mutagenic activity (from Ref. 16)

	rev/m³ +S9	SO_2	NO_2	NO	O_3	CO	Temp.
rev/m³ -S9	0.984	0.733	0.708	0.512	-0.617	0.638	-0.400
rev/m³ +S9		0.701	0.678	0.703	-0.554	0.822	-0.450
SO_2			0.699	0.526	-0.501	0.582	-0.208*
NO_2				0.514	-0.560	0.638	-0.058*
NO					-0.411	0.836	-0.340*
O_3						-0.515	0.538
CO							-0.354*

n = 48   * not significant ($p > 0.05$)

sonal difference in generating sources such as combustion processes for space heating and to meteorological factors such as the height of the boundary layer. This suggestion is supported by the negative correlation of temperature with mutagenic activity, which is presented in Tab. 1. Differences between summer and winter have also been found at other locations (1,8,9,30). Mutagenicity may vary markedly not only within a sampling period, but also between sampling periods. It seems likely that mutagenic activity follows a yearly cycle. However, this cycle is not the same every year: the average direct mutagenicity of samples collected from October to March 1979/1980, 1982/1983, and 1984/1985 was 13, 4, and 51 revertants/m³, respectively.

Table 1 shows that a very strong correlation was found between direct and indirect mutagenicity ($p < 0.01$). The addition of metabolic enzymes gives variable effects. Increases or decreases, if found, are mostly relatively small. Fluctuations from day to day are not found, but between certain periods of time the effect of S-9 mix changes. These effects were neither correlated with season nor with meteorological or air pollution factors. The effect of liver microsomes on the mutagenic potential of airborne particles may be due to the changing composition of the extracts with time. This is supported by data which show a changing action of liver S-9 in the second part of the one-year sampling period (7).

Figure 2 shows the number of direct revertants for samples collected simultaneously at Terschelling and in Wageningen. No differences between mutagenicity of particulate matter in Wageningen and Terschelling were found (Wilcoxon sign rank test, $p > 0.05$). Mutagenic activity at both locations was highly correlated ($r = 0.93$, $n = 27$, $p < 0.01$). On days when there was relatively high mutagenicity the values at Terschelling were generally higher than those in Wageningen. This correlation between Wageningen and Terschelling samples suggests that the mutagenic potential of suspended matter originates not from local source, but rather depends on a large-scale process. It was suggested that mutagenicity at nonindustrial locations is due to long-range transport of mutagenic particulate matter (1). This is supported by the observation that wind direction is of importance for the mutagenic burden (7). The influence of industrial areas in Belgium and Germany was documented by long-range transport studies of air pollution in The Netherlands (11).

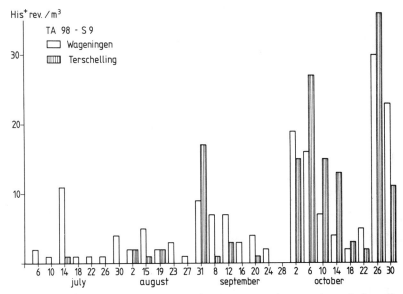

Fig. 2. Mutagenic activity (rev/m³) in Wageningen (rural location) and Terschelling (background location) in Salmonella typhimurium TA98 (-S-9) (taken from Ref. 16).

The origin of air masses can be studied by trajectory studies. In this study, 24-hr trajectories, arriving at 12.00 GMT in De Bilt, were used. Comparing mutagenic activity with those trajectories has limitations. The distance from De Bilt to Wageningen is a minor limitation; comparison of a certain sampling duration with air arriving at 12.00 GMT is a serious one. From Fig. 3 it is clear that actual differences within the summer period may be explained by the passage of the sample air mass over industrial areas, particularly in Belgium and Germany. The same was found in winter. It may be concluded that mutagenic potential of suspended matter in Terschelling and Wageningen originates not from local sources but depends on large-scale processes. This is supported by data which show that the mutagenic potential mainly occurs on the smallest particles which, as a result of their long residence time, may be transported over distances of thousands of kilometers (20).

As long-range transport is important for SO_2 concentrations in The Netherlands (11), the questions therefore arises whether correlations may be found between mutagenicity and commonly registered air pollution parameters such as SO_2, NO_2, NO, NO_x, CO, and O_3. As shown in Tab. 1, all air pollution parameters appear to be strongly correlated with mutagenicity data. However, significant correlations are also found between air pollution parameters. Multiple regression (Tab. 2) shows that the air pollution parameters SO_2, NO_2, NO, CO, and O_3 together account for 70% of the variation in direct mutagenicity and for 80% of the variation in indirect mutagenicity. SO_2 and NO_2 as well as SO_2, NO_2, and CO were significantly associated. For these reasons it can be concluded that, as a general air pollution parameter, monitoring the mutagenic burden of aerosols does not contribute to our knowledge obtained by monitoring SO_2, NO_2, and CO.

Fig. 3. 12.00 GMT trajectories (24 hr, 1,000 mb) and corresponding
         mutagenicity of airborne particles (rev.m³) in Salmonella typhi-
         murium TA98 (-S-9) on days in the summer of 1980. Industrial
         areas are indicated (taken from Ref. 16).

Tab. 2. Multiple regression matrix of dependent variables, direct and
         indirect mutagenicity, and independent variables, temperature
         and pollution levels of SO_2, NO_2, NO, CO, and O_3 (taken from
         Ref. 16)

Dependent variable	rev/m^3 -S9		Dependent variable	rev/m^3 +S9	
	β	p		β	p
Temp.	-0.247	0.043	Temp.	-0.262	0.011
NO_2	0.350	0.028	NO_2	0.300	0.079
SO_2	0.362	0.008	SO_2	0.250	0.024
NO_2	-0.124	0.444	NO_2	-0.009	0.949
O_3	-0.060	0.654	O_3	0.090	0.419
CO	0.189	0.304	CO	0.491	0.002

n = 48 R^2 = 0.706   n = 48 R^2 = 0.797

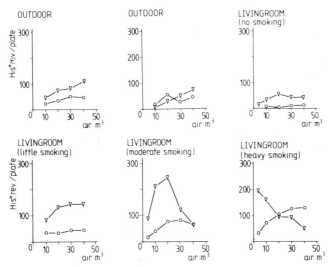

Fig. 4.   Mutagenic activity of extracts of airborne particulate matter, collected simultaneously in different types of living rooms, compared to outdoor mutagenicity.   During collection of indoor samples, respectively 0, 29, 90, and 160 cigarettes were smoked.   Salmonella typhimurium TA98 with (- ▽ -) and without (-o-) liver S-9 (taken from Ref. 14).

Indoor Air

Comparison of the mutagenic activity of particulate matter from indoor and outdoor samples indicates that the direct mutagenic activity of indoor suspended particulate matter in homes is lower than the direct mutgenic activity of outdoor suspended particulate matter sampled at the same time.   Only when samples were taken in living rooms of heavy smokers did the direct indoor mutagenicity exceed the outdoor mutagenicity. It was found that the indirect mutagenic activity is generally larger in indoor samples than outdoor samples.   Moreover, in indoor extracts cytotoxic effects are more pronounced.   From our studies it is obvious that cigarette smoking is the predominant source of airborne genotoxicity in homes.   Wood combustion appeared to be a second important factor producing genotoxic compounds as sometimes a two- to three-fold increase of mutagenic activity is found.   Volatilization of cooking products represents a less important source of mutagens.

Effects of smoking.   In Fig. 4, an example of six samples, collected simultaneously, is presented.   Indirect mutagenicity of indoor suspended-particulate matter was equal to or higher than indirect mutagenicity of outdoor suspended-particulate matter; indirect mutagenicity was high primarily in smokers' living rooms.   This is in agreement with the observation that smoking causes an increase of the mutagenic response of suspended-particulate matter inside an office building (26).   In addition, cytotoxic effects were observed in smokers' living rooms.   This is in agreement with observations of cytotoxicity in cigarette smoke condensate (22).

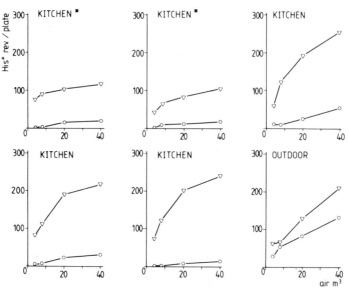

Fig. 5.   Mutagenicity in kitchens of nonsmoking families and the corresponding outdoor sample in <u>Salmonella typhimurium</u> TA98 with
(-▽-) and without S-9 (-o-). *Kitchen with ventilation hood
(taken from Ref. 14).

<u>Effect of cooking</u>. From Fig. 5, a very weak direct activity is ob-
served in air collected in kitchens, with an increase in the number of
revertants after addition of liver S-9. It is not likely that the indirect
mutagenic activity in kitchens is produced by outdoor sources, because in
that case direct indoor activity should also be stronger, and because some
kitchen samples showed a much higher indirect mutagenic activity than
outdoors. It is known that extracts of volatile components produced during cooking show mutagenic activity. A considerable amount of these
components can be produced during cooking (35).

<u>Effect of wood combustion</u>. In homes in which wood combustion took
place, mutagenic activity, adjusted for outdoor effects, is presented in
Fig. 6. The figure shows that the use of wood stoves gives an increase
of airborne mutagenicity in eight of 12 homes (p > 0.05), while using
open fireplaces gives an increase in all homes (p < 0.01). Although the
use of wood stoves (8 hr/da) is much more frequent than open fireplaces
(1.5 hr/da), it is very remarkable that the increase of airborne mutagenicity during wood combustion is only significant when open fireplaces are
used.

<u>Effect of outdoor sources</u>. One of the sources of indoor mutagens
may be penetration of outdoor particles. In the winter 1982/1983, no
contribution of outdoor sources to mutagenic activity indoors was observed. In the winter of 1984/1985, the extremely high levels of outdoor
mutagenic activity were strongly correlated (p < 0.01) with indoor activity. However, in agreement with earlier studies (14), differences in composition of the extracts manifest themselves by a lower ratio -S-9/+S-9 (p
< 0.01) and mild cytotoxic effects indoors. This pehnomenon justifies the

conclusion that penetration of outdoor particles may be one, but certainly not the only source of airborne mutagenic activity indoors. Mostly the contribution of outdoor sources to indoor mutagenicity is only small.

RESULTS, PART II: BIOLOGICAL FATE OF
PARTICLE-BOUND ORGANIC COMPOUNDS

Exposure is a function of concentration and time. In The Netherlands, people spend most of the time at home. As all indoor samples show a strong indirect mutagenic activity, it may be concluded that exposure to genotoxins will be determined to a large extent by the level of pollution inside homes, which implies that exposure to indirect-acting mutagens is quantitatively of far greater concern than exposure to direct-acting mutagens (20).

The relation between the concentration and mutagenicity of airborne particulate matter in breathing air, detected by <u>Salmonella typhimurium</u> TA98 to the resulting toxic doses and potential hazards, depends greatly on the patterns of deposition, the rates and pathways of clearance from the deposition site, solubility in physiological fluids, and biotransformation (20). Respirable airborne particles to which most potential mutagenic compounds detected in the Ames assay, are adsorbed may deposit in various parts of the respiratory tract. One of the defense mechanisms with regard to possible harmful action of these particles is clearance, as it reduces residence time on potentially sensitive epithelial surfaces. Alveolar macrophages provide the initial defense against particulate matter of the lower respiratory tract (20).

The Toxicity of Alveolar Macrophages

The toxicity of indoor and outdoor air samples to rat alveolar macrophages as determiend by studying the phagocytic activity of thes cells in vitro is presented in Fig. 7. Clean air samples did not affect phagocyto-

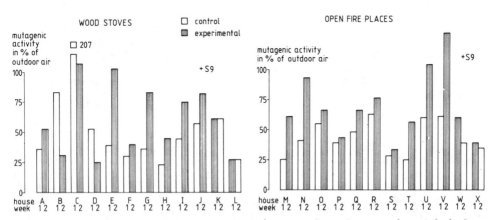

Fig. 6.   Mutagenic activity (+S-9) of control and experimental indoor samples in homes with wood stoves and open fireplaces, expressed as percentage of outdoor mutagenicity (taken from Ref. 15).

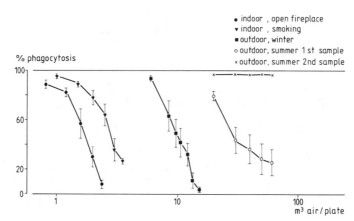

Fig. 7.   Toxicity of extracts of indoor and outdoor airborne particles to
           rat alveolar macrophages in vitro. Results are expressed as
           the percentage of cells that contained yeast per count of 5 x
           100 (mean and standard error) (taken from Ref. 17).

sis at concentrations up to 60 m³/plate, but polluted outdoor air samples
caused a dose-dependent inhibition of the phagocytosis at concentrations
varying from 6 to 60 m³/plate. Indoor air samples, especially when polluted with wood smoke, affected phagocytosis at concentrations below 2
m³/plate. Although more pronounced, these results are in agreement with
toxicity data obtained by mutagenicity testing (15,16).

     Damage to alveolar macrophages by inhaled material may cause a re-
lease of particles and macrophage breakdown products. As a result
clearance of particles may be decreased, which facilitates the transepi-
thelial passage of toxic material to the interstitium (3,21). Conversely,
the recruitment of an unusually large number of cells to the free alveolar
surface, due to the release of macrophage breakdown products, can pro-
mote lung clearance. However, this accumulation of cells may inhibit
drainage, causing dust to be retained (21). It was reported that cigarette smokers had an impaired responsiveness to macrophage migration
(28). An additional risk may arise from biochemical activity by alveolar
macrophages, which may be induced by environmental pollutants (12):
macrophages may metabolize premutagens on airborne particles (5,13),
leading to the release of direct-acting mutagenic metabolites in the
alveolar spaces.

Solubility in Physiological Fluids

     So far studies on the solubility of airborne particles in serum
collected indoors and outdoors in rural areas have not been performed.
Investigations of interactions between mutagens and physiological fluids
were carried out for diesel emissions, coal fly ash, and airborne particulate matter collected in urban areas. For diesel particulate emissions,
serum proved to be less efficient in removing mutagens than organic solvents (4,23). This was also reported for particles collected in urban air
(33). For indirect activity, the same phenomenon was reported. However, in this study, direct activity of serum and DMSO extracts was
about the same (39). Regarding serum extractability of coal fly ash,

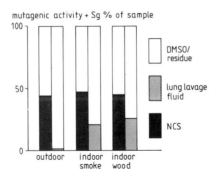

Fig. 8. Distribution in percentages of mutagenic activity of airborne particulate matter in various extracts. Salmonella typhimurium TA98 with S-9 (taken from Ref. 19).

both a 25 times greater mutagenic activity than cyclohexene extracts (6) and a strong decrease related to DMSO solubility were observed (24).

Our results (Fig. 8) show that serum extracts of indoor and outdoor air have a lower mutagenic activity than methanol extracts. Indications were found that serum is more effective in removing direct-acting mutagens from the tar than indirect-acting mutagens. Mutagenic compounds from indoor air are not only soluble in newborn calf serum, lung lavage fluid, although less efficient, is also able to remove mutagens from the tar. In contrast to serum, in lung lavage fluid the percentage of dissolved indirect-acting mutagens exceeds the percentage of direct mutagenicity. Indirect-acting mutagens from outdoor air are almost completely insoluble in lung lavage fluid. The same phenomenon was found for diesel particles (23).

From these experiments it may be concluded that physiological fluids are not as efficient as organic solvents in removing mutagens from particles. Lung lavage fluid, containing a high concentration of phosphatidylcholine (89 mg/ml), is only able to remove about 10% of the total mutagenic burden from airborne particles. Although from our results it is clear that physiological fluids, especially when obtained from lung lavage, are less efficient in removing mutagens than organic solvents, the suggestion seems to be justified that a certain elution of environmental chemicals into body fluids takes place.

Metabolic Activation

Drug-metabolizing enzyme systems may also be seen as a defense mechanism towards chemicals invading the body. These enzyme systems enable the organism not only to convert lipid soluble harmful drugs into harmless water soluble metabolites but also more toxic or mutagenic metabolites may be formed.

In Tab. 3, cytochrome P-450 content in pmol/mg protein is presented for tissue homogenates of rat and mouse and for human lung homogenates. In uninduced liver and lung homogenates, P-450 content is in the same order of magnitude. In contrast to lung homogenates, P-450 content in liver homogenates (38) is strongly enhanced by Aroclor induction. P-450

Tab. 3.   Cytochrome P-450 content (in pmol/mg protein and pmol/ml S-9)
          of various lung and liver tissue homogenates (taken from Ref.
          18)

	P450 pmol/mg protein		P450/ml S9	
	Uninduced	Induced	Uninduced	Induced
Rat liver	159	1300	410	10300
Rat lung	27	14	150	110
Mouse liver	1112	1800	9340	10915
Mouse lung	65	504	370	2360
Human lung	15-43		80-120	

content in rat lung and its decrease after Aroclor induction are in agreement with other data (27,37,40).   The increase in P-450 content in mouse
lung found after Aroclor induction was, although less pronounced, previously found (10,31).   In human lungs, P-450 content varied between 15
and 43 pmol/mg protein, one to five times more than data reported in the
literature (34).   In Fig. 9, the mutagenicity of three air samples was

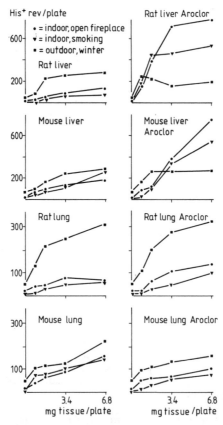

Fig. 9.   Mutagenic activity of extracts of airborne particulate matter in
          Salmonella typhimurium TA98 with different amounts of unin-
          duced and Aroclor-induced liver and lung homogenates of rat
          (Wistar) and mouse (Swiss) (taken from Ref. 18).

assessed by the Salmonella/microsome assay, using different lung and liver homogenates. It is clear that Aroclor-induced liver homogenates gave the strongest metabolic activation in contrast to other homogenates which generally gave lower effects. Aroclor induction produces no increase in activation capacity of the lung homogenates. Despite differences in P-450 content, comparison of the uninduced lung and liver homogenates shows that only minor differences in metabolic activation are found. In contrast to mouse homogenates, a rather strong activation of outdoor air was observed after adding rat homogenates. However, a positive relation ($p < 0.05$) between amount of tissue and number of revertants was found for all homogenates.

Our results show that in addition to liver, lung homogenates of rat (Wistar) and mouse (Swiss) are also able to activate extracts of airborne particulate matter in a comparable way. In certain parts of the lungs, especially in the terminal bronchioles in which Clara cells are located, a rather high metabolic activation may take place. Therefore, these results suggest that the respiratory system may be an important site for in vivo bioactivation of respirable particles.

REFERENCES

1. Alfheim, I., and M. Möller (1979) Mutagenicity of long range transported atmospheric aerosols. Sci. Total Env. 13:275-278.
2. Ames, B.N., J. McCann, and E. Yamasaki (1975) Methods for detecting carcinogens and mutagens with the Salmonella/mammalian microsome mutagenicity test. Mutat. Res. 31:347-363.
3. Bowden, D.H. (1984) The alveolar macrophage. Env. Health Persp. 55:327-342.
4. Brooks, A.L., R.K. Wolff, R.E. Royer, C.R. Clark, A. Sanches, and R.O. McClellan (1979) Biological availability of mutagenic chemicals associated with diesel exhaust particles. In EPA International Symposium on Health Effects of Diesel Engine Emissions, Cincinnati, Ohio.
5. Centrell, E.T., G.A. Warr, D.L. Busbee, and R.R. Martin (1973) Induction of aryl hydrocarbon hydroxylase in human pulmonary alveolar macrophages by cigarette smoking. J. Air Invest. 52:1884-1991.
6. Chrisp, C.E., G.L. Fisher, and J.E. Lammert (1978) Mutagenicities of filtrates from respirable coal fly ash. Science 199:73-75.
7. Commoner, B., P. Madayastha, A. Bronsdon, and A.J. Vithayathil (1978) Environmental mutagens in urban air particulates. J. Tox. Env. Health 4:59-77.
8. Daisey, J.M., T.J. Kneip, I. Hawryluk, and F. Mukai (1980) Seasonal variations in the bacterial mutagenicity of airborne particulate organic matter in New York City. Env. Sci. Tech. 14(12):1487-1490.
9. Dehnen, W., R. Tomingas, and R. Pitz (1981) Studies on the mutagenic effect of extracts derived from airborne particulates within one year of differently burdened areas using the Ames test. Zbl. Bact. Hyg. I. Abt. Orig. B. 172:351-366.
10. Digiovanni, J., D.L. Berry, T.J. Slaga, A.H. Jones, and M.R. Juchau (1970) Effects of pre-treatment with 2,3,7,8-tetrachlorodibenzo-p-dioxin on the capacity of hepatic and extrahepatic mouse tissues to convert procarcinogens to mutagens for Salmonella typhimurium auxotrophs. Tox. Appl. Pharm. 50:229-239.

11. Egmond, N.D. van, O. Tissing, D. Onder de Linden, and C. Bartels (1978) Quantitative evaluation of mesoscale air pollution transport. Atmos. Env. 12:2279-2287.
12. Gardner, D.E. (1984) Alterations in macrophage functions by environmental chemicals Env. Health Persp. 55:343-358.
13. Harris, C.C., I.C. Hsu, G.D. Stoner, B.T. Trump, and J.K. Selkirk (1978) Human pulmonary alveolar macrophages metabolise benzo(a)pyrene to proximate and ultimate mutagens. Nature 272:633-634.
14. Houdt, J.J. van, W.M.F. Jongen, J.S.M. Boley, and G.M. Alink (1984) Mutagenic activity of airborne particles inside and outside homes. Env. Muta. 6:861-869.
15. Houdt, J.J. van, C.M.J. Daenen, J.S.M. Boley, and G.M. Alink (1986) Contribution of wood stoves and fire places to mutagenic activity of airborne particulate matter inside homes. Mutat. Res. 171:91-98.
16. Houdt, J.J. van, G.M. Alink, and J.S.M. Boley (1987) Mutagenicity of airborne particles related to meteorological and air pollution parameters. Sci. Total Env. 61:23-36.
17. Houdt, J.J. van, and I.M.C.M. Rietjens (1988) Toxicity of airborne particulate matter to rat alveolar macrophages: A comparative study of five extracts, collected indoors and outdoors. Tox. In Vitro 2(2):121-123.
18. Houdt, J.J. van, P.W.H.G. Coenen, G.M. Alink, J.S.M. Boley, and J.H. Koeman (1988) Organ specific metabolic activation of five extracts of indoor and outdoor particulate matter. Arch. Tox. 61:213-217.
19. Houdt, J.J. van, L.H.J. de Haan, and G.M. Alink (1989) The release of mutagens from airborne particles in the presence of physiological fluids. Mutat. Res. 222:155-160.
20. Houdt, J.J. van (1989) Mutagenic activity of airborne particulate matter in indoor and outdoor environments--A review. Atmos. Env. (in press).
21. Katnelson, B.A., and L.I. Privalova (1984) Recruitment of phagocytizing cells into the respiratory tract as a response to the cytotoxic action of deposited particles. Env. Health Persp. 55:313-326.
22. Kier, L.D., E. Yamasaki, and B.N. Ames (1974) Detection of mutagenic activity in cigarette smoke condensate. Proc. Natl. Acad. Sci., USA 714010:4159-4163.
23. King, L.C., M.J. Kohan, A.C. Austin, L.D. Claxton, and J.L. Huisingh (1981) Evaluation of the release of mutgens from diesel particles in the presence of physiological fluids. Env. Muta. 3:109-121.
24. Kubitschek, H.E., and L. Venta (1979) Mutagenicity of coal fly ash from electric power plant precipitators. Env. Muta. 1:79-82.
25. Liem, H.H., U. Muller-Eberhard, and E.F. Johnson (1980) Differential induction by 2,3,7,8-tetrachlorodibenzo-p-dioxin of multiple forms of rabbit microsomal cytochrome P-450: Evidence for tissue specificity. Molec. Pharm. 18(3):565-570.
26. Lörfroth, G., L. Nilsson, and I. Alfheim (1983) Passive smoking and urban air pollution: Salmonella/microsome mutagenicity assay of simultaneously collected indoor and outdoor particulate matter. In Short-term Bioassays in the Analysis of Complex Environmental Mixtures III, M.D. Waters et al., eds. Plenum Publishing Corp., New York, pp. 515-535.
27. Lowry, O.H., M.J. Rosebrough, A.L. Farr, and R.J. Randall (1951) Protein measurement with the Folin Pehnol reagent. J. Biol. Chem. 193:265.

28. Martin, R.R., and A.H. Laughter (1976) Pulmonary alveolar macrophages can mediate immune responses: Cigarette smoking impairs these functions. Fed. Proc. 35:716.
29. Mason, R.J., M.C. Williams, R.D. Greenleaf, and J.A. Clements (1977) Isolation and properties of type II alveolar cells from rat lung. Am. Rev. Resp. Dis. 115:1015-1026.
30. Möller, M., and I. Alfheim (1980) Mutagenicity and PAH analysis of airborne particulate matter. Atmos. Env. 14:83-88.
31. Mori, Y., H. Yamazaki, K. Toyoshi, H. Maruyama, and Y. Konoshi (1986) Activation of carcinogenic N-nitrosopropylamines to mutagens by lung and pancreas S9 fractions from various animal species and man. Mutat. Res. 160:159-169.
32. Omura, T., and R. Sato (1964) The carbon monoxide-binding pigment of liver microsomes, II. Solubilization, purification and properties. J. Biol. Chem. 239:2379.
33. Oshawa, M., T. Ochi, and H. Hayashi (1983) Mutagenicity in Salmonella typhimurium mutants of serum extracts from airborne particulates. Mutat. Res. 116:83-90.
34. Prough, R.A., V.W. Patrizi, R.T. Okita, B.S.S. Masters, and S.W. Jacobsson (1979) Characteristics of benzo(a)pyrene metabolism by kidney, liver and lung microsomal fractions from rodents and humans. Cancer Res. 39:1199-1206.
35. Rappaport, S.M., M.C. McCartney, and E.T. Wei (1979) Volatilization of mutagens from beef during cooking. Cancer Lett. 8:139-145.
36. Rietjens, I.M.C.M., L. van Bree, M. Marra, M.C.M. Poelen, P.J.A. Rombout, and G.M. Alink (1985) Glutathion pathway enzyme activities and the ozone sensitivity of lung cell populations derived from ozone exposed rats. Toxicology 37:205-214.
37. Serabjit-Sing, C.J., Ph.W. Albro, I.C.G. Robertson, and R.M. Philpot (1983) Interactions between xenobiotics that increase or decrease the levels of cytochrome P-450 isoenzymes in rabbit lung and liver. J. Biol. Chem. 258:12827-12834.
38. Souhaily-El Amri, H., A.M. Batt, and G. Siest (1986) Comparison of cytochrome P-450 content and activities in liver microsomes of seven animal species, including man. Xenobiotica 16:351-358.
39. Takeda, N., K. Teranishi, and K. Hamada (1983) Mutagenicity in Salmonella typhimurium TA98 of the serum extract of the organic matter derived from airborne particulates. Mutat. Res. 117:41-46.
40. Ueng, T.H., and A.P. Alvares (1985) Selective induction and inhibition of liver and lung cytochrome P-450 dependent monooxigenase by the PCB's mixture, Aroclor 1016. Toxicology 35:83-94.

CHARACTERIZATION OF MUTAGEN SOURCES IN

URBAN PARTICULATE MATTER IN GOTHENBURG, SWEDEN

Göran Löfroth

Nordic School of Public Health
S-402 42 Gothenburg, Sweden

INTRODUCTION

The modern interest in urban air pollution emerged in the early 1950s. Kotin and coworkers (14) were the first to collect samples of airborne particulate matter and use extracts in an animal tumorigenicity assay. These studies seem partly due to a hypothesis that the urban air pollution was a major factor causing the increased lung cancer incidence observed at that time, although data were emerging showing that lung cancer is due largely to tobacco smoking (31). Studies on the carcinogenicity of urban air pollutants continued at a slow pace through the 1960s and early 1970s (9,10). An enhanced interest in the problem started in the latter part of the 1970s when more sophisticated analytical methods became available and the Ames Salmonella mutagenicity assay was shown to be useful for urban air samples (8,20,24,27).

In 1979, the National Swedish Environmental Protection Board initiated a project area "Air pollution in urban areas" and funded a number of separate projects with this framework. The projects cover several aspects, including motor vehicle exhaust, ambient air, toxicology, and risk assessment, but are generally concerned with genotoxicity and genotoxic components.

The initial ambient air studies comprised sample collection, chemical analysis, and short-term tests and have been published (e.g., Ref. 5,17, 23,26). The present report summarizes some of the continued studies undertaken since 1984 which are more fully described elsewhere (1).

MATERIALS AND METHODS

Sampling Sites

The sample collection was performed at several locations within the Gothenburg area. The locations were selected and designated according

Genetic Toxicology of Complex Mixtures
Edited by M. D. Waters *et al.*
Plenum Press, New York, 1990

to the direct impact from traffic. All samplings were made simultaneously at two locations within the series mentioned below:

E   A city rooftop (R) location and a nearby street (S) location on a street with a traffic flow of about 16,000 vehicles per day.

C   A city center (+) street location and a suburban (-) street location, both having a low traffic flow.

D   Street locations with high (+) and low (-) proportion of heavy duty diesel vehicles but with a high traffic flow of >20,000 vehicles per day.

T   Street locations with high (+) and low (-) traffic flow.

B   An open bus terminal (+) and a backyard (-) between the neighboring apartment houses.

Sample Collection

Total suspended particular matter (TSP) was collected by the Swedish Environment Research Institute with high-volume sampling on a glass fiber filter. The sampling periods varied from one to four 24-hour days with two days being the most common period. Filters were always changed each 24 hr. With the exception of the rooftop sampling, all samples were collected at a height of 2 m at the side of the street. For all series, except the E-series, samples were also collected on membrane filters for analysis of elemental composition by X-ray fluorescence and ionic species by ion chromatography; simple aromatic hydrocarbons were also determined with collection on charcoal tubes and gas chromatographic analyses.

Sulfur dioxide, nitrogen oxides, and ozone as well as some meteorological parameters were available from a main station for pollution measurements in the city center (identical to the rooftop location in the E-series).

Extraction and Analysis

The filter-collected particulate material was Soxhlet-extracted with acetone (16) and the extract was subdivided into aliquots for bioassays and chemical analysis. Preselected samples were fractionated on silica gel with a method described by Alsberg et al. (3).

The mutagenicity in Ames Salmonella tester strains was determined with the original plate incorporation method (19) and with a modification (18) of the microsuspension method described by Kado et al. (12).

Cell toxicity was assayed with ascites sarcoma BP8 cells as described elsewhere (6). Binding to the TCDD-receptor was determined with the method described by Poellinger et al. (21).

The determinations of polycyclic aromatic hydrocarbons (PAH) and related compounds, dibenzothiophene, benzo(b)naphtho(1,2-d)thiophene (BNT), and 1-nitropyrene, were made according to the descriptions given by Alsberg et al. (3) and Westerholm et al. (28). Fourteen PAHs in the range from phenanthrene to coronene were analyzed but the results are here condensed to the sum of these PAHs.

RESULTS

Results from the different series are given in Tab. 1-3. The data are condensed from the original data (1) and here presented as averages and ranges. In addition, only a selection of parameters are given as a majority of the elemental composition and ionic species are omitted and only the sum of 14 PHAs is given instead of individual components.

All parameters were not determined for all samples and series. Cell toxicity was only assayed in the E-series with street level and rooftop sampling. Affinity for the TCDD-receptor was only determined for a selected number of samples within the series. All samples have been assayed for mutagenicity with the plate incorporation method with TA98 with the supplementation of TA100 in the E-series, and with the microsuspension method with TA98 in all the other series. Only a few samples in the E-series were subjected to PAH analysis.

Comparison between Street and Rooftop Sampling

There were few apparent differences between the concentrations at the two locations although TSP and lead were much lower at the rooftop than at the street location. The mutagenicity and cell toxicity were, however, greater for rooftop samples when the response is expressed per particulate weight (Tab. 1).

Binding to the TCDD-receptor ranged from 0.12 to 0.0006 m^3/ml for the 12 samples assayed from six sampling periods. With the exception of one highly active street sample (0.0006 m^3/ml) and the corresponding rooftop sample (0.004 m^3/ml), there were no large differences between street and rooftop. The two parallel samples with a large difference in binding affinity were not accompanied by any significant differences in the PAH concentrations or mutagenic effects.

Comparison between Parallel Street Sampling at Two Locations

The comparison between two locations indicated that there were only relatively small differences between the designated (+)- and (-)- locations (Tab. 2). The least differences were found for mutagenic response, whereas other pollution parameters were more or less lower at the site with the lesser direct impact from traffic.

The B-series, with sample collection at an open bus terminal and a neighboring backyard (Tab. 3), showed higher concentrations at the terminal for all parameters associated with motor vehicle exhaust, including (BNT), a potential indicator for diesel exhaust.

All results from the analysis of the C-, D-, T-, and B-series covering 43 sampling periods and 86 samples have been used for various computational analyses (see "Discussion").

Fractionation

Samples from five preselected sampling periods within the C-, D-, and T-series were fractionated with a silica gel column method into five fractions with increasing polarity. The crude extracts, the fractions and samples reconstituted from the fractions, were assayed for mutagenicity and for binding to the TCDD-receptor.

Tab. 1.   Summary table of the E-series measurements with street (S) and
          rooftop (R) samples. Samplers were collected during 20 periods
          from August 1984 to June 1985

Component	Unit	Street level (S)		Rooftop (R)	
		Average	Range	Average	Range
TSP	$\mu g/m^3$	80	29 - 210	50	34 - 120
Soot	"	-[a]		11	1.5 - 33
Lead	ng/m^3	210	65 - 570	110	20 - 400
Mutagenicity					
TA98 -S9	$rev./m^3$	8.7	1.5 - 20	9.0	0.9 - 24
TA98 +S9	"	13	2.0 - 51	11	1.2 - 41
TA100-S9	"	7.5	1.2 - 21	6.9	0.9 - 19
TA100+S9	"	16	2.7 - 61	14	1.6 - 51
TA98 -S9	$rev./\mu g$[b]	130	16 - 260	180	21 - 550
TA98 +S9	"	200	32 - 790	230	28 - 900
Cell toxicity[c]	m^3/ml	8.7	16 - 2.2	9.9	20 - 2.0
"	$\mu g/ml$[b]	530	1000 - 170	440	770 - 130
Sulfur dioxide	$\mu g/m^3$	-		15	3 - 52
Nitrogen oxides	"	-		34	18 - 53
Ozone	"	-		35	8 - 89
Temperature	^{o}C	-		-	-8 - +23

a) - not determined. b) expressed per ug particulate matter (TSP).

c) cell toxicity is expressed in units which cause a 50 % toxicity and the
   higher the toxicity is the lower is the figure.

The average distribution of the mutagenic response is shown in Fig.
1.   No activity is present in fraction I which would contain aliphatic
compounds.   Some activity is present in fraction II which would contain
the major part of the PAH.   The largest activities are found in fractions
IV and V containing relatively polar components.   This distribution of the
mutagenic response is similar to that found previously for other ambient
particulate samples (17), but is dissimilar to the distribution obtained
with exhaust from gasoline cars (3,17) and from diesel vehicles (32) in
which the contribution from the most polar fractions are less.

Tab. 2.   Summary table of the C-, D-, and T-series measurements.
Samples were obtained during the indicated number of periods
from December 1985 to June 1987

Component	Unit	+ Location Average	Range	- Location Average	Range
C-series, 10 periods					
TSP	$\mu g/m^3$	58	22 - 200	46	16 - 160
Soot	"	17	3.5 - 48	17	2.3 - 48
Lead	ng/m^3	160	12 - 640	140	27 - 400
Mutag. TA98-S9	$rev./m^3$	11	1.3 - 47	9.6	1.0 - 38
" TA98+S9	"	8.4	0.8 - 34	8.7	0.6 - 37
PAH	ng/m^3	9.7	0.6 - 42	9.9	1.2 - 33
1-Nitropyrene	pg/m^3	8.6	<1 - 36	3.8	<1 - 11
Benzene	$\mu g/m^3$	3.4	<0.3 - 7.9	2.9	0.3 - 7.0
D-series, 12 samples					
TSP	$\mu g/m^3$	140	29 - 350	86	33 - 190
Soot	"	59	6.6 - 190	26	7.3 - 43
Lead	ng/m^3	730	49 - 1900	700	110 - 1600
Mutag. TA98-S9	$rev./m^3$	13	1.1 - 35	12	2.0 - 30
" TA98+S9	"	13	1.4 - 39	12	2.0 - 28
PAH	ng/m^3	24	1.0 - 57	17	2.8 - 38
1-Nitropyrene	pg/m^3	20	1 - 68	12	2 - 35
Benzene	$\mu g/m^3$	17	0.7 - 50	14	4.2 - 26
T-series, 16 periods					
TSP	$\mu g/m^3$	74	20 - 280	51	15 - 200
Soot	"	28	7.6 - 70	17	5.4 - 55
Lead	ng/m^3	250	60 - 930	140	21 - 380
Mutag. TA98-S9	$rev./m^3$	8.3	2.1 - 18	6.3	0.5 - 18
" TA98+S9	"	8.5	2.0 - 17	6.2	0.3 - 19
PAH	ng/m^3	18	3.3 - 56	9.5	0.7 - 33
1-Nitropyrene	pg/m^3	14	1 - 42	6.8	<1 - 32
Benzene	$\mu g/m^3$	9.6	1.0 - 30	6.7	0.3 - 24

Tab. 3. Summary table of the B-series measurements. Samples were collected during five periods in March 1987

Component	Unit	Bus terminal (B+)		Background yard (B-)	
		Average	Range	Average	Range
TSP	$\mu g/m^3$	140	40 - 230	66	28 - 120
Soot	"	47	14 - 88	29	2.7 - 59
Lead	ng/m^3	380	200 - 700	170	33 - 300
Mutag. TA98-S9	$rev./m^3$	11	5.2 - 16	7.3	2.9 - 13
" TA98+S9	"	15	4.8 - 22	7.3	2.1 - 15
" TA100-S9	"	17	6.9 - 30	9.1	2.9 - 16
" TA100+S9	"	23	8.3 - 35	13	4.1 - 26
PAH	ng/m^3	38	8.8 - 79	14	2.9 - 25
1-Nitropyrene	pg/m^3	17	6 - 32	12	3 - 24
BNT	ng/m^3	0.59	0.06 - 1.2	0.16	0.06 - 0.27
Benzene	$\mu g/m^3$	12	6.3 - 17	7.3	4.2 - 11

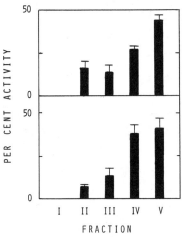

Fig. 1.   The average relative distribution (+SD) of the mutagenic response of ten samples of airborne particulate matter showing that the activity mainly is present in fraction IV and V containing relatively polar compounds. The recovery of the mutagenic activity ranged from 53 to 104% with an average of 85% for the --S-9 condition and from 69 to 99% with an average of 82% for the +S-9 condition.

Binding to the TCDD-receptor ranged from 0.12 to 0.015 m³/ml for unfractionated samples with an average of about 0.04 m³/ml. This binding capacity is of the same magnitude as that found for the E-series samples and similar to those reported earlier for a set of samples collected in Stockholm, Sweden (25). The fractionated samples showed a varied distribution of the binding capacity, although the major part of the activity was found in the PAH-containing fraction II. The recovery of the binding activity varied from 27 to 260% with an average of 120% for six samples from three sampling periods that were fully assayed.

Comparison between Plate Incorporation and Microsuspension Assays

All 86 samples from the C-, D-, T-, and B-series were tested for mutagenicity in both the plate incorporation and microsuspension assay using the same extract from any one of the samples.

The distribution of the responses is illustrated in Fig. 2 with a linear regression slope of about three for the TA98-S-9 condition and about five for the TA98+S-9 condition. The average ratios of the response are given in Tab. 4 together with ratios found for concurrent positive control compounds used in the assays. The differences between the average ratios for all samples and the linear regression slopes are mainly due to the fact that a number of samples with a low response in the plate incorporation assay have a substantially higher enhancement with the mircosuspension method than other samples.

The microsuspension is particularly useful when the amount of sample is limited (13,18), but it should be considered that the increased response in the assay varies between compounds and is, in addition, dependent on the amount of S-9 used in the two assays.

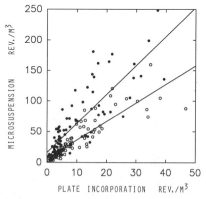

Fig. 2.   The relationship between the mutagenic response with the plate incorporation assay and the microsuspension assay as found for 86 samples of airborne particulate matter. o TA98-S-9, • TA98+5-9.

Tab. 4.   Ratio of the mutagenic response in the microsuspension assay to that in the plate incorporation assay

Sample		Ratio ± S.E.	
		TA98-S9	TA98+S9
Air particles	(n=86)	3.81 ± 0.13	8.18 ± 0.48
Quercetin[a]	(n=33)	4.67 ± 0.34	not tested
Benzo(a)pyrene[a]	(n=33)	-	8.97 ± 0.43

a) The ratio is calculated from the response per unit
   weight obtained with 5 µg quercetin and 0.5 µg
   benzo(a)pyrene in the microsuspension assay and
   25 µg quercetin and 2.5 µg benzo(a)pyrene in the
   plate incorporation assay.

Seasonal Variation of the Mutagenic Response

    Although the samples in the C-, D-, and T-series were collected at different locations, the similarity between simultaneously sampled locations may permit use of the entire set for an inquiry about the seasonal variation.   Figure 3 shows the mutagenic response from 76 samples set against

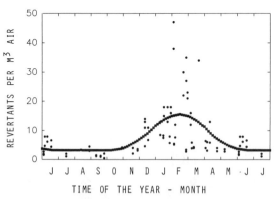

Fig.3.   The annual variation of the mutagenic response of airborne particulate matter collected at several locations in Gothenburg over a period from December 1985 to June 1987. The response per m^3 air in the plate incorporation assay with TA98-S-9 (●) is plotted vs the month of collection and a consecutive moving average has been fitted to the data.

time of sampling. As has been noted in other studies (2,22,30), a peak appears in the colder months. This is probably due to a combination of factors including more combustion, colder vehicle engines, less windy climate, a higher absorption of organic compounds to particles, etc.

DISCUSSION

Short-term Bioassays

The introduction of short-term bioassays as a research tool for studies on airborne particulate matter parallels the increasing interest in the chemical composition and toxicology of air pollutants originating from combustion emissions. Analytical chemistry has made a significant advance in the detection and quantification of nitrosubstituted PAH. It is, however, questionable if a similar advance can be made on other compound classes which remain to be characterized in the still very complex mixture which is present in the urban air. Short-term bioassays, as those used in the present projects or other assays better suited for particular objects, will probably have a part in air pollution studies as long as air pollution problems or interests in air pollution problems exist.

There are some directly observable facts in the present study with respect to the short-term bioassays. Cell toxicity, as assayed in the street and rooftop series, was about the same for both locations. This similarity was also found for the Salmonella mutagenicity and, for a limited number of samples, for the binding to the TCDD-receptor. Although this may seem surprising, it should be considered that the sampling periods comprised entire 24-hr days and that the contribution from a few hours with intense traffic at the street location thus is averaged over a longer time.

A related observation applies to the fact that the mutagenic response is about the same whether the sampling is made in the city or suburb (C), on a street with much or little diesel impact (D), and on a much or less trafficked street (T). The mutagenic components, as well as some other pollutants, are apparently effectively dispersed over a larger area causing similar average concentrations over a longer time period irrespective of the temporary contribution from a particular street. It should be considered that the street locations were typical for the Gothenburg area and that they were not of a canyon-type in which pollutants can be trapped causing a readily detected difference between the street and a neighboring location (2).

Partial Least Square (PLS) Regression

PLS regression, a multivariate data analysis method (29), was used for predicting the mutagenic response from a number of chemical variables (4). The D-series with 24 objects was used as a training set and the remaining 52 objects from the C- and T-series were used as the test set. The best regression was obtained by using the simple aromatic hydrocarbons, individual PAH, dibenzothiophene, BNT, and lead as independent variables giving the result presented in Fig. 4. It is also concluded that the mutagenic response from diesel vehicles is described by low molecular weight PAH and BNT and that high molecular weight PAH, simple aromatic hydrocarbons, and lead described the mutagenic response from gasoline vehicles (1,4).

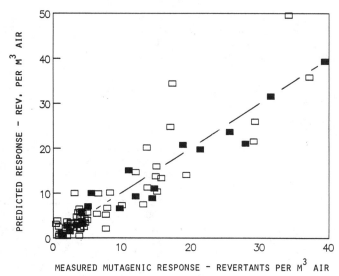

MEASURED MUTAGENIC RESPONSE - REVERTANTS PER M³ AIR

Fig.4.   Partial least square prediction of the mutagenic response with
         TA98+S-9 based on simple hydrocarbons, individual PAH and
         S-PAH, and lead.   ■ Training set, n = 24 from the D-series;
         □ test set, n = 52 from the C- and T-series. The slope is
         0.99, r = 0.92, and the standard error of prediction is 4
         rev./m³ (from Ref. 1 and 4).

Source Apportionment

Two methods of source apportionment have been applied to the entire
sample set for assigning the mutagenic response, a chemical mass balance
and a multilinear regression method (1,7). The overall results are given
in Tab. 5.

The chemical mass balance method indicates an equal contribution
from gasoline and diesel vehicles with a small contribution for oil combustion. The latter result is reasonable, as most space heating in
Gothenburg comes from district heating. The mass balance method is,
however, sensitive to the values assigned to the specific emission from
the different sources and the result would need confirmation by other
methods.

The multilinear regression method used lead as the sole indicator for
gasoline exhaust and BNT for diesel exhaust. The result, however, gave
a nonsignificant linear regression coefficient, although an apportionment
to the two sources could be calculated.

The source apportionment of the mutagenic response has recently
been made for a case with woodsmoke and motor vehicle exhaust (15)
using soil-corrected potassium and lead as indicators. The differentiation
between diesel and gasoline exhaust seems to be a substantial problem and
may only be possible under an experimental situation with a fuel labeling
technique, as has been performed in Vienna, Austria, in order to determine the diesel contribution to the airborne particle concentration (11).

Tab. 5. Calculated fractional source contributions to the TA98-S-9 mutagenic response using all samples from C-, D-, T-, and B-series (from Ref. 1 and 7)

Source	Chemical mass balance[a]	Multilinear regression[b]
Gasoline	0.43	0.40
Diesel	0.42	0.22
Oil	0.15	not used
Roaddust	0.00	not used

a) Based on benzene, lead, vanadium and titanium as source indicators and using literature data on specific emission.

b) Based on lead and BNT as specific indicators for gasoline and diesel exhaust, respectively.

ACKNOWLEDGEMENTS

These projects have been supported by research grants from the National Swedish Environmental Protection Board. Other participants and contributors to the ambient air projects are Eva Agurell and Lennart Romert (Department of Genetic and Cellular Toxicology, University of Stockholm), Tomas Alsberg and Susanne Zackrisson (Special Analytical Laboratory, National Swedish Environmental Protection Board), Eva Brorström and Bengt Steen (Swedish Environmental Research Institute), and Rune Toftgård (Department of Medical Nutrition, Karolinska Institute).

REFERENCES

1. Agurell, E., T. Alsberg, E. Brorström, G. Löfroth, L. Romert, R. Toftgård, B. Steen, and 5. Zakrisson (1989) Chemical and biological characterization of urban particulate matter. National Swedish Environmental Protection Board Report (ms. in prep.).

2. Alfheim, I., G. Löfroth, and M. Moller (1983) Bioassay of extracts of ambient particulatre matter. Env. Health Persp. 47:227-238.

3. Alsberg, T., U. Stenberg, R. Westerholm, M. Strandell, U. Rannug, A. Sundvall, L. Romert, V. Bernson, B. Petterson, R. Toftgard, B. Franzen, M. Jansson, J.Å. Gustafsson, K.E. Egeback, and G. Tejle (1985) Chemical and biological characterization of organic material from gasoline exhaust particles. Env. Sci. Tech. 19:43-50.

4. Alsberg, T., S. Håkansson, M. Strandell, and R. Westerholm (1989) Profile analysis of urban air pollution. Chem. Intellig. Lab. Syst. 7:143-152.
5. Brorström-Lunden, E., and A. Lindskog (1985) Characterization of organic compounds on airborne particles. Env. Intern. 11:183-188.
6. Curvall, M., C.R. Enzell, T. Jansson, B. Pettersson, and M. Thelestam (1984) Evaluation of the biological activity of cigarette-smoke condensate fractions using six in vitro short-term tests. J. Tox. Env. Health 14:163-180.
7. Dahlberg, K., and B. Steen (1987) Computation of the contribution from different sources to air pollutants in central Gothenburg. IVL-Rapport B 860 (in Swedish).
8. Dehnen, W., N. Pitz, and R. Tomingas (1977) The mutagenicity of airborne particulate pollutants. Cancer Lett. 4:5-12.
9. Epstein, 5.5., 5. Joshi, J. Andrea, N. Mantel, E. Sawicki, T. Stanley, and E.C. Tabor (1966) Carcinogenicity of organic particulate pollutants in urban air after administration of trace quantities to neonatal mice. Nature 212:1305-1307.
10. Gordon, R.J., R.J. Bryan, J.S. Rhim, C. Demoise, R.G. Wolford, A.E. Freeman, and R.J. Huebner (1973) Transformation of rat and mouse embryo cells by a new class of carcinogenic compounds isolated from particles in city air. Intern. J. Cancer 12:223-232.
11. Horvath, H., I. Kreiner, C. Norek, 0. Preining, and B. Georgi (1988) Diesel emissions in Vienna. Atmos. Env. 22:1255-1269.
12. Kado, N.Y., D. Langley, and E. Eisenstadt (1983) A simple modification of the Salmonella liquid-incubation assay. Increased sensitivity for detecting mutagens in human urine. Mutat. Res. 121:25-32.
13. Kado, N.Y., G.N. Guirguis, C.P. Flessel, R.C. Chan, K.I. Chang, and J.J. Wesolowski (1986) Mutagenicity of fine (<2.5 μm) airborne particles: Diurnal variation in community air determined by a Salmonella micro preincubation (microsuspension) procedure. Env. Muta. 8:53-66.
14. Kotin, P., H.L. Falk, P. Mader, and M. Thomas (1954) Aromatic hydrocarbons. I. Presence in the Los Angeles atmosphere and the carcinogenicity of atmospheric extracts. AMA Arch. Ind. Hyg. Occupat. Med. 9:153-163.
15. Lewis, C.W., R.E. Baumgardner, R.K. Stevens, L.D. Claxton, J. Lewtas (1988) Contribution of woodsmoke and motor vehicle emissions to ambient aerosol mutagenicity. Env. Sci. Tech. 22:968-971.
16. Löfroth, G. (1981) Comparison of the mutagenic activity in carbon particulate matter and in diesel and gasoline engine exhaust. In Short-term Bioassays in the Analysis of Complex Environmental Mixtures II, M.D. Waters, S.S. Sandhu, J. Lewtas-Huisingh, L. Claxton, and 5. Nesnow, eds. Plenum Publishing Corp., New York, pp. 319-336.
17. Löfroth, G., G. Lazaridis, and E. Agurell (1985) The use of the Salmonella/microsome mutagenicity test for the characterization of organic extracts from ambient particulate matter. Env. Intern. 11:161-167.
18. Löfroth, G., P.I. Ling, and E. Agurell (1988) Public exposure to environmental tobacco smoke. Mutat. Res. 202:103-110.
19. Maron, D.M., and B.N. Ames (1983) Revised methods for the Salmonella mutagenicity test. Mutat. Res. 113:173-215.
20. Pitts Jr., J.N., D. Grosjean, T.M. Mischke, V.F. Simmon, and D. Poole (1977) Mutagenic activity of airborne particulate organic pollutants. Tox. Lett. 1:65-70.

21. Poellinger, L., J. Lund, E. Dahlberg, and J.Å. Gustafsson (1985) A hydroxylapatite microassay for receptor binding of 2,3,7,8-tetrachlorodibenzo-p-dioxin and 3-methylcholanthrene in various target tissues. Analyt. Biochem. 144:371-384.

22. Reali, D., H. Schlitt, C. Lohse, R. Barale, and N. Loprieno (1984) Mutagenicity and chemical analysis of airborne particulates from a rural area in Italy. Env. Muta. 6:813-823.

23. Steen, B.A. (1985) Sampling of airborne particles for biological testing within the Swedish urban air project. Env. Intern. 11:105-109.

24. Talcott, R., and E. Wei (1977) Airborne mutagens bioassayed in Salmonella typhimurium. JNCI 58:449-451.

25. Toftgård, R., G. Löfroth, J. Carlstedt-Duke, R. Kurl, and J.Å. Gustafsson (1983) Compounds in urban air compete with 2,3,7,8-tetrachlorodibenzo-p-dioxin for binding to the receptor protein. Chem. Biol. Interact. 46:335-346.

26. Toftgård, R., B. Franzen, J.Å. Gustafsson, and G. Löfroth (1985) Characterization of TCDD-receptor ligands present in extracts of urban air particulate matter. Env. Intern. 11:369-374.

27. Tokiwa, H., K. Morita, H. Takeyoshi, K. Takahashi, and Y. Ohnishi (1977) Detection of mutagenic activity in particulate air pollutants. Mutat. Res. 48:237-248.

28. Westerholm, R., T. Alsberg, M. Strandell, Å. Frommelin, V. Grigoriadis, A. Hantzaridou, G. Maitra, L. Winquist, U. Rannug, K.E. Egebäck, and T. Bertilsson (1986) Chemical analysis and biological testing of emissions from a heavy duty diesel truck with and without two different particulate traps. SAE Paper 860014:73-83

29. Wold, S., and M. Sjöström (1977) Simca: A method for analyzing chemical data in terms of similarity and analogy. In Chemometrics: Theory and Application. ASC Symposium Series No, 52, B.R. Kowalski, ed. American Chemical Society, pp. 243-282.

30. Wullenweber, M., G. Ketseridis, L. Xander, and H. Ruden (1982) Seasonal variations in the mutagenicity of urban aerosols, sampled in Berlin (West), with Salmonella typhimurium TA98 (Ames test). Staub-Reinhalt. Luft 42:411-415 (in German).

31. Wynder, E.L., and E.A. Graham (1950) Tabacco smoking as a possible etiologic factor in bronchiogenic carcinoma. A study of six hundred and eighty-four proved cases. J. AM. Med. Assoc. 143:329-336.

32. Zakrisson, S., T. Alsberg, and L. Winquist (1989) Chemical composition and biological activity of particulate polycyclic aromatic hyrocarbons (PAH) in the exhaust of passanger cars (ms. in prep.).

ASSESSMENT OF THE MUTAGENICITY OF VOLATILE ORGANIC AIR
POLLUTANTS BEFORE AND AFTER ATMOSPHERIC TRANSFORMATION

Larry D. Claxton,[1] T.E. Kleindienst,[2] Erica Perry,[3]
and Larry T. Cupitt[4]

[1]Health Effects Research Laboratory
U.S. Environmental Protection Agency
Research Triangle Park, North Carolina 27711

[2]Northrop Services International
Research Triangle Park, North Carolina 27709

[3]Environmental Health Research and Testing, Inc.
Research Triangle Park, North Carolina 27709

[4]Atmospheric Research and Exposure Assessment Laboratory
U.S. Environmental Protection Agency
Research Triangle Park, North Carolina 27711

INTRODUCTION

During the past decade, renewed efforts have emerged to examine
the extent to which hazardous compounds (particularly mutagens and
carcinogens) are found in the urban atmosphere (20). The majority of
these studies examined the organic material associated with particles emitted from specific sources (11). In contrast, only a few studies have examined volatile airborne organic compounds before and after they undergo
atmospheric transformation (13). Some studies have shown that the
mutagenicity of organic material from combustion sources was altered dramatically by photooxidation processes (1). Such studies also demonstrated that a variety of atmospheric hydrocarbons can be transformed into
mutagenic species through these same processes (18). In this chapter,
we present an overview of efforts used to assess the effect of atmospheric
transformation upon the mutagenicity of airborne compounds.

RELATIONSHIP OF ATMOSPHERIC TRANSFORMATION TO
ENVIRONMENTAL TOXICOLOGY

Historically, environmental toxicologists have focused their attention
on major air pollutants (e.g., sulfur dioxide, carbon monoxide, and

Genetic Toxicology of Complex Mixtures
Edited by M. D. Waters *et al.*
Plenum Press, New York, 1990

nitrogen dioxide), total suspended particulates, and a few well-established carcinogens [e.g., benzo(a)pyrene]. Although the carcinogenic effects of organic matter extracted from ambient urban air have been known for nearly 50 years (10), Pitts (12) stated that little and sometimes misleading information was published about the role of atmospheric transformation. As illustrated in Fig. 1, however, chemical reactions in the atmosphere have the potential either to increase or decrease the exposure and/or toxicity of airborne compounds. Atmospheric reactions, for example, may alter the formation and size of aerosols thus altering the residence time of toxic materials in the atmosphere. Some of our early work (1) showed that increases and decreases in the mutagenicity of engine emissions could occur after atmospheric reaction had occurred. Whether or not an increase or decrease occurred was dependent upon the amount of light, presence of other gases such as NO_x and ozone, and the length of exposure. Because atmospheric reactions can alter the chemical profiles of emitted chemicals, it appears obvious that the uptake, bodily distribution, metabolism, and toxicity of ambient airborne pollutants may not be the same as that of the pollutants emitted from the source. Public health research efforts, therefore, must determine how to incorporate the role of atmospheric transformation into its evaluations of airborne pollution.

Evaluating the role of atmospheric reactions is complicated by the observation that airborne toxicants can exist as gases, as aerosol-bound components, or can be distributed between the gas and aerosol phases. The association of a toxicant within a particular phase is dependent upon the nature of the compound itself (e.g., molecular weight, size, functionality, charge, etc.), climatic conditions (e.g., temperature, humidity, etc.), and other conditions (e.g., concentration of soot particles). When airborne toxicants are collected, they may not distribute upon collection media in a manner that fully represents ambient air

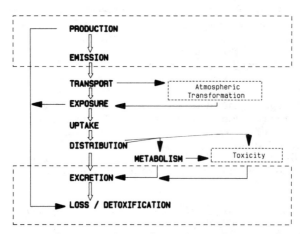

Fig. 1.   The generation and fate of ambient air pollutants. The upper section represents pollutants before ambient exposure. The center section represents processes occurring during exposure periods. The bottom section represents the postexposure fate.

conditions. For example, some of the semivolatile compounds associated with collected particles may disassociate from the particles due to the gas flow across the particles in the collection media. If another collection media (e.g., XAD) is in series after the collection device for particles, the "particle-bound toxicant" may actually be collected separately from the particles. Conversely, gas phase, semivolatile compounds passing across collected particles may absorb onto the particles. Upon collection, these semivolatile compounds may or may not be associated with the same phase. This is important because the distribution and retention of particles and gases within the respiratory tract is different. Also, many airborne toxicant studies have examined compounds associated with only one of the two phases. For example, most of the published genotoxicity studies of ambient air toxicants concentrate upon particle-bound organic compounds (7). However, Duce (4) states that the average global concentration of vapor-phase organic pollutants is approximately 50×10^{11} grams/day while the particulate organic matter concentration is 4×10^{11} grams/day. This is a major difference, especially when one realizes that much of the particulate matter is inert carbonaceous matter. Clearly, toxicological efforts should attempt to evaluate vapors and aerosols both before and after undergoing atmospheric reaction. Bioassay efforts are complicated by the need to have different protocols for gases, semivolatile compounds, and aerosols (7).

Because of the many inherent complexities described, researchers must limit the design of their research to examine rather specific questions. The remaining portion of this chapter will illustrate six types of studies designed to understand atmospheric transformation processes and the effect that these processes have upon the mutagenicity of airborne organic compounds. The six categories of studies to be examined are: (i) environmental inference studies, (ii) sampling site studies, (iii) source versus ambient concentration studies, (iv) complex-mixture reaction-chamber studies, (v) single-compound reaction-chamber studies, and (vi) media-based mechanistic studies.

EXAMPLE OF AN ENVIRONMENTAL INFERENCE STUDY

Although many types of inference studies are possible, most tend to combine a comparison of information from the literature with the analysis of actual environmental and/or source samples. One example is the work of Ramdahl et al. (15). Examination of variously collected ambient air samples from California; Missouri; Washington, D.C.; and Norway showed the presence of 1-nitropyrene, 2-nitropyrene, and 2-nitrofluoranthene in all of the samples. They conclude that these compounds are "ubiquitous in ambient particulate organic matter (POM)." Examining published data, they found no evidence for 2-nitropyrene or 2-nitrofluoranthene being emitted into ambient air by any examined source. They concluded, therefore, that 2-nitropyrene and 2-nitrofluoranthene are formed by chemical reaction from their parent polycyclic aromatic hydrocarbon (PAH) during atmospheric transport.

EXAMPLE OF A SAMPLING SITE STUDY

Using both chemical and bioassay information, Gibson et al. (6) used ambient monitoring to provide evidence for the formation of mutagenic aerosols from secondary atmospheric reactions. In this study, they collected ambient air particles in Delaware and on the southwest coast of Bermuda. At the Bermuda site, sampling was based upon prevailing wind trajectories. Six different wind trajectory categories were used. Figure 2 illustrates the results that were seen. When the collections were influenced mainly by local pollution, as at the Delaware site, the 1-nitropyrene and mutagenicity results corrected by either the levels of lead or elemental carbon were more than an order of magnitude lower than when collections were influenced by pollution undergoing long-distance transport (as in the Bermuda samples collected during times of northerly wind directions). Therefore, after adjusting a specific genotoxicant or bioassay result according to the level of pollution, the particle-bound organics coming from longer distances were more mutagenic. Atmospheric reactions apparently increased the mutagenicity of these organics.

EXAMPLE OF A SOURCE VERSUS AMBIENT CONCENTRATION STUDY

When one combines ambient monitoring with source monitoring, it is possible to examine the contribution of a source to the ambient concentrations of mutagenic species. Driver et al. (3) simultaneously collected stack fly ash from both a medical/pathological waste incinerator and a boiler plant while collecting ambient air particulate samples at upwind and downwind directions. The ambient air samples were collected at relatively short distances (less than 500 feet) from the sources. During the period

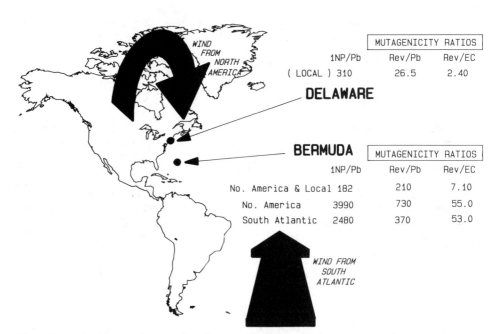

Fig. 2.   Ambient air study illustrating atmospheric transformation. Excerpted from the work of Gibson et al. (6).

Tab. 1.   Comparing the Salmonella typhimurium TA98 mutagenicity of source and ambient air samples

Sample	Revertants/Cubic Meter Air		Revertants/Microgram Organic	
	−S9 (No.)[a]	+S9 (No.)	−S9 (No.)	+S9 (No.)
Upwind	2.4 (7)	2.2 (7)	0.5 (7)	0.5 (7)
Incinerator	70.9 (18)	30.2 (15)	0.6 (18)	0.5 (15)
Boiler	54.4 (8)	28.3 (8)	0.7 (8)	0.4 (8)
Downwind	10.9 (10)	8.0 (10)	1.2 (10)	0.8 (10)

[a]Average slope value (number of replicate samples used).   Extracted from Driver (2).

of sampling, the incinerators were burning medical/pathological waste and the boilers were burning Number 6 residual fuel oil.   Table 1 is a summary of the mutagenicity results normalized to cubic meter of air.   It is readily apparent that the mutagenic activity per cubic meter of air is more concentrated at the sources and that these sources have a direct impact upon the downwind samples.   When examining the revertant values per cubic meter of air (2), one notes that the downwind values (1.16 revertants per μg organic matter without activation and 0.78 revertants per μg organic matter with activation) is higher than upwind, incinerator, and boiler values.   This apparently indicates that even within a very short distance atmospheric reactions altered the mutagenicity of the organic species.

EXAMPLE OF COMPLEX MIXTURE REACTION CHAMBER STUDIES

It is very difficult to study mechanisms and specific interactions in ambient air studies, because one cannot duplicate conditions from day to day and cannot control reaction rates.   Reaction chamber studies, therefore, are used to understand reactions, to identify the important constituents and products of reactions, and to aid in the identification of specific genotoxicants formed during atmospheric reaction.   In order to understand the effect of photooxidation upon wood smoke emissions, Kleindienst et al. (9) irradiated dilute mixtures of wood combustion emissions (with and without additional NO_x) in a 22.7 m^3 Teflon smog chamber.   Although the particle-bound organics were mutagenic in Salmonella typhimurium (in both TA98 and TA100) before irradiation, the gas phase component was not mutagenic.   After irradiation, the mutagenicity of both the gas-phase and particulate-phase organics increased.   Without added NO_x, irradiated wood smoke showed a measurable increase in mutagenic activity only for gas-phase products.   For strain TA100, the mutagenic activity of the organics from the gas and particulate phases were within approximately an order of magnitude (0.3 to 8.5 revertants per μg organic material).

Tab. 2.   Comparison of reactant and product mutagenicity using the Salmonella typhimurium TA100 bioassay without an exogenous activation system

	Substance	Mutagenicity
Reactant	Toluene	Negative
Photooxidation products	Complex mixture	0.069 revertants/hr·ppb
Identified products	NO_2	Negative
	Ozone	Negative
	Acetaldehyde	Negative
	Cresols	Negative
	Nitric acid	Negative
	Formaldehyde	12 revertants/micromole
	Methylglyoxal	200 revertants/micromole
	PAN	∿650 revertants/micromole
	Allyl chloride	400 revertants/micromole

However, due to the relative amounts of gas-phase and particulate-phase material, the mutagenic activity on a cubic meter basis was much greater for the gas-phase component (>12,000 revertants/m^3) than for the particulate-phase component (<200 revertants/m^3). Not only did these studies demonstrate that atmospheric reactions increased the mutagenicity of wood smoke emissions, the study showed that the effect was much greater for the gaseous emissions.

EXAMPLE OF A SINGLE COMPOUND/REACTION CHAMBER STUDY

When a complex mixture is injected into a reaction chamber, the products of photooxidation may have emanated from many of the starting species. Single organic species, therefore, are used in transformation studies so one can understand the relationships between starting species and final products. This information allows the delineation of the mechanisms involved in atmospheric reactions and aids in identifying the classes of compounds that are important precursors to mutagenic species. Using these types of studies, we have examined the production of mutagenic species from several nonmutagenic compounds including toluene (5,16), allyl chloride (17), and propylene (8). When toluene was irradiated in the presence of NO_x and ozone, a number of mutagenic and nonmutagenic products were produced. After identification of the measurable products, the testing of the individual species was performed. Table 2 summarizes the results from these series of experiments. Because 50-70% of the total mutagenicity produced was not accounted for by the identified products, any estimation of ambient air mutagenicity calculated from known concentrations would provide a similar underestimate of potential genotoxity potential. By examining the mutagenicity produced from specific precursors (e.g., toluene, propylene, etc.), however, we can ask "What is the relative importance of each precursor species in terms of the rate of production of mutagenic activity under atmospheric reaction

Tab. 3. Contribution of the mutagenic transformation products of organic precursor compounds to ambient air

	Toluene	Propylene	Allyl Chloride
Revertants/Hr·ppb	0.069	0.036	1.40
Urban Concentration	15.0 ppb	4.0 ppb	0.005 ppb
K(OH)·[OH]·[HC]ppb/Hr	0.34	0.38	0.0003
Revertants/Hr	0.023	0.014	0.0004

Calculated from data contained in Ref. 8, 16-19. Hr, hours; ppb, parts per billion; [], concentrations.

conditions?" This calculation is done with the aid of the typical concentration data (19) and the rate constants for their reaction with the OH radical (18). The results in Tab. 3 illustrate that although the production of mutagenic potential (revertants/hr·ppb) resulting from the irradiation of each precursor is important, the total mutagenicity found in ambient air is determined to a large extent by the urban concentration of the precursor compound.

EXAMPLE OF A MEDIA BASED MECHANISTIC STUDY

By passing reactive gases [e.g., ozone, NO_2, peracetylnitrate (PAN)] over filters impregnated with specific precursor materials [e.g., benzo(a)pyrene and perylene], Pitts et al. (14) demonstrated the generation of products that are direct-acting Salmonella mutagens. For example, benzo(a)pyrene reacted with NO_2 (1 ppm) containing traces of nitric acid (approximately 10 ppb) produced a mixture of 1-nitro-, 3-nitro-, and 6-nitro-benzo(a)pyrene. Likewise, perylene exposed to NO_2 produced the direct-acting mutagen 3-nitroperylene. They also demonstrated that the drawing of ambient photochemical smog across filters impregnated with benzo(a)pyrene yielded products that are mutagenic with S. typhimurium strain TA98 in the absence of exogenous metabolic activation. Because benzo(a)pyrene is an indirect-acting mutagen, the presence of direct-acting mutagenicity demonstrated the production of other mutagenic species. These studies demonstrated two important possibilities. First, atmospheric transformation reactions that affect the mutagenicity of airborne organic matter could occur in the atmosphere. Second, with the passing of reactive gases across organic matter collected upon filters, direct-acting mutagens that are artifacts may be formed.

CONCLUSIONS

Although the carcinogenic potential of ambient urban air, organic matter has been known for 50 years, relatively few studies have dealt

with the generation and/or removal of genotoxicants within ambient air
due to atmospheric reactions. This chapter illustrates that both laboratory experiments and ambient monitoring studies provide evidence that
mutagens can be produced by photooxidation processes. These mutagens
can be produced by known photochemical pathways involving nongenotoxic
hydrocarbons (e.g., toluene) and common atmospheric components such as
NO_x, ozone, and sunlight. Therefore, products emitted into the atmosphere in many cases will not be the toxic products to which human populations are exposed. Because routine monitoring efforts do not quantitate
many of the generated genotoxic products (e.g., 2-nitropyrene, 2-nitro-
fluoranthene, PAN, etc.) and because many of the generated mutagens
have not yet been identified, these studies also illustrate the usefulness
of mutagenicity bioassays in ambient air monitoring. These studies also
show that more effort is needed to identify and measure the vapor-phase
organic components with ambient air. In the final analysis, public health
efforts must determine the most appropriate methods to understand atmospheric transformation and to incorporate the role of atmospheric transformation processes into its assessment of airborne pollution.

ACKNOWLEDGEMENTS

We would like to thank the organizing committee, especially Dr. Mike
Waters and Dr. Joellen Lewtas, for encouraging us to do this review.

The research described in this article has been reviewed by the
Health Effects Research Laboratory, U.S. Environmental Protection
Agency, and approved for publication. Approval does not signify that
the contents necessarily reflect the views and policies of the Agency, nor
does the mention of trade names or commercial products constitute
endorsement or recommendation for use.

REFERENCES

1. Claxton, L.D., and M. Barnes (1981) The mutagenicity of die-
    sel-exhaust particle extracts collected under smog-chamber conditions
    using the Salmonella typhimurium test system. Mutat. Res.
    88:255-272.
2. Driver, J.H. (1988) Risk Assessment of Medical/Pathological Waste
    Incineration: Mutagenicity of Combustion Emissions. Ph.D. Thesis,
    University of North Carolina, Chapel Hill, North Carolina, 187 pp.
3. Driver, J.H., H.W. Rogers, and L.D. Claxton (1989) Mutagenicity of
    combustion emissions from a biomedical waste incinerator. Waste
    Management (in press).
4. Duce, R.A. (1978) Speculations on the budget of particulate and va-
    por-phase nonmethane organic carbon in the global troposphere.
    Pure Appl. Geophys. 116:244-273.
5. Dumdei, B.E., D.V. Kenny, P.B. Shepson, T.E. Kleindienst, C.M.
    Nero, L.T. Cupitt, and L.D. Claxton (1988) MS/MS analysis of the
    products of toluene photooxidation and measurement of their
    mutagenic activity. Env. Sci. and Tech. 22:1493-1498.
6. Gibson, T.L., P.E. Korsog, and G.T. Wolff (1985) Evidence for the
    transformation of polycyclic organic matter in the atmosphere.
    Atmos. Env. 20:1575-1578.
7. Hughes, T.J., D.M. Simmons, L.G. Monteith, and L.D. Claxton
    (1987) Vaporization technique to measure mutagenic activity of vola-

tile organic chemicals in the Ames/Salmonella assay. Env. Mutagen. 9:421-441.

8. Kleindienst, T.E., P.B. Shepson, E.O. Edney, L.T. Cupitt, and L.D. Claxton (1985) The mutagenic activity of the products of propylene photooxidation. Env. Sci. Tech. 19:620-627.

9. Kleindienst, T.E., P.B. Shepson, E.O. Edney, L.D. Claxton, and L.T. Cupitt (1986) Wood smoke: Measurement of the mutagenic activities of its gas- and particulate-phase photooxidation products. Env. Sci. Tech. 20:493-501.

10. Leiter, J., M.B. Shimkin, and M.J. Shear (1942) Production of subcutaneous sarcomas in mice with tars extracted from atmospheric dusts. JNCI 3:155-165.

11. Lewtas, J. (1988) Genotoxicity of complex mixtures: Strategies for the identification and comparative assessment of airborne mutagens and carcinogens from combustion sources. Fund. and Appl. Tox. 10:571-589.

12. Pitts, J.N., Jr. (1979) Photochemical and biological implications of the atmospheric reactions of amines and benzo(a)pyrene. Phil. Trans. R. Soc. Lond. A. 290:551-576.

13. Pitts, J.N., Jr. (1983) Formation and fate of gaseous and particulate mutagens and carcinogens in real and simulated atmospheres. Env. Health Persp. 47:115-140.

14. Pitts, J.N., Jr., K.A. Van Cauwenberghe, D. Grosjean, J.P. Schmid, D.R. Fitz, W.L. Belser, Jr., G.B. Knudson, and P.M. Hynds (1978) Atmospheric reactions of polycyclic aromatic hydrocarbons: Facile formation of mutagenic nitro derivates. Science 202:515-519.

15. Ramdahl, T., B. Zielinska, J. Arey, R. Atkinson, A.M. Winer, and J.N. Pitts, Jr. (1986) Ubiquitous occurrence of 2-nitrofluoranthene and 2-nitropyrene in air. Nature 321:425-427.

16. Shepson, P.B., T.E. Kleindienst, E.O. Edney, G.R. Namie, J.H. Pittman, L.T. Cupitt, and L.D. Claxton (1985) The mutagenic activity of irradiated toluene/NO$_x$/H$_2$O/air mixtures. Env. Sci. Tech. 19:249-255.

17. Shepson, P.B., T.E. Kleindienst, C.M. Nero, D.N. Hodges, L.T. Cupitt, and L.D. Claxton (1987) Allyl chloride: The mutagenic activity of its photooxidation products. Env. Sci. Tech. 21:568-573.

18. Shepson, P.B., T.E. Kleindienst, E.O. Edney, L.T. Cupitt, and L.D. Claxton (1987) The production in the atmosphere of mutagenic products from simple hydrdocarbons. In Short-term Bioassays in the Analysis of Complex Environmental Mixtures, V, S.S. Sandhu, D.M. DeMarini, M.J. Mass, M.M. Moore, and J.L. Mumford, eds. Plenum Publishing Corp., New York, pp. 277-290.

19. Singh, H.B., L.J. Salas, A. Smith, R. Stiles, and H. Shigeishi (1981) Atmospheric Measurements of Selected Hazardous Organic Chemicals. U.S. Environmental Protection Agency, Report Number EPA-600/3-81-032.

20. Tice, R.R., D.L. Costa, and K.M. Schaich (1982) Genotoxic Effects of Airborne Agents, Plenum Publishing Corporation, New York, NY.

CARCINOGENICITY OF COMPLEX MIXTURES IN AIR USING

IN VITRO AND IN VIVO ASSAYS

I. Chouroulinkov

Cancer Research Institute
CNRS
94802, Villejuif Cédex, France

INTRODUCTION

The relationship between the environmental chemicals and the increase of lung cancer in man is now well established. In fact, the first occupational cancer reported by Pott in 1775 (23,24) was caused by a complex mixture of coal soot. In this case the soot in contact for long periods with the scrotal skin of chimney sweeps induced local irritation and squamous cell carcinoma. We can also cite the epidemiological evidence for a correlation between the increase of lung cancer and exposure to cigarette smoke and industrially polluted atmospheres (12-14).

In 1915-1918, Yamagiwa and Itchikawa demonstrated the carcinogenicity of coal-tar in rabbits and mice (26). The identification by Kennaway's group of benzo(a)pyrene (BaP) in coal tar (15) started a major scientific endeavor to understand the role of this carcinogen and the other polycyclic aromatic hydrocarbons (PAHs) in the potential carcinogenicity of PAH mixtures in the air. Based on the carcinogenicity of BaP in animals, it was proposed as an indicator for air quality in terms of cancer hazard. Subsequent studies on the mechanism of tumor initiation (1) and the relationship between its mutagenicity and carcinogenicity (22), led to the utilization of short-term mutagenicity bioassays for detection of the carcinogenic potential of complex environmental mixtures.

The objectives of the studies described here were to evaluate the carcinogenic potential of complex mixtures and to elucidate lung cancer etiophatogenesis. At the beginning our research was concerned essentially with the cigarette smoke condensates (CSCs), but later this work was extended to diesel exhaust extracts, city air pollutant extracts (7), and water organic micropollutants. This chapter summarizes the in vitro and in vivo studies of cigarette smoke condensates and diesel exhaust extracts.

Genetic Toxicology of Complex Mixtures
Edited by M. D. Waters *et al.*
Plenum Press, New York, 1990

STUDIES WITH CIGARETTE SMOKE CONDENSATES

About 1960, it was accepted that tobacco smoke is a cancer hazard. As it was impossible to stop people from smoking, the elaboration of a safer cigarette appears obvious. The problem to us was how to evaluate rapidly and with some certainty the quality of the various tobaccos and cigarettes in terms of carcinogenicity. At the time, the choice was very limited. However, Guerin and Cuzin (11) developed an epidermal hyperplasia induction (EHi) test combined with the already-existing sebaceous gland destruction (SGd) test (2,25), and two short-term skin tests for carcinogenic potential, detection, and evaluation (17). So, these tests were adopted for screening investigations on carcinogenic potential of pure and complex chemicals. Simultaneously, to verify the correlations, long-term skin tests and inhalation experiments for carcinogenicity were performed.

As the inhalation experiments about 1960 were negative, we assumed that the smokers were also exposed to some other carcinogens playing the role of initiator, implicitly tobacco tar playing the role of promoter, we used for long-term skin tests on normal mice and BaP-pretreated (initiated) mice.

Results

We studied more than 1,000 CSCs in short-term skin tests, more than 40 CSCs in long-term skin tests, and some cigarettes in long-term inhalation experiments. One part of the data was published (3,19,20). Briefly, the two short-term skin tests, SGd and EHi, showed a dose-response relationship opening the possibility of comparative evaluations, the results of the two tests are correlated in between; finally they correlated well with the results from long-term skin tests and particularly with the promoting activity of the CSCs. Actually we know that EHi is related to the promoting effect through epigenetic mechanisms.

Association of Skin Tests and Inhalation Experiments

The results from short- and long-term skin tests with two CSCs (A and B) showed that CSC-A, compared to CSC-B, induced higher levels of epidermal hyperplasia, and exhibited higher carcinogenic and promoting activities in the same manner as 12-0-tetradecanoylphorbol-13-acetate (TPA) and phorbol, 12,13-didecanoate (PDD) (5). At the section entitled "Discussion," I will cite the results from direct cell exposure to diesel exhaust. To verify if such a difference should appear in an inhalation experiments, Syrian hamster cells were exposed to the smoke of cigarettes A and B using Borgwald (Hamburg) smoking machines according the Dontenwill's method (9). The results expressed as total proliferative lesions in the respiratory tract (larynx, trachea, lung) (Tab. 1) and tumors (Tab. 2)--clearly confirmed the results obtained with the respective CSCs in short- and long-term skin tests. The smoke of cigarette A induced more local (respiratory tract) proliferative lesions than the smoke of cigarette B. The smoke of cigarette A also significantly increased the incidence of other tumors. Thus, it appears that cigarette B is less hazardous based on the lower inactivity in these different tests for carcinogenicity. Cigarette B also had three times less tar per cigarette.

Tab. 1.   Localization and type of proliferative lesions in the respiratory tract of Syrian hamsters exposed to cigarette smoke

Localization - Type of lesions	Cigarette A	Cigarette B	Control (machine)
Larynx			
- hyperplasia	2	1	2
- epidermal papilloma	9	5	0
- epidermoid carcinoma	2	0	0
Trachea			
- hyperplasia	7	2	2
- papillary hyperplasia	20	8	5
- adenoma	2	0	0
- epidermoid papilloma	2	1	0
Lung			
- papillary hyperplasia	4	0	1
- adenoma	7	2	2
- glandular carcinoma	0	1	1
Total	55	20	13
♂	38	12	2
♀	17	8	11

[a]Exposure: 5 da/wk, 10 min (30 cig.) twice a day, during 20 mo, sacrificed at 24 mo. Borgwald (Hamburg) smoking machines. 100 (50 0 + 50 0) animals per group.

Tab. 2.   Malignant and benign tumors, other than respiratory tract tumors in Syrian hamsters exposed to cigarette smoke

	Cigarette A	Cigarette B	Machine control
Malignant	13	7	5
Benign	18	10	7
Total	31	17	12

[a]Exposure: 5 da/wk, 10 min (30 cig.) twice a day, during 20 mo, sacrificed at 24 mo. Borgwald (Hamburg) smoking machines. 100 (50 0 + 50 0) animals/group.

Tab. 3.   Sister chromatid exchanges induced in V79 cells with cigarette
          smoke condensates (CSCs) I, II, and III.   Data from 30 meta-
          phases

Dose (μg/ml)	SCEs per metaphase + S.D.		
	CSC-I	CSC-II	CSC-III
Acetone	6.20 + 0.36	5.2 + 0.24	6.27 + 0.22
0.5%			
2	10.37 + 0.46	6.65 + 0.32	8.03 + 0.32
4	13.47 + 0.38	6.67 + 0.27	8.67 + 0.43
8	12.12 + 0.33	8.83 + 0.22	10.4 + 0.47

Comparison of Mutagenicity Assays and Skin Tests

As the use of mutagenicity bioassays increased, we evaluated a series
of CSCs in sister chromatid exchange (SCE) tests and the gene mutation
(HPRT locus) using V79 cells simultaneously with the short- and long-
term skin tests.   The results from the SCE assay, presented in Tab. 3,
indicate that all CSCs exhibited positive results with dose-response relationship without an exogenous metabolic system.   The CSC-I showed the
highest activity, while CSC-II showed the lowest.   We may then classify
these condensates according to the activity in decreasing order as
follows: CSC-I > CSC-III > CSC-II.   According to the results from the
gene mutation assay (HPRT locus, V79 cells), CSC-II is inactive with or
without metabolic activation, CSC-I is active only with metabolic activation, and CSC-III is active with or without metabolic activation.   The
activity of these samples in decreasing order was CSC-III > CSC-I >
CSC-II (inactive).

The results from the short- and long-term tests with the same CSCs
(Tab. 4) showed a dose-response relationship, but the difference between
the activities of the three CSCs in all tests is very narrow.   However, we
may also classify them according to the activities in decreasing order, as
indicated in Tab. 5.   As we can see from Tab. 5, there is a good agreement between the results of all skin tests, as CSC-II appears the most
active and CSC-I less active.   In mutagenicity tests the inverse was
observed with CSC-II as the less active.   That suggests that CSC-II
contains more epigenetic carcinogens (promoters) than the other CSCs,
and fewer mutagens.   Moreover, these results indicate that each type of
bioassay may detect specific components which justify the association of
mutagenicity and carcinogenicity bioassays for biological characterization
of complex mixtures.

Table 6 is the last illustration of parallel studies with CSCs in short-
and long-term skin tests associated with the cell growth inhibition test
using an established cell line of rat lung epithelial cells.   The range of
CSC activity determined in all tests is always the same.   The cell growth
inhibition test gave excellent results, with a dose-response relationship.

Tab. 4. Activity of cigarette smoke condensates I, II, and III in short-
         and long-term skin tests for carcinogenicity in mice

CSC	Tests		Dose/Application (mg) 14	22	30	
	− Seb. glands (destruction %)		0.00	0.00	49.58	
I	− Hyperplasia (control = 100)		109.95	131.59	203.23	
	− Mice with	− initiated	15	37	50	= 102
	tumors	− normal	1	15	22	= 38
		Total	16	52	72	= 140
	− Seb. glands (destruction %)		0.00	24.05	68.45	
II	− Hyperplasia (control = 100)		111.94	157.21	228.36	
	− Mice with	− initiated	14	32	44	= 90
	tumor	− normal	12	27	30	= 69
		Total	26	59	74	= 159
	− Seb. glands (destruction %)		0.00	3.44	52.85	
III	− Hyperplasia (control = 100)		100.00	140.55	199.00	
	− Mice with	− initiated	13	43	50	= 106
	tumor	− normal	5	14	31	= 50
		Total	18	57	81	= 156

Long-term skin test: applications three times per week, 80 wk, 80
Swiss female mice per group.

However, the significance of this test with respect to carcinogenicity is
not known.

STUDIES WITH DIESEL EXHAUST EXTRACTS

The increase of diesel-powered light-duty vehicles and medium- and
heavy-duty truck and buses raises problems of particle emissions into the
ambient air. Several studies using various systems have demonstrated
mutagenic, transforming, and initiating (in vivo) activity of diesel particle
extracts or fractions (3,4). Studies and data are needed to provide a
better quantitative evaluation of the biological effects of automobile
emissions.

Tab. 5.   Classification of CSC I, II, and III in the function of the
           results in different tests

Tests	Classification
SCE	I > III > II
Gene mutation (HGPRT)	III > I > II
Sebaceous glands	II > III > I
Epidermal hyperplasia	II > III = I
Carcinogenicity	II ≥ III > II
Promoting activity	III > I > II
Total carcinogenic activity	II ≥ III > I

The aim of this work is to quantitatively evaluate the carcinogenic
potential of diesel exhaust extracts and fractions, and to try to
understand the mechanism as either genotoxic or epigenetic (promoter).
The bioassays used were short-term skin tests and Syrian hamster embryo
cell transformation systems.

Results

Short-Term Skin Tests

As for CSCs, we observed with diesel particulate extract (DPE) a
dose-response relationship in the two short-term skin (SGi and EHi tests)
(Fig. 1). For comparative evaluation the activity is expressed per mg
(Fig. 2). From these data we can draw the following conclusions:

Tab. 6.   Classification of four cigarette smoke condensates (CSCs) in
           relation to the endpoint (CSC, a,b,c,d)

Endpoint	Classification (decreasing activity)
- Cell growth inhibition (lung epithelial cells)	a > b ≥ c > d
- Sebaceous gland destruction	a > b ≥ c ≥ d
- Hyperplasia induction	a > b > c ≥ d
- Papillomas	a ≥ b > c ≥ d
- Squamous cell carcinomas	a > b > c > d
Total tumors	a > b > c > d

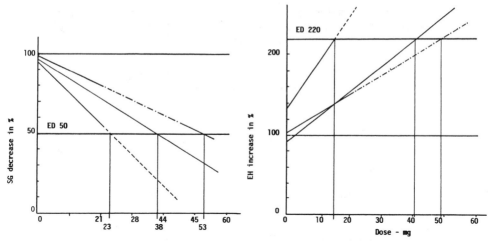

Fig. 1. Dose-response relationship for sebaceous gland (SG) test and epidermal hyperplasia (EH) test. Mouse skin treated with three different diesel exhaust extracts.

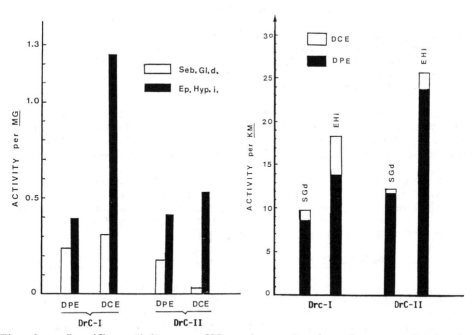

Fig. 2. Specific activity per MG and per km in sebaceous glands destruction (SGd) and epidermal hyperplasia induction (EHi) tests of diesel particulate (DPE) and diesel condensate (DCE) extracts from driving cycles type I and II.

Tab. 7.   Activity in sebaceous glands (SG) and epidermal hyperplasia
          (EH) tests of the diesel exhaust extract (Driving cycle II)
          fractions; (-) negative, (+) positive

Fraction	SG Test	EH Test
Insoluble	-	-
Acidic	-	+
Basic	-	-
Paraffinic	-	+
Aromatic	+++++	+++
Transitional	++++	++++
Oxygenated	+	++

Number of (+) = relative activity

- For driving cycle type I (DrC-I), the diesel condensate extract
  (DCE) showed higher activity than the DPE.

- For driving cycle type II (DrC-II), generally the activity of the
  extracts was lower compared to DrC-I. DCE (DrC-II) is also less
  active than DCE (DrC-I), but more active than DPE(DrC-II). We
  should note a lower range of activities in the SGs test of the two
  extracts from DrC-II compared to those of DrC-I.

The specific activities (expressed per km) are a function of the
quantity emitted per km and the activities of the extracts. The higher
emission quantity from DrC-II resulted in the specific activity per km
being higher, even though specific activity per mg was less.

In conclusion, the results from short-term skin tests indicate that
mixtures from diesel exhaust have carcinogenic potential, activity of the
type consistent with tumor promotion (epigenetic) mechanism. It is also
possible to evaluate this activity quantitatively as a function of the
driving cycle, and of the diesel engine type.

Diesel Particulate Extract Fraction Activities in the SGd and EHi Tests

The chemical fractions from DPE-DrC-II were studied in both short-
term skin tests. The same quantity of the fraction in the corresponding
doses of DPE were used in the same test to provide dosing equivalents.
From the summarized results presented in Tab. 7, it appears that insoluble and basic fractions are inactive; the acidic and paraffinic fractions
are positive only in the EHi test; the aromatic, transformational, and
oxygenated fraction are positive in both tests. The aromatic fraction is
more active in the SGd test like benzo(a)pyrene (BaP), while the others

Tab. 8.   Study of the initiating activity of diesel particulate extracts:
concentrations, combinations, and results (TPA-twice)

	DPE (µg/ml)	TPA (µg/ml)	No. of dishes	CE (%)	No. of transformed med. colonies	Transformation frequency
1	0	0	10	31.3	0	0.0
2	0	0.025	10	26.6	10	1.3
3	0	0.05	10	28.1	28	3.7
4	0	0.1	10	25.2	36	4.8
5	1.25	0	10	30.4	2	0.2
6	1.25	0.025	10	26.1	2	0.3
7	1.25	0.05	10	24.8	3	0.4
8	1.25	0.1	10	25.5	20	2.6
9	2.5	0	10	28.9	0	0.0
10	2.5	0.05	10	28.8	1	0.1
11	2.5	0.05	10	25.8	7	0.9
12	2.5	0.1	10	26.9	10	1.2
13	5.0	0	10	30.7	2	0.2
14	5.0	0.025	10	25.2	4	0.5
15	5.0	0.05	10	25.9	6	0.8
16	5.0	0.1	10	26.8	19	2.4
17 BaP	0.1	0.025	10	12.3	25	5.1
18 BaP	0.1	0.05	10	16.6	25	5.0
19 BaP	0.1	0.1	10	16.3	45	9.2
20 BaP	0.1	–	10	18.8	3	0.5
21 BaP	0.1 DPE	2.5	10	18.2	5	0.9

are predominantly active in the EHi test, expressing its promoting potential.

The Transformation Effect of DPE in Syrian Hamster Embryo (SHE) Cells

This study was conducted to explore the initiating and the promoting
activities of DPE in SHE cells according to the classical method for transformation but using the orthogonal design (21).

Table 8 presents the modalities and the results from the experiment
where DPE was used as an indicator at three concentrations, and 12-0-
tetradecanoxlphorbol-13-acetate (TPA), also at three contractions, as
promoter. TPA alone transforms SHE cells with a dose-response relationship. DPE alone, at the concentrations used, did not induce transformed
colonies. The DPE significantly reduced ($p < 0.05$) the TPA-transforming
effect. We concluded that DPE at the concentrations used does not
exhibit initiating activity in this assay.

In Tab. 9, the results from an experiment where DPE is used as
promoter are presented. The cells were treated once with BaP at three
concentrations, then followed by DPE (twice) also at three concentrations.

Tab. 9.   Study on the promoting activity of diesel particulate extract
           (DPE) in Syrian hamster embryo (SHE) cells: concentrations,
           combinations, and results (DPE-twice)

	BaP (µg/ml)	DPE (µg/ml)	No. of dishes	C.E. (%)	No. of transformed med. colonies	Transformation frequency
1	0	0	10	22.8	0	0.0
2	0	5	10	20.2	6	1.0
3	0	10	10	14.8	8	2.0
4	0	15	10	0	0	0.0
5	0.025	0	10	22.5	4	0.6
6	0.025	5	10	19.4	6	1.0
7	0.025	10	10	16.4	7	1.4
8	0.025	15	10	0	0	0.0
9	0.05	0	10	20.9	6	1.0
10	0.05	5	10	18.1	8	2.5
11	0.05	10	10	15.0	18	4.4
12	0.05	15	10	0	0	0.0
13	0.1	0	10	19.7	4	0.7
14	0.1	5	10	16.4	5	1.0
15	0.1	10	10	12.6	2	0.5
16	0.1	15	10	0	0.0	0.0

The higher DPE concentrations were toxic in all groups.   At 5 and 10
µg/ml DPE showed a promotion-like effect, particularly evident on cells
pretreated with 0.05 µg/ml of BaP.

In conclusion, the carcinogenic potential of mixtures from diesel
exhaust can be detected and evaluated by short-term skin tests and cell
transformation systems.   These studies suggest that diesel exhaust
extracts could act as epigenetic carcinogens with promoting effects.

DISCUSSION

The data from the chemical analyses and mutagenicity bioassays
clearly indicated the presence of genetoxic compounds in the studied
mixtures of CSCs and DPEs.   Probably significant for heritable genetic
effects, these compounds seem to play a very marginal role in the carcinogenicity of such mixtures.   Enrichment of CSC with BaP (32 times the
normal level) did not increase the carcinogenicity.   But, the same quantity
of BaP alone (in acetone) induced skin tumors (18).   It appears that CSC
and DPE contain many promoters (nongenotoxic or epigenetic carcinogens).   In fact, epidermal hyperplasia (EH) is a specific indicator for
promoting potential and it is inhibited by dexamethasone or by peanut oil
when used as a solvent without inhibition of adduct formation when BaP
or DMBA are used (16).   In long-term skin tests, CSCs are carcinogenic
alone.   This is not in contradiction with its characterization as a
promoter. TPA is both a strong promoter and carcinogen by itself (4).

As the promoting (epigenetic) effects are reversible, a promoter--to be effective as a carcinogen--should be applied frequently for a long period and at an elevated dose (6). This is in fact the nature of the exposure of smokers to tobacco smoke.

We do not have convincing epidemiological data on the relationship between increased incidence of lung cancer and exposure to diesel exhaust. This is due to irregular human exposure and low concentrations of diesel particles. We know that rats exposed regularly to diesel exhaust at a high concentration of particles (2,000 to 7,000 $\mu g/m^3$) and for a period of 24 mo developed lung tumors with a latency period of more than 20 mo. These experimental conditions are in agreement with the conditions necessary for promoters to induce tumors.

One of the criticisms of the skin tests is the very high doses used (10). However, a simple calculation will show that this criticism is not justified. If a man smokes 20 cigarettes containing 20 mg/cigarette (old cigarettes contained more than 30 mg/cigarette) during 20 yrs, and 60% of the smoke is deposited in the respiratory tract, he will expose his lungs to 1,750 g (1.750 kg) of tar. Assuming the lung surface is 1 m^2, each cm^2 will receive 175 mg (some area will in fact receive two or three times more). Our lower effective dose used in skin tests was 14 mg/application on about 10 cm^2 of skin. That means, for 80 weeks mice received 336 mg/cm^2 of skin. Under these conditions of skin applications, the tumor incidence is higher than the lung-tumor frequency in smokers. This suggests that the doses used in skin tests for carcinogenicity of complex mixtures are unusually high.

CONCLUSION

In conclusion, these complex mixtures of cigarette smoke and diesel particles do have carcinogenic potential. The biological characterization of the potential carcinogenicity of these extracts should include bioassays for mutagenicity, such as the Ames test (e.g., cell transformation) and/or short-term skin tests and, only in some cases, long-term skin tests. This combination of assays will provide information on both the genotoxins and epigenetic carcinogens (promotors) in these mixtures.

REFERENCES

1. Berenblum, I. (1974) Carcinogenesis as a Biological Problem. North Holland Co., Amsterdam.
2. Bock, F., and R. Mund (1958) A survey of compounds for activity in the suppression of mouse sebaceous glands. Cancer Res. 18:887-892.
3. Chouroulinkov, I., Ph. Lazar, C. Izard, C. Liberman, and M. Guerin (1969) "sebaceous glands" and "hyperplasia" tests as screening methods for tobacco tar carcinogenesis. JNCI 42:981-985.
4. Chouroulinkov, I., and P. Lazar (1974) Action cancérogène et cocancérogène du 12-0-tétradécanoy-phorbol-13-acétate (TPA), sur la peau de souris. C.R. Acad. Sci. (Paris) 278D:3027-3030.
5. Chouroulinkov, I., and C. Lasne (1978) Two-stage (initiation-promotion) carcinogenesis in vivo and in vitro. Bulletin du Cancer 65:254-264.

6. Chouroulinkov, I., C. Lasne, R. Lowy, J. Wahrendord, H. Becher, N.E. Day, and H. Yamasaki (1989) Dose and frequency effect in mouse skin tumor promotion. Cancer Res. 49:1964-1969.

7. Coulomb, H., Y. Courtois, F.B. Callais, B. Festy, and I. Chouroulinkov (1983) Short-term Bioassays in the Analysis of Complex Environmental Mixtures, III, M.D. Waters et al., eds. Plenum Publishing Corp., New York, pp. 485-497.

8. DiPaolo, J.A. (1980) Quantitative in vitro transformation of Syrian golden hamster embryo cells with the use of frozen stored cells. JNCI 64:1485-1489.

9. Dontenwill, W., H.J. Chevalier, H.P. Harke, et al. (1973) Investigation on the effects of chronic cigarette smoke inhalation in Syrian golden hamsters. JNCI 51:1781-1832.

10. Feron, V.J., R.A. Griesemer, S. Nesnow et al. (1986) Testing of complex chemical mixtures. IARC Scientific Publ. 83:483-492.

11. Guerin, M., and J. Cuzin (1961) Tests cutanés chez la souris pour déterminer l'activité cancérogène des goudrons de fumée de cigarettes. Bull. Cancer 48: 111-121.

12. IARC (1984) Polynuclear aromatic compounds, part 3, industrial exposures in aluminum production, coal gasification, coke production, and iron and steel founding. IARC Monographs on the Evaluation of the Carcinogenic Risk of Chemicals to Humans, Vol. 34. Lyon, France.

13. IARC (1985) Polynuclear aromatic compounds, part 4, bitumens, coal-tars and derived productions, shale-oils and soots. IARC Monographs on the Evaluation of the Carcinogenic Risk of Chemicals to Humans, Vol. 36. Lyon, France.

14. IARC (1986) Monographs on the Evaluation of the Carcinogenic Risk of Chemical to Humans, Vol. 38, Tobacco Smoking. Lyon, France.

15. Kenaway, E.L. (1955) The identification of a carcinogenic compound in coal-tar. Brit. Med. J. 11:749-752.

16. Lasne, C., G. Nguyen-Ba, and I. Chouroulinkov (1988) Inhibition of chemical skin carcinogenesis in mice by arachis oil in an excipient. Proc. of the AARC 29:163.

17. Lazar, P., C. Liberman, I. Chouroulinkov, and M. Guérin (1963) Test sur la peau de souris por la détermination des activitiés carcinogènes: Mise au point méthodologique. Bull. Cancer 50:567-577.

18. Lazar, Ph., I. Chouroulinkov, C. Liberman, and M. Guerin (1966) Benzo(a)pyrene content and carcinogenicity of cigarette smoke condensate-results of short-term and long-term tests. JNCI 37:573-579.

19. Lazar, Ph., and I. Chouroulinkov (1974) Validity of the sebaceous gland test and the hyperplasia test for the prediction of the carcinogenicity of cigarette smoke condensate and their fractions. In Experimental Lung Cancer, Carcinogenesis and Bioassays, Springer-Verlag, Heidelberg, Germany.

20. Lazar, Ph., I. Chouroulinkov, C. Izard, P. Moree-Testa, and D. Hemon (1974) Bioassays of carcinogenicity after fractionation of cigarette smoke condensate. Biomedicine 20:214-222.

21. Lu, Y.P., C. Lasne, and I. Chouroulinkov (1986) Use of an orthogonal design method to study two-stage chemical carcinogenesis in BALB/3T3 cells. Carcinogenesis 7:893-898.

22. McCann, J., and B.N Ames (1976) Detection of carcinogens as mutagens in the Salmonella/mircosome test: Assay of 300 chemicals: Discussion. Proc. Natl. Acad. Sci., USA 73:950-954.

23. Potter, M., and Percival Potts (1963) Contribution to cancer research. In NCI Monograph No. 11, pp. 1-13.

24. Shimkin, M.B., and V.A. Triolo (1969) History of chemical carcino-
     genesis: Some prospective remarks.   Progr. Exp. Tumor Res.
     11:1-20.
25. Simpson, W.L., and W. Cramer (1943) Sebaceous glands and experi-
     mental skin carcinogenesis.   Cancer Res. 4:236-240.
26. Yamagiwa, K., and K. Ichikawa (1918) Experimental study of the
     pathogenesis of carcinoma.   J. Cancer Res. 3:1-21.

CONTRIBUTION OF POLYCYCLIC AROMATIC HYDROCARBONS AND

OTHER POLYCYCLIC AROMATIC COMPOUNDS TO THE CARCINOGENICITY

OF COMBUSTION SOURCE AND AIR POLLUTION

G. Grimmer,[1] H. Brune,[2] G. Dettbarn,[1] J. Jacob,[1]
J. Misfeld,[3] U. Mohr,[4] K.-W. Naujack,[1] J. Timm,[5]
and R. Wenzel-Hartung[2]

[1]Biochemical Institute for Environmental Carcinogens
Grosshansdorf

[2]Vaselinwerk, Biological Laboratory, Hamburg

[3]Institute of Mathematics, University, Hannover

[4]Institute of Experimental Pathology, Medical School
Hannover

[5]Department of Mathematics, University, Bremen

Federal Republic of Germany

INTRODUCTION

From epidemiological studies, it is obvious that various cancer diseases are caused by exogenous factors. However, the existence of these factors has been proven only for a few types of cancers. For example, a definite correlation is well established between smoking habits and formation of malignant tumors of the respiratory tract. The Department of Epidemiology and Statistics of the American Cancer Society has calculated from the epidemiological studies so far that 82.7% of the male lung cancer cases in the United States can be attributed to cigarette smoking. Accordingly, only 17% of lung cancers remain, possibly caused by occupational impact, which may be correlated to ubiquitously occurring environmental carcinogens or which cannot be attributed to any factors. However, only 35% of the cancer mortality can be attributed to smoking, if all kinds of cancers occurring in males in the U.S. are regarded.

There are several classes of environmentally occurring compounds which may cause malignant neoplasms in animal experiments. Regarding lung cancer--the main cancer disease in males--one has to look for carcinogenic compounds in the gaseous and the particle phase of various emissions. However, these emissions are very complex mixtures of various classes of compounds. The logical approach for the identification of carcinogenic constituents in these mixtures is to fractionate them such as the emission condensate from coal combustion or automobile exhaust, into chemical classes and then to compare the biological activity of the fractions with that of the total emission.

All emissions from incomplete combustions contain polycyclic aromatic hydrocarbons (PAHs) which are a well-known class of carcinogens. The question whether there are additional carcinogenic compounds present in this emission can be answered by separating the condensate into PAH-containing and PAH-free parts. These are subsequently tested in an animal experiment by means of a carcinogen-specific test system such as implantation into the lung of Osborne-Mendel rats (2,3,15) or the topical application of the chemical to the skin of mice.

CONTRIBUTION OF POLYCYCIC AROMATIC HYDROCARBONS TO THE CARCINOGENIC POTENTIAL OF DIESEL EXHAUST PARTICLES

Figure 1 shows a fractionation scheme of diesel exhaust particles. In a first step, the toluene extract of the diesel exhaust loaded glass fiber filters together with the rinsing of the glass cooler is separated into a water soluble and lipid soluble fraction. Subsequently, the lipid soluble part is separated into a PAH-free, a PAH-containing, and a polar fraction by column chromatography on Sephadex LH 20. Finally, the PAH-containing mixture is fractioned into PAH and nitro-PAH by chromatography on silica (9).

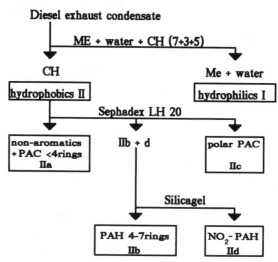

Fig. 1.   Separation scheme of diesel exhaust condensate (ME = methanol, CH = cyclohexane).

The above fractions then are tested in a carcinogen-specific animal test system, dosed proportionately, i.e., in ratios as they appear in the original diesel exhaust. Figure 2 shows the various fractions which have been tested by implantation into the lungs of Osborne-Mendel rats (9).

On the right side of the Fig. 2 illustration, the carcinoma incidence is given in percent. The fractions tested are indicated on the abscissa. The left part of Fig. 2 compares the carcinogenic potency of the water- and the lipid-soluble part. Only the latter causes lung carcinoma in the test animals.

After further chromatographic separation, only the PAH-fraction and, to a minor extent, the nitro-PAH exhibit carcinogenic potentials. Although only 1% by weight of the total extract, the PAH-fraction represents the carcinogenic activity of the total toluene extract. The fraction IIa+b+c+d has been obtained by reconstituting the total lipid-soluble parts.

The potency of benzo(a)pyrene (BaP) is shown at the right side of the figure. About 0.06 mg of pure BaP simulate the effect of 26.7 mg of diesel particle extract which, however, contains only 0.0024 mg BaP. Accordingly, the BaP content of the exhaust accounts for only 4% of the carcinogenic effect, when the lungs of Osborne-Mendel rats are used as the target organ.

The composition of the carcinogenic PAH-fraction is given in Fig 3. This fraction contains PAH exclusively. The PAH-profile is characteristic of emissions from incomplete combustion processes of sulphur-poor fuels.

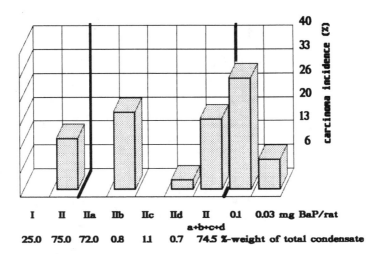

I	II	IIa	IIb	IIc	IId	II a+b+c+d	0.1	0.03 mg BaP/rat
25.0	75.0	72.0	0.8	1.1	0.7	74.5 %-weight of total condensate		

I = hydrophilic fraction (single dose: 6.7 mg)
II = hydrophobic fraction (20.0 mg)
IIa = non-aromatics and PAC 2+3 rings (19.22 mg)
IIb = PAH 4-7 rings (0.21 mg)
IIc = polar PAC (0.29 mg)
IId = nitro-PAH (0.19 mg)
IIa+b+c+d = reconstituted hydrophobics

Fig. 2. Carcinoma incidence of diesel exhaust condensate and its fractions as evaluated from implantation into the lungs of rats.

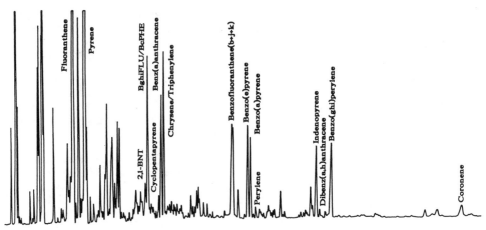

Fig. 3.   Gas chromatographic separation of the most potent carcinogenic
           fraction (fraction llb) from diesel exhaust condensate containing
           PAH.

CONTRIBUTION OF POLYCYCLIC AROMATIC HYDROCARBONS TO THE CARCINOGENIC POTENTIAL OF THE PARTICLE PHASE OF EXHAUST FROM GASOLINE-DRIVEN ENGINES

When the particle phase of gasoline-engine exhaust is fractionated analogously and fractions are proportionately administered, by far the most part of the carcinogenic potential is also located in the PAH-fraction. These findings are nondependent on the biodetector used (rat lung implantation or topical application onto the skin of mice). This is demonstrated with one dose each in Fig. 4.

As shown in the upper part of Fig. 4, the effect of 10 mg of exhaust condensate may be compared with that of 0.28 mg of its PAH-fraction in the case of a rat lung implantation (5). The lower part of the figure shows results obtained from the topical application onto the skin of mice. In this series the vehicle exhaust condensate has been applied twice weekly using a dose of 1.752 mg which has been compared with the carcinogenic effect provoked by fractions thereof which have been administered proportionately. In this model, again 0.061 mg of the PAH-containing fraction causes the same effect as 1.752 mg of the original condensate, whereas all other fractions exhibit only a very minor effect (4).

To calculate the portion of the biological effect caused by the various fractions, it is recommended to use several different doses. Figure 5 demonstrates this with vehicle exhaust condensate in doses of 5, 10, and 20 mg, respectively, administered to the lungs of Osborne-Mendel rats. After transformation to a Probit net, the dose-incidence relationship becomes linear. The three doses of pure BaP and of the PAH-containing fraction result in a linear curve being parallel to that obtained from the original material within the margin of error (5).

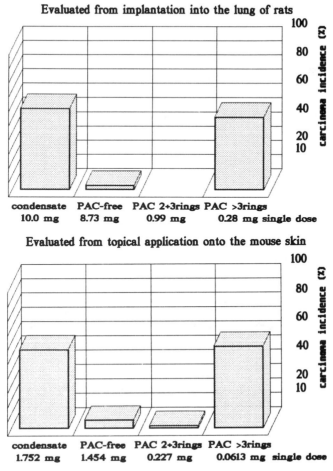

Fig. 4.   Comparison   of   carcinoma   incidence   of   automobile   exhaust
condensate and its fractions.

CONTRIBUTION   OF   POLYCYCLIC   AROMATIC   COMPOUNDS   TO   THE
CARCINOGENIC   POTENTIAL   OF   PARTICLES   FROM   HARD--COAL
COMBUSTION

      The   results   obtained   with   particulates   of   hard-coal   combustion
effluents   resemble   very   much   the   aforementioned   ones.   They   are
presented   in   Fig   6.   The   upper   part   of   the   figure   shows   the   results
obtained   after   implantation   of   the   filter   extract   and   fractions   thereof   into
the   lungs   of   rats   (8).   The   lower   part   presents   the   results   of   the
experiment   using   topical   administration   to   the   skin   of   mice.   It   also   can
be   recognized   that   in   this   case   (hard-coal   combustion)   that   the
carcinogenic   activity   is   caused   by   PAC   (6,7).   Apart   from   PAH   this
fraction contains predominantly thiaarenes and azaarenes (13,14).

      Table   1   presents   a   comparative   identification   and   quantification   of
the   carcinogenic   potencies   of   vehicle   exhaust   and   hard-coal   combustion
flue   gas   using   the   implantation   into   the   rat   lung   and   the   topical   application   to   the   mouse   skin   as   bioindicators.   In   the   case   of   gasoline-engine

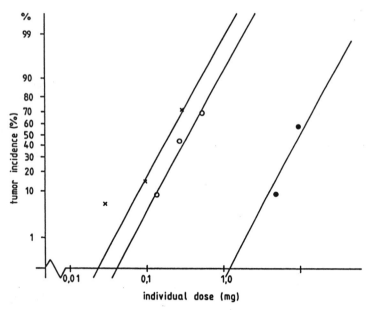

Dose-response relationship of the condensate (●) (doses: 5 and 10 mg,
20 mg was not soluble homogeneously and the PAH-fraction with > 3
rings (o) (doses: 0.14, 0.28 and 0.56 mg). BaP (x) (doses: 0.03, 0.10
and 0.30 mg).

Fig. 5.   Comparison of carcinoma incidence of automobile exhaust
condensate and its fractions as evaluated from implantation into
the lungs of rats.

exhaust, the PAC-free fraction causes a minor carcinogenic effect in both
biological systems. A contribution of 12.4% and 8.3%, respectively, to the
total effect may be calculated by Probit analysis. Similarly owner contributions, namely 6.0% and 16.2%, are found for the PAC-fraction
containing 2- and 3-ring systems. The main part of the carcinogenic
potential is caused by PAC with four and more rings. They contribute
by 84% and 81% to the carcinogenicity of the total exhaust condensate in
the two biosystems.

Once again, in the case of hard-coal flue gas condensate, the
PAC-free fraction and the fraction containing 2- and 3-ring systems
exhibit a very low, hardly detectable effect in both of the aforementioned
biosystems, whereas the PACs with 4-, 5-, 6-, and more rings cause the
carcinogenic effect. The contribution of BaP to the total carcinogenic
effect was found to be different in both systems in case of vehicle
exhaust with 7.6% and 2.4% and in the case of hard-coal flue gas with 12%
and 1.4%. This might be explained by the sensitivity of some PAHs
against oxygen resulting in their destruction during the mouse skin
experiment. Accordingly, these reactive PAHs contribute to the total
carcinogenic effect only in the case of lung implantation.

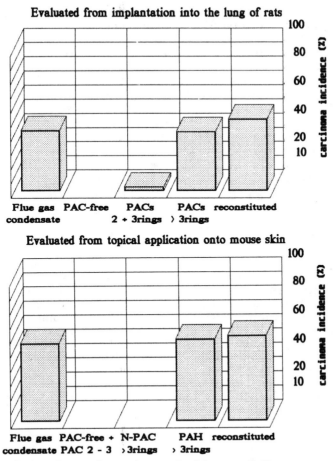

Fig. 6.   Comparison of carcinoma incidence of flue gas condensate from
coal combustion, its fractions, and the reconstituted flue gas
condensate.

CONTRIBUTION OF POLYCYCLIC AROMATIC COMPOUNDS TO THE
CARCINOGENIC POTENTIAL OF CIGARETTE SMOKE (SIDESTREAM
SMOKE)

The separation of sidestream cigarette smoke (11) leads to similar
results when the carcinogenic effect of semivolatiles is compared with that
of PAC-free and PAC-containing fractions from the particle phase. For
the collection of the particle and the gaseous phase (semivolatiles), a
filter combination is used, consisting of a glass fiber for the particles and
a subsequent sorption filter (silanized polystyrene beads) which collects
the semivolatiles with boiling points higher than 130°C quantitatively.

The sidestream semivolatiles of a cigarette do not cause lung cancer
carcinoma following lung implantation in Osborne-Mendel rats. The PAC-
free fraction of the particle phase which contains also 2- and 3-ring PACs
exhibit only a minor carcinogenic potency, whereas the main effect of the
sidestream is caused by the PAC-containing fraction (three rings),
although only 3% of the total cigarette smoke by weight (Fig. 8).

Tab. 1.   Comparison of the percentage contribution from the PAH-
          fractions to the total carcinogenic potency of automobile exhaust
          and flue gas condensate from coal combustion evaluated from
          implantation into the lung of rats or application onto the skin of
          mice (% of total potency)

	automobile exhaust skin/mice	automobile exhaust lung/rats	flue gas condensate skin/mice	flue gas condensate lung/rats
PAC-free fraction	12.4	8.3	0	1.9
PAC, 2 and 3 rings	6.0	16.2	0	2.0
PAC, 4 and more rings	84.0	81.0	ca.100	
- PAC, 4 and 5 rings			56	68.2
- PAC, 6 and more rings			50	54.6
Benzo(a)pyrene	7.6	2.4	12	1.4

A linear dose-response relationship in the range used provided a
dose of 0.06 mg BaP corresponds to the carcinogenic effect of the
sidestream of one cigarette which, however, contains only 0.0001 mg BaP.
Assuming that the carcinogenic potential of the total PAH-fraction
represents the sum of the effects caused by the individual PAH and PAC,
this means that only about 0.17% of the carcinogenic potential may be
explained by the BaP content in the case of sidestream cigarette smoke. A
more detailed investigation of the three sidestream fractions shows that
the carcinogenic fraction of the PAC-containing particle phase (1 mg)
exclusively consists of PAH, carbazoles, acridines, and aromatic amines
with more than three rings.

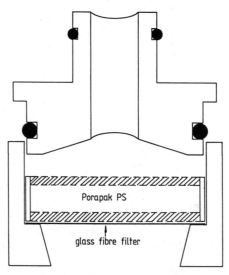

Fig. 7.   Filter system for collection of particles (silanized glass fiber
          filter) and semivolatiles (Porapak PS).

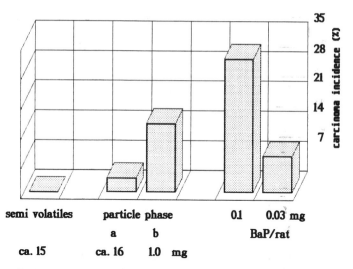

a: non-PAC and 2-3 ring PAC
b: PAC > 3 rings

Fig. 8.   Carcinoma incidence of fractions from sidestream cigarette smoke
          (semivolatiles and particulates) as evaluated from implantation
          into the lungs of rats.

The main fraction of the particle phase (about 16 mg) contains all of
the nonaromatic smoke constituents, 2- and 3-ring PAC, as well as all of
the tobacco-specific nitrosamines such as N-nitroso-nornicotine and
4-methylnitrosamino-1-(3-pyridyl)-1-butanone (NNK), a locally potent
carcinogen. Aromatic amines, such as 2-aminonaphthalene, are found in
both fractions of the particle phase.

Dimethylnitrosamine, the main part of aniline, toluidines, and a
lesser part of NNK, are found in the gaseous phase of the sidestream
smoke (about 15 mg) which contains the semivolatiles. As a result of the
comparison of the three cigarette smoke fractions, it may be stated that
the fraction which contains only PAH and azaarenes exhibits the most
pronounced carcinogenic potential.

EXTRAPOLATION OF THE RESULTS TO HUMANS

From occupational studies, it may be concluded that PAC-containing
emissions can provoke cancer in human skin and lung, such as the
formation of scrotum cancer in chimney sweeps, skin tumors in paraffin
workers, or increased lung cancer rates in coke plant workers. A
further possibility to compare biological reactions of animal and human
tissues is to investigate the metabolism of carcinogenic xenobiotics in cell
cultures of these tissues. This is demonstrated with chrysene as an
example in Fig. 9. It shows the gas chromatographic separation of
chrysene metabolites formed after incubation of epithelial hamster lung
cells with chrysene after a period of eight days. Four isomeric
hyroxychrysenes and two trans-dihydrodiols are formed (12).

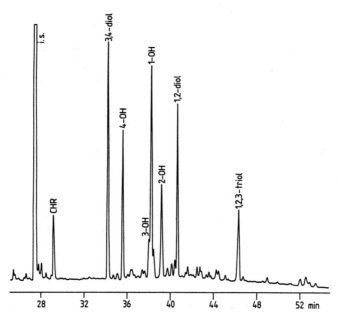

Abbreviations: 1-OH, 2-OH, 3-OH, 4-OH = 1-, 2-, 3-, 4-hydroxychrysene;
1,2-diol and 3,4-diol = trans-1,2-dihydroxy-1,2-dihydrochrysene and trans-3,4-
dihydroxy-3,4-dihydrochrysene; 1,2,3-triol = 1,2,3-trihydroxy-1,2-dihydrochry-
sene; i.s. = internal standard (benzo(c)phenanthrene); CHR = chrysene.

Fig. 9.   Gas chromatographic separation of chrysene metabolites formed
           during eight days incubation with epithelial hamster lung cells.

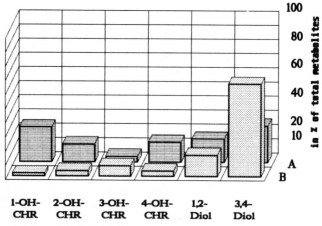

A: hamster lung epithelial cells

B: fetal human bronchial epithelial cells

Fig. 10.   Comparison of the profiles of chrysene metabolites after an
           eight-day incubaion with epithelial human and hamster lung cells
           (%).

Tab. 2.   Comparison   of   urinary   PAH-metabolite   concentration   of
          unexposed volunteers and coke plant and road workers (µg/l)

	1	2	3
1-OH-PHE	1.39	35.12	15.20
2-OH-PHE	0.80	15.71	8.71
3-and 9-OH-PHE	1.15	18.96	10.61
4-OH-PHE	0.11	0.52	0.39
ΣOH-PHE	3.45	70.31	34.91
ΣOH-FLU	‹0.0001	8.99	1.28
1-OH-PYR	0.25	7.59	2.63
5-OH-CHR	‹0.0001	0.79	0.05
6-OH-CHR	‹0.0001	0.11	0.003
4-OH-CHR	‹0.0001	0.28	0.01
3-OH-CHR	0.014	0.61	0.013
1-OH-CHR	0.015	0.71	0.01
ΣOH-CHR	0.029	2.50	0.086
3-and 9-OH-BaP	0.006	0.28	0.012
6-OH-BaP	‹0.0001	0.09	0.007
ΣOH-BaP	0.006	0.37	0.019

1: volunteers, unexposed
2: coke plant workers
3: road workers

As shown in Fig. 10, the metabolism is qualitatively similar in the
case of human epithelial lung cells.   Figure 10 compares the chrysene
metabolites formed with epithelial lung cells from hamster (A) and human
(B) within eight days.   The same metabolites are formed in both cell
systems.   However, quantitative differences can be recognized.   In
hamster lung cells, significantly higher concentrations of 1- and 2-hydro-
xychrysene   are   formed,   whereas   the   formation   of   trans-3,4-
dihydroxy-3,4-hydroxychrysene predominates in the human cells.

In summary, it may be stated that the metabolism of chrysene is
qualitatively similar in hamster and human lung tissue, which indicates an
analogous enzyme equipment in these tissues.   No basic differences of the
metabolic potential were found.   In the meantime, this has been confirmed
with various other PAHs.   The metabolites formed react either with cell
constituents such as DNA or protein, or are conjugated and excreted via
urine or feces.   Table 2 compares the urinary metabolite profiles of
different PAH-exposed people.

Comparing 14 urinary PAH-metabolites, significant differences between PAH-exposed and less exposed people can be recognized.   In
addition, the PAH-composition of the working place atmosphere is reflected in the urinary metabolite profile.   More volatile PAHs, such as
phenanthrene occurring in about the same concentrations in coke plants
and in road paving atmospheres, are excreted in the urine of the workers
as hydroxy-phenanthrenes also in about the same concentrations.   The
concentrations of the hydroxy-BaP, however, are found to be 370 ng/l
and 19 ng/l and hence differ by a factor of 20.   From this it may be
concluded that road workers are significantly less exposed to higher-boil-
ing carcinogenic PAHs (14).

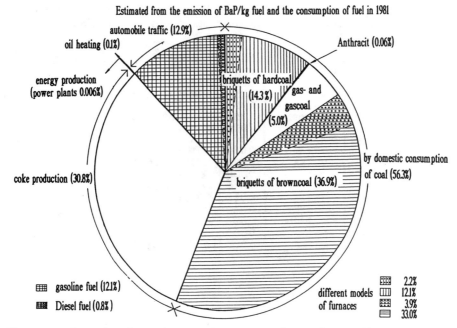

Fig. 11.   Contribution of benzo(a)pyrene from five different sources to the air pollution of Federal Republic of Germany (in total, about 18 t benzo(a)pyrene/year).

Tab. 3.   Percentage contribution of the PAH-fraction and benzo(a)pyrene to the carcinogenic potency of various emission condensates (evaluated from implantation into the lung of rats by probit analysis)

	PAH-fraction	BaP
DIESEL exhaust (0.8 %)[x]	ca. 80 %[xx]	4 %
automobile exhaust (2.8 %)[x]	81 %	2.4 %
flue gas of coal fired residential furnaces:		
a. PAC 4 - 7 rings (17.3 %)[x]	68.2 %	1.4 %
b. PAC > 7 rings and polar PAC (11.7 %)[x]	54.6 %	
cigarette smoke sidestream (3.5 %)[x]	ca. 75 %[xx]	0.17%

[x] weight-% of the PAH-fraction related to the total emission extract

[xx] estimated assuming that the dose-incidence-relation corresponds to that of BaP

CONTRIBUTION OF VARIOUS SOURCES OF EMISSIONS TO THE TOTAL BENZO(A)PYRENE BURDEN IN THE FEDERAL REPUBLIC OF GERMANY

Figure 11 shows the contribution of five different sources of emissions to the total BaP burden of the Federal Republic of Germany (1). The main contribution—more than one-half of the BaP emitted—originated in 1981 from residential household furnaces fired with hard-coal- or lignitte-briquets. Coke production also contributes about 31% to the total BaP-emission. Vehicle exhaust contributes about 13%, whereas coal-fired power plants and domestic oil heatings contribute only very little to the total BaP-emission.

SUMMARY

The investigations on the contributions of PAC-fractions and of BaP to the carcinogenic potential of four different pyrolytic condensates are summarized in Tab. 3. The carcinogenic effect of the particle phase is predominantly caused by polycyclic aromatic compounds containing four or more rings in the cases investigated, accounting for 75-100% of the carcinogenic potential evaluated from the implantation into the lungs of rats or from topical application onto mouse skin. In the case of hard-coal flue gas, the fractions containing PAC with four to six rings and PACs consisting of more than six rings have been tested separately and both fractions were found to equally contribute to the carcinogenic effect of the flue gas condensate. The contribution of BaP to the carcinogenic potency of various condensates, however, is minor in all cases investigated and accounts for only 0.17% to 4% of the total carcinogenicity as evaluated from implantation into the lungs of rats.

REFERENCES

1. Ahland, E., J. Borneff, H. Brune, G. Grimmer, M. Habs, U. Heinrich, P. Hermann, J. Misfeld, U. Mohr, F. Pott, D. Schmähl, J. Thyssen, and J. Timm (1985) Luftverschmutzung und Krebs--Prüfung von Schadstoffen aus verschiedenen Emissionsquellen auf ihre krebserzeugende Wirkung. Münch. Med. Wschr. 127:218-222.
2. Deutsch-Wenzel, R., H. Brune, and G. Grimmer (1983) Experimental studies on the carcinogenicity of five nitrogen containing polycyclic aromatic compounds directly injected into rat lungs. Cancer Lett. 20:97-101.
3. Deutsch-Wenzel, R., H. Brune, G. Grimmer, G. Dettbarn, and J. Misfeld (1983) Experimental studies in rat lungs on the carcinogenicity and dose response relationships of eight frequently occurring environmental polycyclic aromatic hydrocarbons. JNCI 71:539-544.
4. Grimmer, G., H. Brune, R. Deutsch-Wenzel, K.-W. Naujack, J. Misfeld, and J. Timm (1983) On the contribution of polycyclic aromatic hydrocarbons to the carcinogenic impact of automobile exhaust condensate evaluated by local application onto mouse skin. Cancer Lett. 21:105-113.
5. Grimmer, G., H. Brune, R. Deutsch-Wenzel, G. Dettbarn, and J. Misfeld (1984) Contribution of polycyclic aromatic hydrocarbons to the carcinogenic impact of gasoline engine exhaust condensate evaluated by implantation into the lungs of rats. JNCI 72:733-739.

6. Grimmer, G., H. Brune, R. Deutsch-Wenzel, G. Dettbarn, J. Misfeld, U. Abel, and J. Timm (1984) The contribution of polycyclic aromatic hydrocarbons to the carcinogenic impact of emission condensate from coal-fired residential furnaces evaluated by topical application to the skin of mice. Cancer Lett. 23:167-176.
7. Grimmer, G., H. Brune, R. Deutsch-Wenzel, G. Dettbarn, J. Misfeld, U. Abel, and J. Timm (1985) The contribution of polycyclic aromatic hydrocarbon fractions with different boiling ranges to the carcinogenic impact of emission condensate from coal-fired residential furnaces as evaluated by topical application to the skin of mice. Cancer Lett. 28:203-211.
8. Grimmer, G., H. Brune, R. Deutsch-Wenzel, G. Dettbarn, and J. Misfeld (1987) Contribution of polycyclic aromatic hydrocarbons and polar polycyclic aromatic compounds to the carcinogenic impact of flue gas condensate from coal-fired residential furnaces evaluated by implantation into the rat lung. JNCI 78:935-941.
9. Grimmer, G., H. Brune, R. Deutsch-Wenzel, G. Dettbarn, J. Jacob, K.-W Naujack, U. Mohr, and H. Ernst (1987) Contribution of polycyclic aromatic hydrocarbons and nitro-derivatives to the carcinogenic impact of diesel engine exhaust condensate evaluated by implantation into the lung of rats. Cancer Lett. 37:173-180.
10. Grimmer, G., G. Dettbarn, K.-W. Naujack, and J. Jacob (1988) Excretion of hydroxy-derivatives of polycyclic aromatic hydrocarbons of the masses 178, 202, 228 and 252 in the urine of coke and road workers. 18th International Symposium on Environmental Analytical Chemistry, Barcelona, Spain (Sept. 5-8).
11. Grimmer, G., H. Brune, G. Dettbarn, K.-W. Naujack, U. Mohr, and R. Wenzel-Hartung (1988) Contribution of polycyclic aromatic compounds to the carcinogenicity of sidestream-smoke of cigarettes evaluated by implantation into the lung of rats. Cancer Lett. 43:173-177.
12. Jacob, J., G. Grimmer, G. Raab, M. Emura, M. Riebe, and U. Mohr (1987) Comparison of chrysene metabolism in epithelial human bronchial and Syrian hamster lung cells. Cancer Lett. 38:171-180.
13. Schmidt W., G. Grimmer, J. Jacob, and G. Dettbarn (1986) Polycyclic aromatic hydrocarbons and thiaarenes in the emission from hard-coal combustion. Toxi. and Env. Chem. 13:1-16.
14. Schmidt, W., G. Grimmer, J. Jacob, G. Dettbarn, and K.-W. Naujack (1987) Polycyclic aromatic hydrocarbons with mass number 300 and 302 in hard-coal flue gas condensate. Fres. Z. Analyt. Chem. 326:401-413.
15. Stanton, M.F., E. Miller, C. Wrench, and R. Blackwell (1972) Experimental induction of epidermoid carcinoma in the lungs of rats by cigarette smoke condensate. JNCI 49:867-877.

OLD AND NEW CARCINOGENS

Lorenzo Tomatis

International Agency for Research on Cancer
Lyon, France

In populations of developed countries, 60% of all cancer cases occur in people over 65 years of age, confirming that the older one gets, the greater is the risk of dying of cancer. Age, in this sense, is indeed the most important risk factor for cancer, as it would permit a disease that is genetically determined to appear late in life to manifest itself, and it certainly also extends the duration of exposure to the variety of factors that directly or indirectly increase the risk of developing a clinical cancer.

The hypothesis that the occurrence of cancer is predominantly related to genetic factors can hardly be supported, except for very rare instances. It is clear instead that there is a genetic component, as there is in every disease, of very varied importance in the etiology of most cancers.

It is unknown if and to what extent the common and often lethal infectious diseases prevalent in early life in previous centuries exerted a severe selection, favoring not only individuals resistant to them but also to the development late in life of chronic degenerative diseases such as cancer. It is a fact that the sharp increase of the proportion of the total population reaching an age older than 65 has coincided with an increase in number and in exposure levels of carcinogenic agents in our environment, and it is at least theoretically possible that among the many individuals today reaching old age, there is a considerable proportion that is particularly susceptible to chronic diseases like cancer.

It is also possible that improved living conditions and health care may partly compensate for a marginal decrease in fitness, possibly related to the accumulation of slightly deleterious mutations (8) due to exposure to environmental mutagens. If this compensation were only temporarily adequate, then the growing proportion of individuals reaching old age having escaped hard selection in infancy, and the combined effect of the growing pollution of our environment and of hazardous behaviors, would eventually lead us to a situation where an increasing fraction of the aging

Genetic Toxicology of Complex Mixtures
Edited by M. D. Waters *et al.*
Plenum Press, New York, 1990

world population will need health care, turning our much-cherished progress into a nightmare. At present, however, the benefits that mankind
has obtained from the industrialized society--among which are primarily
the lowering of infant mortality, the decrease of infectious diseases, and
the extension of the life span--are still more evident than the possible
adverse effects.

There is, in fact, an apparent incongruity between the rather pessi-
mistic concern expressed above and the present evidence that, for the
fraction of the world population in industrialized countries, living conditions have never been better during the entire history of mankind.

Another incongruity is that the considerable decrease in mortality
rates became apparent in cohorts born around the end of last century and
the beginning of this century. As was noted by Brody (2), more than
half of the decline in mortality for individuals of 65 years or older was
achieved in the late 1940s. The decline in mortality continued thereafter,
at least until the end of the 1970s. One cannot help but observe that the
cohorts that are now, or have recently been, around and over 80 years
of age and for whom the decline in mortality has been most conspicuous
have probably had little time to follow the recommendations we presently
make for a better and healthier life. These cohorts have instead been
through some of the most difficult times, including two wars and a de-
pression and, as far as most of Europe is concerned, the long postwar
period which was characterized by little food of poor quality, no ideal
housing, and, in general, little of the conditions that make life pleasant.

There is presently some disagreement on whether the human population has already reached the longest attainable average life expectancy or
whether it could be further extended. In both instances, however, the
fraction of the population aged between 65 and 85 years cannot but
considerably increase in the near future. If the rate of morbidity remains at today's level, and even more so if it increases, the health care
structures already inadequate at present will be completely swamped by
the demand.

Focusing on the extension of active life expectancy in an aging
population is certainly very appealing, as it puts emphasis on personal
autonomy rather than on dependency on paternalistic and expensive care.

Whether Fries is right with his paradigma on the compression of
morbidity, which also implies that life expectancy cannot be further
extended (4,5) or whether others are (2) who claim that life expectancy
can be further extended, there is a general consensus that most of the
gain in years of life has consisted of years of poor health requiring
continued care. The health problems that increase after age 65 certainly
do not only relate to cancer. To cite two examples taken by Brody (2)
the projected number of hip fractures and of demented persons in the
U.S. shows a sharp increase in the next decades. If we want to prevent
a worsening of the recently observed trend of a substantial decline in
active life expectancy between ages 65 and 85, we have clearly to decide
what can be done to reduce the morbidity burden in the elder population.

Whatever the role of genetic predisposition to chronic diseases may
be, it is reasonable to expect that an intervention on environmental

exposures will definitely lead to a containment of a disease like cancer that in most cases becomes clinically apparent only late in life. Tuberculosis, for instance, a disease for which the genetic component is important, underwent a sharp decrease in its prevalence following favorable environmental measures, and not, or at least not mainly, because there was a decrease in the proportion of susceptible individuals (9). For one type of cancer, namely gastric cancer, the decrease we presently observe is more likely to be attributable to changes in eating habits and in the preservation of food, than to a decrease in the proportion of individuals with blood group A, who have a greater risk for such cancer.

Industrialization and a series of other events, mainly occurring during the 19th Century, have contributed to the addition of new environmental hazards and new environmental carcinogens to those that existed well before industrialization began and with which humans have possibly been exposed since the origin of the species. Among the "new" carcinogens can also be considered certain naturally occurring substances, such as several metals and asbestos since they became serious threats to human health only following their massive exploitation which began in the last century. Similarly, even if tobacco as a natural plant is present on our planet since immemorial times, cigarettes and tobacco smoke are rather recent products that could hardly be defined as natural; like many other industries, chemical industries in particular, the industrial production of cigarettes began in the middle of the last century (6).

The "old" carcinogens, that is, those to which humans had been exposed long before industrialization began, are ionizing and nonionizing radiations, mycotoxins, certain combustion products, and probably certain viruses. There are different levels of uncertainty with regard to these "old" carcinogens: (i) What proportion of cancer cases can be attributed to them? (ii) How many more "old" carcinogens are there and remain to be uncovered?

Similar uncertainties exist for the "new" carcinogens which represent the large majority among the 57 agents carcinogenic to humans firmly identified until now (10). We have until now been unable, however, to establish satisfactorily the proportion of cancer cases that are causally related to them. Moreover, we also ignore if there are other carcinogens, and how many more, among the thousands of chemicals to which we are exposed in our environment and for which there are not data on their possible carcinogenicity.

While accepting the limitations of our knowledge, we may, however, agree on the fact that we have at least been successful in identifying certain categories of risk factors, most of which are amenable to primary prevention. The limited knowledge that we presently have of the etiology of human cancer is no excuse for not implementing prevention wherever this is possible. We can, for instance, prevent the further spread of smoking among youngsters and, in developing countries, prevent occupational risks, prevent pollution of the environment by the carcinogens we have already identified, as well as prevent new carcinogens from being produced and used.

A different type of uncertainty relates to the role that certain risk factors could play, for which a causal relationship with human cancer has not been firmly established, but which are usually considered to be relevant in the origin of cancers occurring at certain sites. The most important among these are certainly those related to nutrition. Dietary factors are generally classified among the "old" carcinogens since, by definition, humans must have always been exposed to food. The composition of food, however, changes conspicuously with time. During the 19th century, for instance, eating habits of the majority of the population underwent considerable change. The average European consumed about 18 ounces of meat per year until the beginning of the last century, whereas in the middle of the century the average consumption of meat and fish was estimated to have risen in Britain, Belgium, and Germany to about 100 pounds a year (7). At the same time, other products rather unusually consumed before became more common, among the last being the dairy products, the wide consumption of which could begin only after the introduction of pasteurization. More recent has been the availability of fresh vegetables and fruits all year 'round practically everywhere in the industrialized countries, while the rapid expansion of the so-called "fast food" is introducing another element of change in the eating habits of today's youngsters.

With the caveat dictated by the uncertainties mentioned above, we may note that: (i) most of the agents so far firmly identified as being causally related to human cancers are relatively new, and mankind began to be exposed to them for less than two centuries, and more intensely for less than one century; and (ii) the vast majority of these carcinogenic agents are chemicals or chemical mixtures.

This may be due to various reasons, some of them interrelated. (i) It may be that cancer is mainly a chemical disease, that the majority of carcinogens are indeed environmental chemicals and that the preponderance of such chemicals among the identified carcinogenic agents reflects the reality of the situation. (ii) The agents identified until now may be potent carcinogens and/or exposure to them may have been particularly high. Their effect has been, therefore, easily detected by the reports of cases, in many instances, even before an epidemiological survey was carried out. Even if these hypotheses are true, it would not necessarily imply that most human cancer cases are due to the chemicals so far identified, nor that most cases must by necessity be attributed to chemicals. (iii) The preponderance of chemicals among the carcinogenic agents so far identified may be the result of a selective bias related to the type of approaches used for their identification, that may be more apt to identify chemical carcinogens than other carcinogenic agents. Some inherent characteristics and limitations of the epidemiological and experimental approaches, at least in the way they have been used until now, could, therefore, be the limiting factors for the type of carcinogenic agents that have been so far identified. (iv) The experimental approaches used for the identification of carcinogens have been largely based on the mechanistic assumption that most human cancers are causally associated with exposures to agents that directly damage DNA and cause mutations. This assumption could be at least partly wrong (3), in which case it would be reasonable to assume that agents that increase the risk of cancer by different mechanisms could remain undetected. The preponderant proportion of environmental chemicals among the identified human carcinogens would, therefore, be explained by the fact that the agents operating through mutational mechanisms were preferentially detected.

The sensitivity of long-term animal tests to detect human carcinogens, as it can be measured by using results on the limited number of human carcinogens for which a comparison can be made, is higher than that of short-term tests. in particular if data with sufficient and limited evidence of carcinogenicity in animals are pooled together. The decline that has been reported recently in the degree of concordance between carcinogenicity in laboratory animals and the results in short-term tests is probably to be attributed to the increasing proportion, among the chemicals included in validation assays, of those which may act through mechanisms other than those involving mutagenic/electrophilic intermediates (1). To what extent this reflects the actual situation in relation to the etiology of human tumors is not at all clear.

The sensitivity of short-term tests with different end-points is between 67% and 88%; if tests with different phylogenetic orders are combined with regard to end-points, the sensitivity varies between 54% and 92% (10).

Among the most important recognized risk factors for human cancer are complex mixtures within which it has been difficult or impossible in some instances to identify precisely the agent(s) responsible for the carcinogenic effect. In fact rarely, if ever, have humans been exposed to pure, single agents. Even if not necessarily always in the form of a mixture, human exposures generally involve a plurality of agents which act either simultaneously or in sequence, by the same or by different routes. The inhalation of tobacco smoke, for example, may involve the exposure to a variety of carcinogenic agents that may act simultaneously as well as in sequence following the same route of exposure; rhinopharyngeal carcinomas and primary liver carcinomas instead may be the result of an action of different agents acting in sequence and following different routes of exposure, as it would, for instance, be the case for the Hepatitis B virus (HBV) and aflatoxins for liver cancer.

There are two conspicuous limitations in our present knowledge of the etiology of human cancer. (i) The agents so far identified as human carcinogens are much more relevant to tumors occurring in men than in women. This imbalance may be related to the fact that many more men than women have been exposed to occupational hazards, that men took up the habit of smoking earlier than women, and consumed greater quantities of alcoholic beverages than women. It is unclear to which extent, in addition, our male-dominated society has influenced the choice of priorities in research, favoring matters that involved predominantly males. (ii) The etiological agents so far firmly identified are mainly associated with tumors occurring at certain sites, the most frequent targets being the lung, bladder, emolymphopoietic system, skin, liver, nasal sinuses, and pleura. Notably absent among these are cervix and breast, the most frequent cancer sites in females [diethystilbestrol (DES) and other estrogens are the only chemicals which have been shown to be causally related to the breast cancer], colon-rectum, stomach, ovary, brain, and prostate. There are 18 agents or exposures causally related to lung cancer, 12 to bladder cancer and emolymphopoietic systems, and 11 relate to skin cancer.

CONCLUSIONS

(i)   The identification of a number of human carcinogens, in parti-
       cular the "new" carcinogens, has been rather successful.

(11)   Most of the identifed human carcinogens are chemicals and/or
       chemical mixtures and are amenable to primary prevention; they
       are mainly associated with tumors occurring at the lung, blad-
       der, emolymphopoietic system, skin, liver, nasal sinuses, and
       pleura.

(iii) While there are etiological hypotheses concerning cancers occur-
       ring at some of the most frequent target sites--breast, cervix,
       and ovary for women; stomach, colon, and rectum for both
       sexes; and prostate in males--no agents have yet been clearly
       identified that could be amenable to primary prevention; the
       observance of sexual hygiene for cervical cancer, and the
       regular consumption of fruits and vegetables for cancer of the
       gastrointestinal tract are, however, correlated with a lower
       risk.

(iv)   Priorities in primary prevention are, therefore, of two types:
       (a) the efficient implementation of measures concerning the
       carcinogens we have already identified, and the avoidance of
       "new" carcinogens entering our environment; and (b) a re-
       search effort identifying agents involved in the development of
       cancers occurring at the sites mentioned above, and clarifying
       the mechanisms underlying the various steps of the process
       leading to the clinical appearance of cancer.

A better knowledge of the different combinations of events that may
be at the origin of certain tumors may, in the future, influence the type
and efficacy of preventive measures that should be taken.

REFERENCES

1. Bartsch, H., and C. Malaveille (1989) Prevalence of genotoxic
    chemicals among animal and human carcinogens evaluated in the IARC
    Monograph series. Cell Biol. Virol. 2:115-127.
2. Brody, J.A. (1985) Prospects for an aging population. Nature
    315:463-466.
3. Cairns, J. (1981) The origin of human cancer. Nature 289:383-387.
4. Fries, J.F. (1988) Aging, illness and health policy: Implications of
    the compression of morbidity. Persp. Biol. Med. 31:407-428.
5. Fries, J.F., L.W. Green, and S. Levine (1989) Health promotion and
    the compression of morbidity. The Lancet 1:481-483.
6. International Agency for Research on Cancer (1986) IARC Mono-
    graphs on the Evaluation of Carcinogenic Risks to Humans, Vol. 38.
    Tobacco Smoking, Lyon, France.
7. Knapp, V.J. (1988) Major dietary changes in nineteenth-century
    Europe. Persp. Biol. Med. 31:188-193.
8. Kondrashov, A.S. (1988) Deleterious mutations and the evolution of
    sexual reproduction. Nature 336:435-440.

9. McKeown, T. (1988) The Origins of Human Disease. Basil Blackwell, Oxford and New York.

10. Tomatis, L., A. Aitio, J. Wilbourn, and L Shuker (1989) Human carcinogens so far identified. Japanese J. of Cancer Res. 80:795-807.

EXPERIMENTAL AND EPIDEMIOLOGIC APPLICATIONS

TO CANCER RISK ASSESSMENT OF COMPLEX MIXTURES

Marja Sorsa,[1] Harri Vainio,[1] and Anthony J. McMichael[2]

[1]Institute of Occupational Health
Helsinki Finland

[2]Department of Community Medicine
University of Adelaide
Adelaide, Australia

INTRODUCTION

Both the testing and the risk evaluation of human exposure to complex mixtures present difficult scientific problems. To better define the toxicity of complex mixtures, and thus to give a more precise and reliable risk assessment, there is a need for extensive experimental research as well as development of the risk assessment strategies. In the international workshop held in Espoo, Finland, in May 1989, the lines of exploration that are likely to yield the most relevant and practically useful information were discussed. The approaches for improved risk assessment include more effective use of epidemiology (including molecular epidemiology), health surveillance and biological monitoring (in vivo data on exposed humans), as well as in vivo toxicology studies supplemented with the relevant in vitro experiments and knowledge of chemistry.

The workshop on risk assessment with complex mixtures was designed to give scientists from different disciplines the opportunity to talk to each other in order to facilitate a multidisciplinary approach to solving difficult problems. The papers presented and the topics discussed, as examples of human exposure situations with complex mixtures, covered typical cases from both environmental and occupational settings in an interdisciplinary spirit (2).

Most instances of environmental human exposures involve concurrent or sequential exposures to mixtures of compounds in air, water, and food. Many of the individual components in typical complex mixtures are carcinogenic in humans and/or in animals (1). A complex mixture is considered to be a combination that is either deliberately formulated (e.g., gasoline, paint) or has a common source (e.g., diesel exhaust, tobacco smoke). The distinguishing feature of such an exposure is that it entails

Genetic Toxicology of Complex Mixtures
Edited by M. D. Waters *et al.*
Plenum Press, New York, 1990

149

Tab. 1.   Particular problems in studying complex mixtures

- Individual components are unknown or insufficiently specified

- The composition may change over time and differ according to the source of emission

- Samples are difficult to collect, extract, and prepare for testing

- Quantitative estimates of cumulative exposure at the individual level are difficult to make

- Identification of the appropriate biomarkers(s) is difficult when the mixture contains a number of potential carcinogens

- Interactive effects may complicate the assessment of risk

- No mechanistically relevant dose-response models are available for carcinogenesis of complex mixtures

an essentially indivisible "package" of exposures to a combination of agents. Mixtures may be of widely different chemical complexity, ranging from technical chemical products with a few defined major components (e.g., some pesticides and dyes), to highly complex mixtures or series of similar components [e.g., polychlorobiphenols (PCBs)], to mixtures that can be best defined only by their sources (e.g., coal tars).

PARTICULAR PROBLEMS WITH COMPLEX MIXTURES

Certain characteristics of complex mixtures may lead to special problems concerning cancer risk assessment as noted in (Tab. 1).

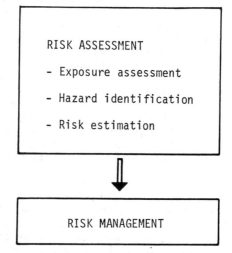

Fig. 1.   Definition of risk assessment, as a prerequisite for risk management, used in the context of this discussion.

COMPONENTS OF RISK ASSESSMENT

There is no generally agreed definition of the term "risk assessment." In relation to human cancer, it typically entails the use of qualitative and quantitative data, directed at the identification of carcinogenic hazards and the estimation of carcinogenic risks.

The components of risk assessment and its relationship, as a prerequisite, to risk management, are shown in Fig. 1. Risk assessment is a process consisting of exposure assessment, risk identification (hazard identification), and risk estimation (dose-response assessment).

EPIDEMIOLOGICAL AND EXPERIMENTAL APPROACHES
TO RISK ASSESSMENT

Quantitative risk assessment has conventionally estimated the risk of cancer incidence in humans in relation to measured levels of external environmental exposure. Such estimation has generally depended on extrapolation from animal bioassay data, since epidemiological data are infrequently available. This applies to both simple and complex exposures. This reliance upon animal data, in the absence of human biomarkers of known carcinogenic predictivity, is necessary to anticipate and minimize time awaiting definitive human evidence. However, use of predictive biomarkers in the exposed population is warranted in order to save time from compiling epidemiological data, and still be able to prevent unnecessary end effects (i.e., cancer) (Fig. 2).

Risk assessment of complex mixtures can begin either from epidemiological studies in relation to actual exposures, or from laboratory studies of an existing complex mixture for which no human observations have yet been made. In case human exposure has already occurred, the data relevant for risk assessment may be obtained from studying the people exposed, supplemented with the data obtained by examining the complex mixture, its chemistry, and its toxicology. If no human exposure has yet occurred, the data for risk assessment can be obtained only by studying the experimental toxicological characteristics of the mixture, and

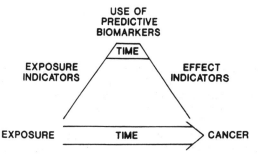

Fig. 2. The general logic in using biochemical/molecular changes as biomarkers is their early predictiveness of late effects, such as cancer.

Tab. 2. Suggested outline for a risk assessment strategy with complex mixtures

---

1. Epidemiological studies of populations exposed to complex mixtures

      1.1 Causality inference (i.e., risk identification)
      1.2 Study of interactions specific to humans
      1.3 Quantitative considerations

2. Assessment of the human exposure and relevant biological effects to complex mixtures

      2.1 "Nonspecific" indicators (i.e., chromosomal aberrations, micronuclei, SCEs, point mutations, oncogene activations, etc.)
      2.2 Specific" indicators: DNA adducts
      2.2.1 Broad "nonchemical specific" spectrum adducts (^{32}P-post-labelling)
      2.2.2 Chemical-class specific adducts (e.g., PAH-adducts)
      2.2.3 Specific constituents of complex mixtures (e.g., BaP-adducts)
      2.3 Exposure indicators
      2.3.1 "Nonspecific" broad spectrum indicators (e.g., urinary mutagenicity, etc.)
      2.3.2 Specific constituents of the complex mixtures (e.g., I-OH-pyrene)

3. Experimental approaches to risk assessment of complex mixtures; study of complex mixtures in vivo and in vitro

      3.1 In vivo studies
      3.1.1 Animal bioassays of complex mixtures
      3.1.2 Animal bioassays of components of complex mixtures
      3.2 In vitro studies
      3.2.1 Activity-guided fractionation of complex mixtures
      3.2.2 Study of individual components of complex mixtures
      3.3 Structure-activity relationships

4. Chemical characterization of the complex mixture

      4.1 Activity-guided fractionation: chemical analysis
      4.2 Search for active principles; macromolecular adducts as a guide
      4.3 Traditional chemical characterization of the mixtures

---

then extrapolating the results to humans. Whenever biological markers, as intermediate outcomes measured in either animals or humans, are used in the quantitative modelling of carcinogenic risk, it is important that their toxicological validity be already known.

The quality and quantity of pertinent information available for risk assessment of complex mixtures varies considerably and is generally less than for pure chemicals. It is usual that only a few components of the complex mixture are known, exposure data are uncertain, toxicological data are limited, and epidemiological data are nonexistent.

Tab. 3. Problem areas, gaps in knowledge, and needs for future research specified in the workshop (2)

Problem Area	Action/Recommendation
Exposure assessment	Individual exposure assessment with improved specificity should be applied in future epidemiological studies of complex mixtures. This must entail reconstruction of retrospective exposure situations, detailed exposure histories, and application of exposure matrices.
	More work needs to be undertaken on total exposure assessment, i.e., integration of exposures arising from multiple routes.
Representativity of samples	To study mechanistic aspects and possible variations in processing of complex mixture sampling, both "real" samples and "artificial" samples should be used. Laboratory-derived mixtures can be controlled in their composition thus allowing replication of experiments.
	There is a need to develop further the in situ environmental assessment technologies and the utility of sentinel surveillance systems.
Interactions of individual components	More work should be targeted towards understanding of additive, synergistic, or antagonistic interactions in complex mixtures not observed with the individual fractions.
	It needs to be confirmed whether the prevailing toxicological evidence showing less than one order of magnitude deviations from additivity also holds true in carcinogenicity studies.
Dose-response relationships	Carefully designed dose-response studies on complex mixtures should be conducted which fully utilize the current knowledge on mechanisms of action and appropriate biomarkers, applying both experimental animals and human exposure situations.
Nongenotoxic components	There is a need to develop reliable molecular markers for exposure to nongenotoxic carcinogens.
Genotoxicity data	Continued use and development of short-term testing for genotoxicity should be encouraged to identify potentially carcinogenic components in complex mixtures.
	Work with computerized databases, such as the Genetic Activity Profiles, should be developed further.
Interindividual variation	Research is needed to more precisely characterize the implications of interindividual variation to risk estimations, including the development and validation of markers of susceptibility for exposure to complex mixtures.
	Repetition of well-designed human studies, including proper exposure assessments in combination with epidemiology, should be encouraged to be performed in different populations and geographic areas.

Tab. 3, continued

Problem Area	Action/Recommendation
Macromolecular adducts	Further validation studies are needed to understand interindividual variability, intraindividual variation, persistence of adducts in different cell types, and relationship between adducts in surrogate and target tissue, as well as adducts and other biomarkers measuring genetic effects. Follow-up studies both in animals and prospectively in humans are needed to validate the final relevance of adducts to be used as predictors of cancer risk.

RECOMMENDATIONS FOR IMPROVEMENT OF RISK ASSESSMENT STRATEGIES WITH COMPLEX MIXTURES

Present Applications

There is no simple and reliable method for assessing risks from complex mixtures. Each situation is unique. Although a formalized approach can be misleading, a series of essential considerations (a checklist) can be defined, which may give the ideas of data needed for risk assessment. In the international workshop, the four categories of presently available methodologies were discussed, from chemical characterization to experimental testing and, further, to human exposure assessment and epidemiology of end effects (Tab. 2).

Much of the discussion was oriented toward application of biological markers in humans. However, great individual variability in response is to be expected because of genetic or acquired host susceptibility factors, and these may change with age. Since epidemiological studies of cancer cannot anticipate risks for agents only relatively recently brought into use, studies of other biological effects on humans are, therefore, needed for predicting potential cancer hazards and to achieve prevention. The development, validation, and more widespread use of biological markers enable more rapid assessment of human cancer risks.

Future Needs of Research

The workshop discussed various lines of exploration likely to yield the most relevant and practically useful information, and identified obvious gaps in knowledge. In the following, some of the specific recommendations are listed (Tab. 3).

In general, international collaboration was emphasized in carrying out prospective follow-up studies using biological indicators in potential risk groups. For future follow-up the necessity of banks preserving biological specimens was stressed, considering that many human tissue banks are already available and can be applicable with modern molecular biology techniques.

REFERENCES

1. IARC monographs on the evaluation of carcinogenic risks to humans (1987) In Overall Evaluations of Carcinogenicity: An Updating of IARC Monographs Volumes 1 to 42, Suppl.7, pp. 1-440. International Agency for Research on Cancer, Lyon, France.
2. Vainio, H., M. Sorsa, and A.J. McMichael, eds. (1990) Complex mixtures and cancer risk. IARC Scient. Publ. (in press).

MUTAGENICITY, CARCINOGENICITY, AND HUMAN CANCER RISK

FROM INDOOR EXPOSURE TO COAL AND WOOD COMBUSTION

IN XUAN WEI, CHINA

Judy L. Mumford,[1] Robert S. Chapman,[1]
Stephen Nesnow,[1] C. Tuker Helmes,[2] and Xueming Li[3]

[1]U.S. Environmental Protection Agency
Research Triangle Park, North Carolina 27711

[2]SRI International
Menlo Park, California 94025

[3]Institute of Environmental Health and Engineering
Beijing, China

ABSTRACT

The residents in Xuan Wei County, China, have been exposed to high levels of combustion emissions from smoky and smokeless coal and wood combustion under unvented conditions in homes. An unusually high lung cancer mortality rate that cannot be attributed to tobacco smoke or occupational exposure was found. The communes using smoky coal, which emits more organics than smokeless coal, generally have a higher lung cancer rate than the communes using smokeless coal or wood. The mutagenicity and carcinogencity of organic extracts of indoor air particles collected from Xuan Wei homes during cooking were investigated. The objectives of this study were: (i) to investigate the characteristics of lung cancer mortality in Xuan Wei, (ii) to determine the genotoxicity and chemical and physical properties of the combustion emissions, and (iii) to link bioassay results to human lung cancer data. The organic extracts of these emission particles were tested for mutagenicity in the Ames Salmonella and the L5178Y TK$^{+/-}$ mouse lymphoma assays, and for skin tumor-initiating activity and complete carcinogencity in SENCAR mice. The two coal samples showed higher activity in both mutagenicity and tumor initiation. When the emission rate of organics was taken into consideration, the smoky coal emission showed the highest potency of the three fuels. The smoky coal sample was also a more potent complete

Genetic Toxicology of Complex Mixtures
Edited by M. D. Waters *et al.*
Plenum Press, New York, 1990

carcinogen than the wood sample. Higher mutagenicity and carcinogenicity of the smoky coal emission compared to wood or smokeless coal emissions are in agreement with the epidemiological data.

INTRODUCTION

Studies have shown that a significant portion of human cancer is caused by environmental agents. Short-term genetic and carcinogenic bioassays have been useful in providing relevant information on mechanisms involved in the induction of cancer and genetic diseases. These bioassays have also been used to assess the potential carcinogenicity of environmental agents, both in the forms of pure compounds and complex mixtures. Humans are exposed daily to thousands of pollutants, mostly in the form of complex mixtures. Most of these complex environmental mixtures are not well characterized. In addition, numerous new chemicals for commercial use are introduced into the environment each year. Due to the cost and time required, it is impossible to conduct long-term chronic animal studies or human studies to assess cancer risk from exposure to all of these environmental agents. There has been great concern about how well these short-term bioassays can be used to predict human cancer risks as a result of exposure to environmental agents (12). The products of incomplete combustion, which contain carcinogenic compounds of polycylic aromatic hydrocarbons (PAHs) and other compounds, have been reported to be human carcinogens (4,5,11). The studies reported here investigated a human population exposed to high concentrations of wood and coal combustion emissions in a rural county in China. Because human cancer data for this population are available, there are unique opportunities to evaluate the applications of short-term genetic and carcinogenic bioassays in assessing human carcinogenicity.

The China National Cancer Survey indicated that the residents in the rural Xuan Wei county, located in Yunnan Province, had lung cancer mortality rates that were about five times the national average during 1973 to 1975 (3,8). Xuan Wei women seldom smoked tobacco, and their lung cancer mortality rates were the highest among women of all counties in China. Xuan Wei men's lung cancer rates were among the highest in men of all counties in China. Xuan Wei residents traditionally burn coal or wood in an unvented pit in homes for cooking and heating. Common fuels used are "smoky" coal, "smokeless" coal, and wood. Smoky coal is similar to medium-volatile, low sulfur (0.2%), bituminous coal with a heating value of 27.1 MJ/kg, whereas smokeless coal has a heating value of 14.5 MJ/kg and contains a high sulfur (1.9%) and ash content (49%). Under the China-United States Protocol for Scientific and Technological Cooperation in the Field of Environmental Protection, Chinese and American investigators jointly conducted interdisciplinary studies to examine the etiology of lung cancer in Xuan Wei, including epidemiology, air monitoring, chemistry and toxicology studies.

The specific objectives of the studies reported here are: (i) to investigate the characteristics of lung cancer mortality in Xuan Wei, (ii) to characterize the indoor air from Xuan Wei homes using short-term genetic and carcinogenic bioassays and chemical and physical analysis data to investigate the etiology of lung cancer in Xuan Wei, and (iii) to compare the results of the bioassays to human lung cancer data.

LUNG CANCER IN XUAN WEI

The lung cancers reported here include carcinomas of lung parenchyma, bronchus, and trachea. A sputum study of lung cancer patients from a high lung cancer commune, Lai Bin, showed that 58% of 115 specimens examined contained cells with squamous carcinomas; 28% contained cells with adenocarcinomas; and the remaining 14% of the specimens were mixed cell, undifferentiated, or alveolar carcinomas (9). The characteristics of lung cancer in Xuan Wei are as follows.

High Lung Cancer Mortality Rate and
Lung Cancer Death at an Early Age

In Xuan Wei, lung cancer is the only cancer for which the mortality exceeds China's national average (7). The lung cancer mortality rate in Xuan Wei county, as a whole, is about five times the China national average. In the three high lung cancer communes in Xuan Wei, the lung cancer rate is about 24 times the national average (8). Table 1 shows the annual lung cancer mortality rates in Xuan Wei. The lung cancer rate in Xuan Wei is also much higher than the Yunnan Province where Xuan Wei county is located. In comparison to the lung cancer mortality rates in the U.S., the Xuan Wei men's rates are similar to the men's rates in the U.S., but the Xuan Wei women's lung cancer mortality rate was four times the American women's rate (when comparing the lung cancer mortality rates adjusted to the 1964 China population). The age of lung cancer patients at death was six years younger than the China national average (9).

Similarity of Lung Cancer Mortality Rates Between Men and Women

Lung cancer mortality rates in men are usually higher than those in women, mostly due to tobacco smoking in men. In the U.S. and China, the ratios of man-to-woman lung cancer mortality are 4.8 and 2.1, respectively (see Tab. 1), whereas the corresponding ratio in Xuan Wei is 1.09, indicating that men and women have a similar lung cancer mortality

Tab. 1. Annual lung cancer mortality rates in China and the United States (8)

Place	Time Period	Unadjusted Males	Unadjusted Females	Unadjusted Combined	Age–adjusted to 1964 China population Males	Age–adjusted to 1964 China population Females	Age–adjusted to 1970 U.S. population Males	Age–adjusted to 1970 U.S. population Females
China	1973-75			5.0	6.8	3.2	12.3	5.7
United States	1970	53.7	12.0		30.0	6.3	53.7	12.0
Yunnan Province	1973-75			2.8	4.3	1.5	6.9	2.5
Xuan Wei County	1973-79	27.0	24.5		27.7	25.3	43.2	38.7
Three high-mortality Xuan Wei communes (Cheng Guan, Lai Bin, Rong Cheng)	1973-79	114.4	120.6		118.0	125.6	186.8	193.4
55- to 59-year age group in three high-mortality communes	1973-79	849.4	904.0					
Three low-mortality Xuan Wei communes (Pu Li, Re Shui, Yang Liu)	1973-79	4.0	2.8		4.3	3.1	5.8	4.3

Column header top: Mortality rate (per 100,000)

[a] Taken from Mumford et al. (1987) Science 235:217-220.

rate. The percentage of Xuan Wei women that smoke is low (0.2%)
compared to the percentage of men that smoke (42.7%). This suggests
that tobacco smoke is not the major cause of lung cancer in Xuan Wei, at
least in women (9). Domestic factors may be associated with lung cancer
in Xuan Wei.

Great Geographic Variation in Lung Cancer Mortality Rates Among Communes in Xuan Wei

There is a great variation among Xuan Wei communes in lung cancer
mortality rates, ranging from 0 to 152 per 100,000. The central
communes-Cheng Guan, Lai Bin, and Rong Cheng--have the highest rates
(see Tab. 1); most peripheral communes showed lower rates.

High Correlation Between Lung Cancer and Smoky Coal Use in Homes

A fuel survey conducted in 11 Xuan Wei communes showed the lung
cancer mortality rate was highly correlated with the percentage of households using smoky coal (8), whereas no correlation was found between
lung cancer and the percentage of households using wood. This finding
suggests that indoor air pollution from domestic use of smoky coal may be
associated with lung cancer in Xuan Wei.

INDOOR AIR MONITORING AND CHARACTERIZATION

Air monitoring, chemical and physical characterization, and toxicological studies of indoor air from Xuan Wei homes were conducted.
High-volume samplers with size-selective inlets for collecting air particles
smaller than 10 μm (PM_{10}) were used to collect large quantities of samples
for detailed chemical analyses and bioassays, as described previously (8).
Air sampling was conducted during cooking in a central commune where
lung cancer mortality is Xuan Wei's highest, and where smoky coal is the
major fuel used. Sampling also was conducted in a southwestern commune
where mortality was low and where households used wood (67%) and
smokeless coal (33%) as fuels. The size of air particles was determined
by electron microscopy, and the air particles were extracted with dichloromethane. The organic extracts were fractionated and analyzed for
chemical composition, including quantitative analysis of PAHS, as reported
previously (8). The organic extracts were bioassayed for (i) mutagenicity in Ames Salmonella/microsome assay, according to the method of Ames
et al. (1); (ii) mutagenicity in the L5178Y TK[+/-] mouse lymphoma assay,
according to the method of Clive et al. (2); and (iii) tumor-initiation
activity and complete carcinogenicity in SENCAR mouse skin (10). In the
skin tumor initiation studies, female mice (40 mice/group) were initiated
with the organic extracts of the combustion emission particles at 1, 2, 5,
10, and 20 mg/mouse and one week later promoted with TPA (2 μg/mouse)
twice a week for 26 weeks and scored for papillomas. In the complete
carcinogenesis studies, female mice (40 mice/group) were treated with the
organic extracts at 1 mg/mouse, twice a week for 52 weeks, and were
scored for carcinomas 25 weeks later.

The detailed characterization of indoor air particles collected from
Xuan Wei homes during cooking has been reported previously (8,9). This
report summarizes the up-to-date findings. Table 2 shows the characteristics of indoor air particles collected from Xuan Wei homes that burn

Tab. 2. Characteristics of indoor air particles from Xuan Wei homes during cooking[a]

Sample	Smoky Coal	Wood	Smokeless Coal
Particulate concentration[b] (mg/m³)	24.4 ± 3.3	22.3 ± 2.0	1.8
Organic mass extractable[b] (percent)	72.3 ± 6.7	55.1 ± 7.0	27.0
Particle size[c] (percent of submicron-particles)	51	6	–
PAH concentrations (µg/m³)			
Benz[a]anthracene[b]	25.1 ± 5.0	4.0 ± 1.1	1.0
5-Methylchrysene[b]	7.3 ± 2.7	0.2 ± 0.06	0.2
Benzo[a]pyrene[b]	14.7 ± 3.0	3.1 ± 1.0	0.6
Dibenz[a,j]acridine[c]	0.7	BDL	BDL
Organic fractions in percent mass[c] (revertants/m³ air)			
Neat	100 (58.9)	100 (11.1)	100 (1.3)
Aliphatics	5.0 (0)	1.5 (0)	3.2 (0)
Aromatics	36.4 (18.1)	7.7 (3.1)	29.7 (0.5)
Moderately polar	35.0 (8.7)	14.4 (4.4)	20.7 (0.3)
Polar	30.8 (24.4)	78.3 (5.0)	34.4 (0.4)

[a] Portions of these data were taken from Mumford et al. (In Press) Recent Results in Cancer Research, Springer-Verlag, NY.
[b] Data from four homes (mean ± SEM) except one home from smokeless coal combustion.
[c] Data from one home.
BDL = Below detection limit.

smoky coal, smokeless coal, or wood. These data show that the Xuan Wei residents who used smoky coal inhaled extremely high concentrations of mostly submicron-sized particles, which can be deposited efficiently in the lung. These fine particles were composed mostly of organic compounds (72%), including mutagenic and carcinogenic organic compounds, especially in the aromatic and polar fractions. The residents were exposed to polycylic aromatic compounds, such as benzo(a)pyrene (BaP), at comparable or higher levels than those measured in coke oven plants and other occupational environments (6). Table 3 shows the comparative potency of organic extracts of the combustion emission particles from burning the three fuel types. The organic extracts from smoky coal and smokeless coal combustion showed higher potency than organic extracts from wood combustion in the Ames Salmonella mutation assay, the mouse lymphoma mutation assay, and skin tumor initiation activity in SENCAR mice. When emission rates of organics were taken into consideration to determine the concentrations of mutagens or tumor initiators in the air, the smoky coal emissions showed the highest concentrations among the three fuels. The organic extract of the emission particles from smoky coal combustion also is a more potent complete carcinogen than that of the particles from wood combustion. At the end of the study, at Week 77, 88% of the mice treated with the organic extract from smoky coal combustion showed carcinomas (average 1.1 carcinomas/mouse) in contrast to only 5% (1 carcinoma/mouse) of the mice treated with the organic extract from wood combustion. Higher mutagenicity and carcinogenicity of smoky coal emissions as compared to wood or smokeless coal emissions were in general agreement with the epidemiological data.

Tab. 3.   Comparative potency of organic extracts from combustion emission particles (PM_{10}) collected in Xuan Wei homes during cooking

Fuel/Lung Cancer Mortality	Salmonella Mutagenicity TA98 (+ S9) Revertants		L5178Y TK^{+}/$^{-}$ Mouse Lymphoma (+ S9) Mutation Frequency		Tumor Initiation in SENCAR Mouse Skin Papillomas	
	Per µg Organics	Per m³ Air	Per µg Organics	Per m³ Air	Per µg Organics	Per m³ Air
Smoky coal High (152/100,000)	2.6	58.9	62.1	1,406.5	0.6	13.8
Smokeless coal Low (2.3/100,000)	2.7	1.3	86.7	42.5	0.4	0.2
Wood Low (2.3/100,000)	0.7	11.1	18.4	292.6	0.1	2.1

[a] Data presented are from a representative home in each commune. A sample was considered positive if a dose-response relation was observed. Each sample was tested at a minimum of five doses. The slope of the initial linear portion of each dose-response curve was calculated by least-squares linear regression ($R^2 \geq 0.85$) to obtain the potency per microgram of organic extract. The potency per cubic meter of air for each type of emission was calculated by multiplying the potency per microgram of organic extract by the micrograms of organic extract per cubic meter of air.

In summary, studies reported here investigated nontobacco-related environmental causes of lung cancer. The study of fuel use and lung cancer mortality in Xuan Wei suggested an association between domestic smoky coal use and Xuan Wei lung cancer. The collaborative studies of physical characterization, chemical analysis, and toxicology studies further substantiated this link.

ACKNOWLEDGEMENTS

The authors wish to thank the investigators of the Institute of Environmental Health and Engineering in Beijing and the staff from the anti-epidemic stations in Yunnan Province, Qu Jing Region, and Xuan Wei County for their assistance in field sampling and data collection. The authors thank Dr. Xingzhou He of the Institute of Environmental Health and Engineering in Beijing for providing epidemiology data, Mr. D. Bruce Harris of the U.S. Environmental Protection Agency for his technical assistance in sampling, Dr. Colette Rudd of SRI International for conducting the mouse lymphoma assay, and Ms. Virginia Houk of the U.S. Environmental Protection Agency for conducting the Ames assay.

REFERENCES

1. Ames, B.W., J.M. McCann, and E. Yamasaki (1975) Methods for detecting carcinogens and mutagens with the Salmonella/mammalian microsome mutagenicity test. Mutat. Res. 31:347–364.
2. Clive, D., K.O. Johnson, J.F.S. Spector, A.G. Batson, and M.M.M. Brown (1979) Validation and characterization of the L5178Y TK^{+}/$^{-}$ mouse lymphoma mutagen assay system. Mutat. Res. 59:61-108.
3. He, X.Z., S.R. Cao, W.Z. Jiang, R.D. Yang, and C.W. Xu (1986) Research on Etiology of Lung Cancer in Xuan Wei. Institute of Environmental Health and Engineering, Chinese Academy of Preventive Medicine, Beijing, China.
4. International Agency for Research on Cancer (1983) Polynuclear aromatic compounds: Chemical, environmental, and experimental data.

In IARC Monographs on the Evaluation of the Carcinogenic Risks of Chemicals to Humans, Vol. 32, Part 1:95-447. International Agency for Research on Cancer, Lyon, France.

5. International Agency for Research on Cancer (1986) Tobacco smoking. In IARC Monographs on the Evaluation of Carcinogenic Risk of Chemicals to Humans, Vol. 38. International Agency for Research on Cancer, Lyon, France.

6. International Agency for Research on Cancer (1984) Polycyclic aromatic compounds, industrial exposures in aluminum production, coal gasification, coke production, and iron and steel founding. In IARC Monographs on the Evaluation of the Carcinogenic Risk of Chemicals to Humans, Vol. 34, Part 3. International Agency for Research on Cancer, Lyon, France, pp. 107-108.

7. LI, J.Y., et al., eds. (1979) Atlas of cancer Mortality in the People's Republic of China, China Map Press, Shanghai, China.

8. Mumford, J.L., X.Z. He, R.S. Chapman, S.R. Cao, D.B. Harris, X.M. Li., Y.L. Xian, W.Z. Jiang, C.W. Xu, J.C. Chuang, W.E. Wilson, and M. Cooke (1987) Lung cancer and indoor air pollution in Xuan Wei, China. Science 235:217-220.

9. Mumford, J.L., X. He, and R.S. Chapman (1989) Human lung cancer risks due to complex organic mixtures of combustion emissions. In Recent Results in Cancer Research, P. Band, ed. Springer-Verlag, New York (in press).

10. Mumford, J.L., C.T. Helmes, X. Lee, L. Seidenberg, and S. Nesnow (1989) Mouse skin tumorigenicity studies of indoor coal and wood combustion emissions from homes of residents in Xuan Wei, China, with high lung cancer mortality. Carcinogenesis (in press).

11. Pott, P. (1775) Chirurgical observations relative to the cataract, the polypus of the nose, the cancer of the scrotum, the different kinds of ruptures, and the modifications of the toes and feet. In The Chirurgical Works of Percivall Pott, Vol. 5, L. Hawes, W. Clarke, and R. Collins, eds. London.

12. Tennant, W.R., B.H. Margolin, M.D. Shelby, E. Zeiger, J.K. Haseman, J. Spalding, W. Caspary, M. Reshick, S. Stasiewicz, B. Anderson, and R. Minor (1987) Prediction of chemical carcinogenicity in rodents from in vitro genetic toxicity assays. Science 236:933-941.

THE ROLE OF NITROARENES IN THE MUTAGENICITY OF

AIRBORNE PARTICLES INDOORS AND OUTDOORS

Hiroshi Tokiwa,[1] Nobuyuki Sera,[1] Mamiko Kai,[1]
Kazumi Horikawa,[1] and Yoshinari Ohnishi[2]

[1]Department of Health Science
Fukuoka Environmental Research Center
Fukuoka 818-01, Japan

[2]Department of Bacteriology
School of Medicine
The University of Tokushima
Tokushima 770, Japan

INTRODUCTION

Nitroarenes are nitrated by-products of the reaction of various environmental agents with polycyclic aromatic hydrocarbons (PAHs); the presence of nitroarenes in the environment appears to result mainly from man's activities. Most nitroarenes are strongly associated with mutagenicity, genotoxicity, and carcinogenicity. Some nitroarenes, which are potent mutagens in bacteria, have been reported to be carcinogenic at various sites in animals [see review by Tokiwa and Ohnishi, (7)], suggesting a correlation between mutagenicity and carcinogenicity, at least in nitroarenes. However, there is no clear evidence that indicates an epidemiological relation with human carcinogenesis. On the other hand, it has been found that most nitroarenes are induced from diesel exhaust emissions, wood burning, kerosene heaters, city gas, and liquified petroleum gas, and are widespread in the environment (8). In relation to cigarette smoking, some nitroarenes were also found to be induced when cigarette smoke condensates were treated with nitric acid or Chinese cabbage pickles (4).

In this paper, the role of nitroarenes in particulates is described from the viewpoint of biological and chemical analyses of airborne particulates and soot of heavy oil burning outdoors, and particulates in kerosene heaters and city gas burner emissions.

Genetic Toxicology of Complex Mixtures
Edited by M. D. Waters *et al.*
Plenum Press, New York, 1990

MATERIALS AND METHODS

Bacterial Strains

The bacterial strains used were Salmonella typhimurium His⁻ strain TA98 which was kindly provided by Dr. B.N. Ames, University of California and strain YG1024, constructed by transfer of the plasmid of the acetyltransferase gene into TA98 (9) which was kindly provided by Dr. M. Watanabe, National Institute of Hygienic Sciences, Tokyo.

Collection of Particulates

Airborne particulates were collected in Sapporo, Hokkaido, in 1989. Particulates from a kerosene heater (Toyo stove RCA-36B) and city gas were collected under the condition of incomplete combustion according to the procedure described previously (5). The particulate matter was collected on a Teflon-coated silica fiber filter and extracted with dichloromethane by ultrasonification three times for 30 min. The composition of city gas was 4-20% methane, 40-50% hydrogen, 10% carbon monoxide, 3-4% oxygen, 5-25% nitrogen, 3-4% butane, and traces of pentane, isopentane, and propane.

Mutagenicity Test

The mutagenicity test was carried out according to the procedure described previously (5).

Silica Gel Column Chromatography

As indicated in a previous report (2), the crude extracts were applied to a column filled with silica gel (Neutral, Sigma Chemical Co.) (0.9 cm x 25 cm), and were eluted stepwise with 30 ml each of, in order, hexane, hexane/benzene (1:1, v/v), benzene, benzene/methanol (1:1, v/v), and methanol as effluents. The mutagens were fractionated and identified by means of high performance liquid chromatography (HPLC) and gas chromatography and mass spectrometry (GC/MS).

High Performance Liquid Chromatography

HPLC was performed on a column of Unisil Q C18 (Gaschro Kogyo, Inc., Japan) according to the procedure described previously (2).

Sample was loaded onto the column (0.46 cm i.d. x 25 cm), and eluted with acetonitril-water (80 :20, v/v) at a flow rate of 1.0 ml/min for HPLC (Toyosoda Manufacturing Co. Ltd., HLC-803D).

Gas Chromatography and Mass Spectrometry

The GC/MS analysis was performed on a DB-5 capillary gas chromatograph interfaced to a Finnigan MAT ITD-80 mass spectrometer. The GC/MS system was equipped with a fused silica capillary column (0.25 mm i.d. x 30 m) coated with SE-30 stationary phase. The column oven temperature was programmed from 50-280°C at 4°C/min. The GC injection and the ion source temperatures were maintained at 250°C.

RESULTS

Fractionation of Mutagen in Airborne Particulates, Kerosene Heater and City Gas Emission Particulates, and Heavy Oil Emission Particulates

Under the condition of incomplete combustion, the particulate materials collected were 17,315 mg for kerosene heater emissions, 2,013 mg for city gas emissions, 801.3 mg for airborne particulates, and 7,618 mg for heavy oil emissions. The neutral fraction of each sample was obtained, and the results of mutagenicity tests in strains TA98 and YG1024 are indicated in Tab. 1. The neutral fraction of each sample was fractionated on a column of silica gel as described in "Materials and Methods" and the benzene fraction, the highest mutagenic fraction containing polar and nonpolar compounds, was analyzed by GC/MS. It was suggested that the benzene fraction contains nitroarenes, which were activated by acetyltransferase, because of the higher mutagenicity for strain YG1024 except with the sample from heavy oil combustion.

HPLC Analysis of Benzene Fractions Separated from Each Sample

The benzene fraction of each material was analyzed on a HPLC column of Unisil Q C18. Figure 1a and b shows the analytic results with airborne particulates and kerosene heater emission materials. Active fractions with direct-acting mutagenicity that eluted on the column appeared at

Tab. 1. Mutagenicity of fractions from samples of kerosene and city gas emissions, heavy oil emissions, and airborne particulates

Fraction	Particulate weight (mg)	TA98 (revertants/mg)		YG1024 (revertants/mg)	
		-S9	+S9	-S9	+S9
Kerosene heater	17,315				
Crude extract	76.1	6,300	11,400	9,200	11,500
Neutral fraction	32.3	12,700	23,200	19,300	24,200
Benzene fraction	2.0	113,100	183,400	189,200	193,100
City gas	2,013				
Crude extract	103	2,200	3,900	3,100	3,700
Neutral fraction	44.2	4,400	7,700	6,200	8,100
Benzene fraction	2.7	37,200	61,100	63,200	64,100
Heavy oil	7,618				
Crude extract	1,261	180	390	110	360
Neutral fraction	48.9	4,200	8,700	2,500	8,900
Benzene fraction	2.4	41,100	95,700	23,400	94,300
Airborne particulates	801.3				
Crude extract	73.2	2,300	4,100	3,600	7,100
Neutral fraction	29.4	5,100	8,600	7,600	16,400
Benzene fraction	1.8	43,600	80,100	61,800	155,300

Tab. 2.   Nitroarenes and polycyclic aromatic hydrocarbons in particulate
           matter

Chemical	μg/g of particulate			
	Kerosene heater	City gas	Airborne particulates	Heavy oil
1-Nitropyrene	8.7	0.057	0.16 (0.068)[a]	0.026
1,3-Dinitropyrene	1.8	0.019	0.024 (0.010)	0.001
1,6-Dinitropyrene	1.2	0.01	0.018 (0.008)	
1,8-Dinitropyrene	0.08	0.002		
3-Nitrofluoranthene	3.2	0.072	0.31 (0.13)	
2-Nitrofluoranthene			0.37 (0.16)	
3,7-Dinitrofluoranthene	0.14		0.012 (0.005)	
3,9-Dinitrofluoranthene			0.009 (0.004)	
2-NItrofluorene		0.021	0.50 (0.21)	0.013
9-Nitroanthracene			0.41 (0.17)	
1-Nitronaphthalene			2.1 (0.90)	
2-Nitronaphthalene			1.3 (0.55)	
Pyrene	141	1.19	17.8 (7.59)	0.173
Fluoranthene	96	1.04	14.1 (6.01)	
Benzo(e)pyrene	63	0.92	6.2 (2.6)	
Benzo(a)pyrene	52	0.61	3.4 (1.4)	0.16
9-Fluorenone			0.1 (0.04)	
Phenanthrene			5.2 (2.21)	
Triphenylene			0.16 (0.07)	
Benzo(a)anthracene	4.3	0.083	0.21 (0.09)	
Chrysene			2.8 (1.19)	
Benzo(k)fluoranthene	37	0.27	1.6 (0.68)	0.14
Benzo(b)fluoranthene	62	0.31		
Perylene			3.2 (1.36)	
Benzo(ghi)perylene			3.8 (1.62)	

[a] (   ), ng/m^3

retention times from 9-17 min, and those with indirect-acting mutagenicity
appeared mainly at retention times from 20-37 min.   The active peaks
were pooled and partially concentrated.   The main active peaks were
analyzed by GC/MS to determine the mutagen.   For airborne particulates
materials, the 14 chemical mutagens identified were 9-fluorenone in Pl,
2-nitrofluorenone in P2, 3,9-dinitrofluoranthene (DNF) in P3, 3,7-DNF and
1,6-dinitropyrene (DNP) in P5, 1,3-DNP in P6, l-nitropyrene (NP) in P7,
3-nitrofluoranthene (NF) in P8, triphenylene in P9, benzo(a)anthracene
(BaA) in P10, chrysene in P11, benzo(e)pyrene (BeP) in P12, perylene in
P13, and benzo(a)pyrene (BaP) in P14.   Eight active peaks appeared from
the kerosene heater emission material (Fig. lb).   The mutagens identified
were 1,8-DNP in Pl, 1,6-DNP and 3,7-DNF in P2, 1,3-DNP in P3, 1-NP
and 3-NF in P4, BaA in P5, benzo(b)fluoranthene (BbF) and BeP in P6,

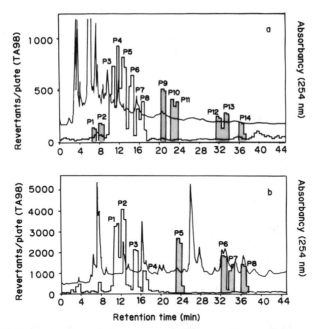

Fig. 1.   HPLC analysis of benzene fractions from airborne particulates
(a) and kerosene heater emission particulates (b). Mutagenicity
is shown as revertant colonies of strain TA98 without (white
column) and with (dark column) S-9 mix.

benzo(k)fluoranthene (BkF) in P7, and BaP in P8. The results with city
gas and heavy oil emissions are shown in Fig. 2a and b. In the city gas
emission material, seven main active peaks appeared on the column of
Unisil Q C18, and each peak was identified by GC/MS. The mutagens
identified were 1,6- and 1,8-DNP in Pl, 1,3-DNP in P2, 3-NF and 1-NP in
P3, BaA in P4, BbF and BeP in P5, BkF in P6, and BaP in P7. Similarly, mutagens identified in heavy oil emission material were determined as
1,3-DNP in P4, 1-NP in P5, BkF in P6, and BaP in P7, but the active
peaks in Pl, P2, and P3 could not be identified. Figure 3a, b, and c
show the gas chromatograms and mass spectra of nitroarenes which were
determined from airborne particulates and kerosene heater emission
particulates.

Measurement of Mutagens in the Materials from Kerosene Heater and City Gas Emission Indoors and Airborne Particulates and Heavy Oil Burning Soot Outdoors

Major nitroarenes, 1-NP, 1,3-, 1,6-, and 1,8-DNP, 3-NF, 3,7- and
3,9-DNF, 2-NF and 2-nitrofluorene, were detected in all samples. In
samples of heavy oil combustion, only 1-NP and 1,3- DNP were detected.
For indirect-acting mutagens that appeared on a HPLC column, 12 species
of PAHs were identified from materials of kerosene, city gas, and heavy
oil combustion, and airborne particulates by the GC/MS system. The
large amount of PAHs detected may be due to collection of the particulates
under the condition of incomplete combustion of kerosene and city gas.

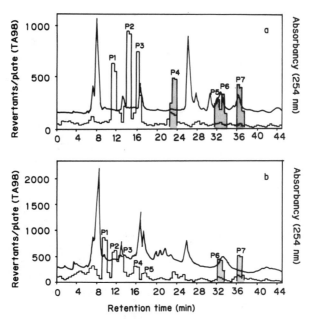

Fig. 2.   HPLC analysis of benzene fractions from city gas (a) and heavy
oil burning particulates (b). Mutagenicity is shown as revertant colonies of strain TA98 without (white column) and with
(dark column) S-9 mix.

DISCUSSION

Toxic agents generated from kerosene heaters and gas burners dis-
played high direct-acting mutagenicity in the Salmonella-microsome test
system. Each of these samples reverted strain YG1024, constructed by
transfer of the acetyltransferase gene, much more strongly than strain
TA98, containing DNPs as polar compounds. Similarly, the increasing
mutagenicity for the new strain was observed in each successive fraction
from airborne particulates. As shown in a previous study (5), the
mutagenicity of kerosene heater and gas and liquified petroleum particulate extracts was markedly reduced for strain TA98/1,8-DNP$_6$, an acetyl-
transferase-deficient strain. These findings suggest that 45-79% of all
mutagenicity observed is due to DNPs and related compounds which are
activated by acetyltransferase. About 48-55% of direct-acting mutagenicity for each sample was concentrated in the benzene fraction when the
neutral fraction was separated by silica gel column chromatography. The
benzene fraction is considered to contain polar and nonpolar compounds.
These facts indicate that, in each sample, this fraction possesses potent
direct-acting mutagenicity, suggesting the presence of nitroarenes.

It is important to know if the accumulation of pollutants indoors
when a kerosene heater is ignited in a closed room poses a risk to human
health because of their mutagenicity. As reported previously (5), the
indoor pollutants in a 211.2 m^3 room after ignition of a kerosene heater
were collected on a silica fiber filter at 1-hr intervals for 5 hr and their

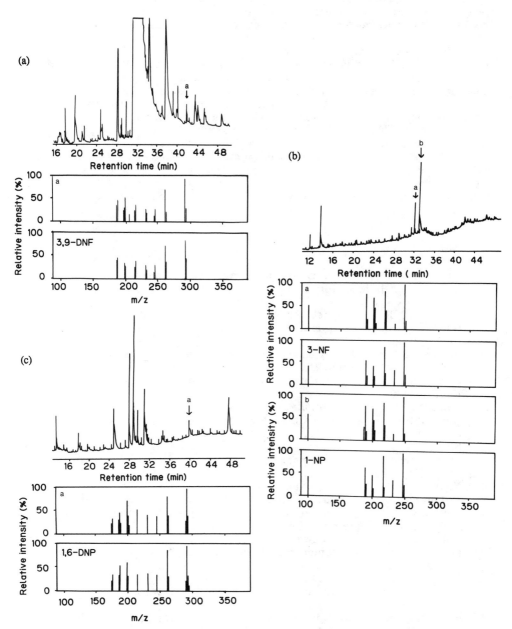

Fig. 3.   a. Gas chromatograms and mass spectra of 3,9-DNF detected in
           airborne particulates.   b.   Gas chromatograms and mass spectra
           of 3-NF and 1-NP detected in kerosene heater emissions.   c.
           Gas chromatograms and mass spectra of 1,6-DNP detected in
           kerosene heater emissions.

mutagenicity was measured. It was found that mutagens in the room air were at their highest concentration during the first hour after ignition of the heater, but were decreased dramatically by ventilation.

It is no exaggeration to say that some airborne particulates are due to mostly diesel engine exhaust. It has also been shown that diesel emission particulates contain a number of nitroarenes. In Japan, diesel passenger cars have dramatically increased since 1975 (6), and had reached as many as 7 x 10 cars in 1987. Recently, a research project in Japan determined the mutagenicity of particulates extracts in air in Hokkaido, Sendai, Tokyo, Nagoya, and Fukuoka. It was reported that direct-acting mutagenicity for strains TA98 and TA100 was much more potent than indirect-acting mutagenicity, although indirect-acting mutagenicity was detected in the period 1960-1970. It has been pointed out that the presence of nitroarenes in the atmosphere will spread; in particular, the presence of l-NP, 2-NF, 3-NF, DNP, and DNF is very important (1,3,8).

ACKNOWLEDGEMENT

This work was supported in part by Grants-in-Aid for Cancer and Scientific Research from the Ministry of Health and Welfare, Japan.

1. Arey, J., B. Zielinska, R. Atkinson, and A.M. Winer (1988) Formation of nitroarenes during ambient high-volume sampling. Env. Sci. Tech. 22:457-462.
2. Nakagawa, R., S. Kitamori, K. Horikawa, K. Nakashima, and H. Tokiwa (1983) Identification of dinitropyrenes in diesel-exhaust particles. Their probable presence as the major mutagens. Mutat. Res. 124:201-211.
3. Ramdahl, T., B. Zielinska, J. Arey, R. Atkinson, A.M. Winer, and J.N. Pitts, Jr. (1986) Ubiquitous occurrence of 2-nitrofluoranthene and 2-nitropyrene in air. Nature 321:425-427.
4. Sera, N., H. Tokiwa, and T. Hirohata (1989) Induction of nitroarenes in cigarette smoke condensate treated with nitrate. Tox. Lett. (in press).
5. Tokiwa, H., R. Nakagawa, and K. Horikawa (1985) Mutagenic/carcinogenic agents in indoor pollutants: The dinitropyrenes generated by kerosene heaters and fuel gas and liquefied petroleum gas burners. Mutat. Res. 157:39-47.
6. Tokiwa, H., R. Nakagawa, K. Horikawa, and A. Ohkubo (1987) The nature of the mutagenicity and carcinogenicity of nitrated, aromatic compounds in the environment. Env. Health Persp. 73:191-199.
7. Tokiwa, H., and Y. Ohnishi (1986) Mutagenicity and carcinogenicity of nitroarenes and their sources in the environment. CRC Critical Rev. 17:23-60.
8. Tokiwa, H., T. Otofuji, R. Nakagawa, K. Horikawa, T. Maeda, N. Sano, K. Izumi, and H. Otsuka (1986) Dinitro derivatives of pyrene and fluoranthene in diesel emission particulates and their tumorigenicity in mice and rats. In Carcinogenic and Mutagenic Effects of Diesel Engine Exhaust, N. Ishinishi et al., eds. Elsevier Science Publishers B.V., pp. 253-270.
9. Watanabe, M., M. Ishidate, Jr., and T. Nohmi (1989) A sensitive method for the detection of mutagenic nitroarenes: Construction of nitroreductase-overproducing derivatives of Salmonella typhimurium strains TA98 and TA100. Mutat. Res. (in press).

CHARACTERIZATION OF MUTAGENIC COMPOUNDS FORMED

DURING DISINFECTION OF DRINKING WATER

Leif Kronberg

Department of Organic Chemistry
The University of Abo Åkademi
SF-20500 Turku/Åbo
Finland

INTRODUCTION

The primary purpose for disinfection of drinking water is to destroy and eliminate pathogenic organisms responsible for waterborne diseases. Commonly used disinfectants are chlorine, chlorine dioxide, chloramine, and ozone. The most widely used disinfectant is chlorine. Chlorine has many advantages as it is toxic to many microorganisms, is cheaper than most other disinfectants, can be prepared in situ, and is not critically sensitive to pH and temperature variations (3). However, the very reactivity of chlorine produces problems. Especially harmful are the reactions with organic compounds present in water resulting in halogenated organics (4,14).

The seriousness of the problem of halogenated organics in drinking water was not recognized fully until Rook in 1974 demonstrated that chlorination of water containing natural humic substances led to the formation of chloroform and other trihalomethanes (36). However, it was soon found out that most of the chlorine bound to organic compounds resides in the nonvolatile fraction of the organic material (33). Generally, the nonvolatile organohalides are more difficult to identify than are the trihalomethanes, partly because the nonvolatiles are more polar and because they are present as a complex mixture of individual compounds in very small quantities. Concern over potential human health hazards associated with drinking water has been heightened by the widespread recognition in recent years of mutagenic activity exhibited by the nonvolatile fraction (5,7,8,26,30,37,42).

This paper gives a review of work on the characterization, isolation, and identification of compounds responsible for considerable proportions of the mutagenicity formed during chlorination of drinking water.

Genetic Toxicology of Complex Mixtures
Edited by M. D. Waters *et al.*
Plenum Press, New York, 1990

REACTIONS OF CHLORINE IN AQUEOUS SOLUTIONS

Chlorine gas hydrolyzes very rapidly in water according to the following reaction:

$$Cl_2 + H_2O \longrightarrow HOCl + H^+ Cl^-$$

At normal conditions of drinking water treatment the hydrolysis is essentially complete. Hypochlorous acid (HOCl) is a weak acid and dissociates to hypochlorite ion:

$$HOCl \longrightarrow H^+ + OCl^-$$

This is an equilibrium reaction which is controlled by pH. At pH 7.5 and 25°C, HOCl and OCl$^-$ are essentially equimolar in concentration (32).

The chlorine added to water is primarily consumed in inorganic oxidation reactions and in reactions with organic compounds (16). The reactions of free chlorine (HOCl, OCl$^-$) with organic compounds in aqueous solutions are oxidation, addition, and substitution. Oxidation is the predominant reaction at conditions for water disinfection. Only addition and substitution reactions produce chlorinated organic compounds.

ISOLATION AND CONCENTRATION OF MUTAGENS

Organic compounds are present in drinking water at low concentrations, typically 1 µg/l or less. To make the compounds available to chemical-analytical and biological-toxicological measurements they have to be isolated from water and concentrated. When a known compound is under study, the concentration method giving maximum recovery with minimum interference is selected. However, when compounds with unknown structures and physicochemical properties are to be recovered, such optimization is impossible. Therefore, the method to be selected should give high recoveries of a wide range of organic compounds. The necessity of a direct relation between chemical and toxicological measurements implies that the measurements have to be carried out in one and the same concentrate, i.e., the organic solvent to be used in the concentration procedure should be compatible with both the intended bioassay and the analytical method, or be easily exchangable to a compatible solvent. A further requirement of the method is that the method should be able to process a large quantity of water in a reasonable time.

Wilcox et al. (43) and Jolley (15) reviewed some of the techniques that have been used to produce concentrated extracts of organic matter in water samples. Methods discussed were freeze drying, reverse osmosis, liquid-liquid extraction, and binding to different adsorbents. The most popular method seems to be the use of the macroreticular XAD resins, held in a column. Amberlite XAD-2, XAD-4, XAD-7, and XAD-8 are nonionic resins: the first two are styrene-divinyl benzene copolymers and are effective in adsorbing nonpolar solutes, while the latter two are copolymers based on methacrylate and give better recoveries for compounds of intermediate polarity.

The recovery of organic carbon by the XAD procedure can be enhanced by passing water acidified to pH 2.0 through the XAD column

(31). Recently, several studies have reported substantially higher activity in extracts of drinking water passed through XAD columns at acid pH compared to extracts of neutral water (19,35,41). The results suggest that a major proportion of the mutagenicity is associated with organic acids. It has been shown, however, that residual chlorine in water can react with XAD resins and thereby produce mutagenic artifacts (39). Nevertheless, the work of Ringhand et al. (35) showed that the acid mutagens were products of water disinfection and not artifacts of the sample acidification step in the XAD concentration procedure. Ringhand and co-workers also observed higher activity for freeze-dried samples of chlorinated surface water compared to the activity of XAD extracts, suggesting the presence of polar mutagens not retained efficiently by the XAD resin.

THE IDENTIFICATION OF MUTAGENS IN CHLORINATED WATER

An effective way of reducing the number of compounds possibly representing a main mutagen is to fractionate the organic material in complex mixtures in regard to mutagenicity. The ideal fractionation procedure leads to a fraction containing only one compound, a significant mutagen. Subsequently, the identity of the compound might be determined by spectrometric studies.

Chemical fractionation and biological testing was successfully used by Holmbom et al. (10) in isolating a major Ames mutagenic compound present in kraft chlorination liquors. A multistep isolation scheme was developed for the separation of the key compound from a complex matrix of other organics. The scheme consisted of sorption/desorption with a XAD-4 resin, liquid-liquid partitioning followed by repeated thin layer chromatographic (TLC) and reversed-phase high performance liquid chromatographic (RP-HPLC) fractionations. On the basis of high-resolution mass spectrometry, and UV and infra red (IR) spectroscopic data the compound was tentatively identified as 3-chloro-4-(dichloromethyl)-5-hydroxy-2-(5/$\underline{H}$)-furanone (MX). Later the compound was synthesized and its structure comfirmed (34).

Kronberg et al. (21) applied the coupled chemical fractionation/bioassay procedure for the study of the mutagens formed during chlorination of a sample of surface water with a high content of humic material (total organic carbon, TOC = 25 mg/l). Since humic substances account for the major portion of the organic material in surface waters and are known to be precursors for trihalomethanes as well as for other volatile and nonvolatile organics formed during water chlorination (6,14,36), the chlorinated humic water was considered an appropriate model solution for chlorinated drinking water. The mutagens were isolated and concentrated by passing water at pH 2 over a column of XAD-4 and XAD-8 (1:1 volume mixture). Adsorbed material was eluted with ethyl acetate. Fractionation of the XAD extract by C_{18} column RP-HPLC showed 76% of the extract activity to be located in one single fraction (Fig. 1). This fraction was subjected to further RP-HPLC fractionation using a C_6 column. This time mutagenicity was collected in two distinct fractions, fraction A accounted for 22% and fraction B for 77% of the injected activity. The fractions were methylated and analyzed by gas chromatography/mass spectrometry (GC/MS). The mass spectrum collected at the retention time of standard MX was consistent with the mass spectrum of MX. Although previously Hemming et al. (9) had obtained strong indications of the presence of MX

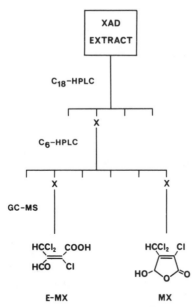

Fig. 1.   Procedure for isolation and identification of mutagens in chlorinated humic water.

in chlorinated humic and drinking water from studies carried out by selected ion monitoring (SIM) mode GC/MS and various techniques of sample ionization, this finding provided final evidence of the presence of MX in chlorine treated humic water.   Fraction A was found to contain a compound having the identical mass spectrum as one of the by-products formed during the synthesis of MX.   Mass spectrometric and ^{1}H nuclear magnetic resonance studies of the isolated by-product led to the identification of the geometric isomer of the open form of MX, (E)-2-chloro-3-(dichloromethyl)-4-oxobutenoic acid (EMX).   Support for correct structural determination was provided by the finding that the compound isomerized slowly in water at pH 2.0 quantitatively to MX (23).

Pure MX generated 5,600 net revertants/nmol in the Ames test with strain TA100-S-9 and its concentration in the original XAD extract and in fraction B was 380 ng/l and 270 ng/l, respectively (21).   MX accounted for all of the mutagenicity observed in the extract and the fraction.   In other samples of chlorine-treated humic waters MX was shown to account for approximately 50% of the observed mutagenicity (1).   The mutagenicity of EMX was determined as 320 net revertants/nmol (TA100-S-9).   It is not known whether this activity is caused by EMX or by MX formed as a result of EMX isomerization during the bioassay.   In any case, at the levels of EMX found in extracts of chlorine-treated humic waters the compound accounted for less than 10% of the overall mutagenicity.

A very similar approach was carried out by Meier et al. (28).   Meier and co-workers prepared an aqueous solution of a commercial soil humic acid at a concentration of 1 g TOC per liter distilled water.   The solution was treated with chlorine at a 1:1 ratio of chlorine equivalents per mol of carbon.   The mutagenicity of the solution was found to be associated with

Tab. 1.   Mutagenicity of tap waters in Finland and the mutagenicity con-
           tributions of MX

Disinfectant	No. of Plants	Mutagenicity	MX Contribution (%)
–	3	nonmutagenic	–
Cl_2	20	+ to +++	15 to 57
ClO_2/Cl_2	2	+	31 and 54
O_3/NH_2Cl	1	(+)	–

(+) = mutagenicity just detectable

+ = weak mutagenicity

+++ = very high mutagenicity

acid components.   Fractionation of the acids by C_{18} column HPLC showed
the activity to be concentrated in two distinct fractions: most of the
activity was collected in the fraction eluted in the middle of the
chromatographic run and minor activity in the more strongly retarded
fraction.   GC/MS analyses of the major peak of activity resulted in the
tentative identification of trichlorohydroxyfuranone isomers.   Recently
Meier et al. (29) reported MX to be one of the furanone isomers.

MX and EMX in Chlorinated Tap Waters

A preliminary investigation of two extracts of tap water indicated MX
to be a significant mutagen in the samples (9).   In order to find out
whether MX and EMX are compounds commonly present in chlorine-treated
tap waters, Kronberg and Vartiainen (22) analyzed water samples collected from 26 localities in Finland.   The mutagens were concentrated by
passing water at pH 2 through columns of XAD-8.   Ethyl acetate was
used for desorption.   The recovery efficiency of the procedure was tested
and it was found that approximately 70% of MX and 100% of EMX was collected in the extract.

Mutagenic activity was found in all drinking waters disinfected with
chlorine (n=20) or a combination of chlorine and chlorine dioxide (n=2)
(Tab. 1).   MX and EMX were detected in all active extracts except for
two extracts which exhibited only marginal activity.   The concentration of
MX ranged from 5-67 ng/l and the mutagen accounted for 15-57% of the
observed activity.   [In one sample the concentration of the compound was
too low (<4 ng/l) to allow for exact quantification and for calculation of
the activity contribution.]   The concentration of EMX was slightly lower
than the concentration of MX in corresponding samples.   EMX accounted
for a few percent or less of the extract mutagenicities.   Linear correlations were found between extract mutagenicity and concentrations of MX
and EMX; (correlation coefficient 0.894 and 0.910, respectively).

The water treatment method employed was found to have a pronounced effect on mutagenicity and on MX and EMX levels.   Although no
strict correlation was observed between chlorine dose and mutagenicity,
drinking water samples of unprecipitated surface water subjected to post-

chlorination at dosages of 1.0-1.5 mg Cl_2/l showed the lowest activity. Slightly more activity was formed in other waters when alum flocculation was included in the treatment and the chlorine dosage was 1.0 mg/l, perhaps because of the increased Cl_2/C dosage resulting from removal of organic material. Waters using postchlorination at dosages of 1.2-2.5 mg/l resulted in still higher activity of the extracts. Some of the waters studied employed both pre- and postchlorination; those with the low chlorine dosages (0.6 mg/l) at both stages did not show enhanced activity, whereas those with higher chlorine dosages, in particular during prechlorination, (1.1-2.8 mg Cl_2/l) generated the most active drinking waters monitored in this study.

Disinfection with chlorine dioxide in combination with chlorine produced tap water of low activity. Marginal activity was noted in water treated with ozone and monochloramine (n=1). Previously it was shown that ozone treatment does not generate mutagenicity (1) and thus the low activity noted was most likely due to the application of monochloramine. The samples of undisinfected drinking water, derived from groundwater or artificial groundwater (n = 3), were not mutagenic, and neither MX nor EMX was detected.

Recently Meier et al. (29) found MX in samples of chlorinated drinking water collected from three localities in the United States. The concentration of MX was estimated to 2-33 ng/l and the activity contribution of the mutagen was calculated to 15-34%. In addition, MX and EMX have been identified in eight samples of chlorinated drinking water collected in the United Kingdom (12). Very recently Backlund et al. (2) observed MX and EMX in chlorinated waters in The Netherlands.

Collectively, these studies show that MX and EMX are compounds commonly present in chlorine-treated waters, and that MX accounts for a considerable portion of the observed mutagenicity.

APPROACHES TO THE IDENTIFICATION OF MUTAGENS RESPONSIBLE FOR THE RESIDUAL ACTIVITY

In the chlorinated drinking waters studies so far the mean activity contribution of MX has been approximately 30%, and of EMX, a few percent at most. In order to achieve a more comprehensive understanding of the impact of mutagens in drinking water, the chemical identity of the compounds responsible for the residual activity should be determined. From previous studies, some information can be found of chemical and physical characteristics of the unidentified mutagens. Several studies have shown that as much as 90% of the mutagenicity is associated with compounds extractable from water only at acid condition (19,35,41). The acid properties of the mutagens were also demonstrated in an experiment where chlorinated water was extracted at various pH conditions using diethyl ether (20). A substantial increase in extract mutagenicity was observed when the extraction pH was lowered from pH 6.0 to pH 4.0. The mutagens have been shown to be very stable in water at pH 2, but at higher pH conditions they are susceptible to nucleophilic attacks by hydroxyl ions (20). Further, it has been found that the mutagens are relatively nonvolatile compounds (18,20).

Although the neutral mutagens may account for a smaller portion of the activity, they might be of considerable importance, especially if it is

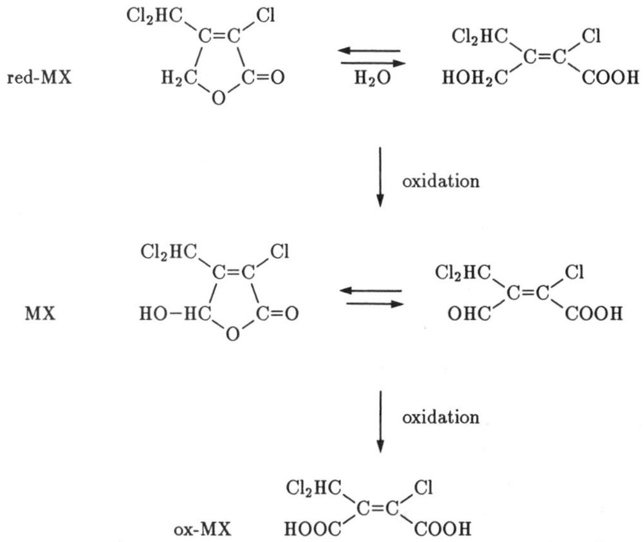

Fig. 2. Hypothetical pathway for the formation of MX and ox-MX through sequential oxidation of the reduced form of MX.

found that they have a large capability of bioaccumulation. However, the findings of previous work with neutral extracts of chlorinated drinking water indicated the presence of numerous weakly or at most moderately active neutral mutagens (11,17,40). The identification of individual neutral mutagens seems to be a laborious task. Work on the structural determination of unidentified mutagens in highly active extracts or acidified water might prove more promising. The following two approaches could be useful in carrying out this task: (i) search for compounds with structural similarities to MX or (ii) use chemical and chromatographic fractionation combined with mutagenicity testing of highly active extracts where the amount of MX is minimized.

MX is one of the most active mutagens ever tested in the Ames assay. In order to determine the structural elements of MX essential for rendering the compound its high level of mutagenic activity, various compounds with structural similarities to MX have been studied in the Ames assay. The effect on mutagenicity of changing the substituent at the 4-position of MX (the dichloromethyl group) by a chloromethyl and a methyl group was investigated by Streicher (38). The data clearly showed the importance of the dichloromethyl substituent since replacement by a chloromethyl substituent resulted in ten-fold less, and the replacement with a methyl substituent in 1,000-fold less active compound than MX. Ishiguro et al. (13) suggested that the cis arrangement of the CHCl$_2$ and Cl substituents on the carbon-carbon double bond might be a structural feature governing the mutagenic response of MX. This suggestion finds support from the observation that the geometric isomer of MX, EMX, where the CHCl$_2$ and Cl substituents are located trans to each other, generates at most 1/10 of the mutagenicity of MX (21). However, since it has been shown that EMX partly isomerizes to MX at pH 7 (the pH of the incubation medium in the Ames assay) the ultimate mutagen tested might in fact have been MX. Another difference between MX and EMX is the ability of MX to undergo ring-chain tautomerism.

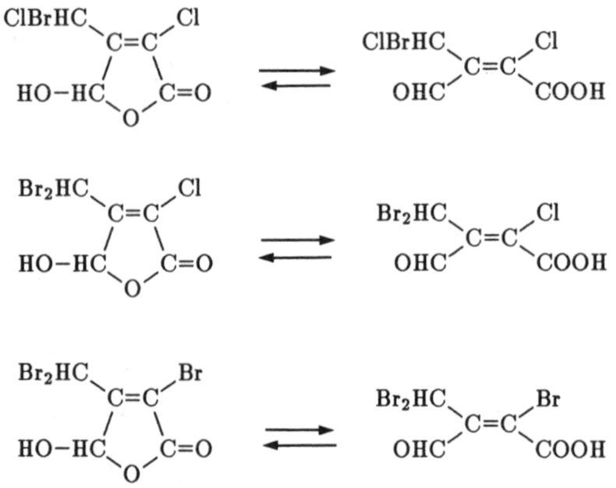

Fig. 3.   Structure of the three synthesized brominated analogues of MX.

During Ames testing, the prevailing form of MX is the chain-form but simultaneously traces of the ring-form are present. It is not known whether the ring-form of MX significantly contributes to the mutagenic response of the compound.

Two MX analogues having the cis arrangement of the $CHCl_2$ and Cl substituents and could theoretically be present in chlorinated waters are shown in Fig. 2. Assuming that MX is an intermediate of oxidation reactions in chlorinated water, (Z)-2-chloro-3-(dichloromethyl)-4-hydroxybut-2-enoic acid (red-MX, chain form) should be a precursor to MX while MX should be a precursor of (Z)-2-chloro-3-(dichloromethyl)-butenedioic acid (ox-MX). Kronberg and Christman (24) have synthesized the reduced and the oxidized analogues of MX and SIM-mode GC/MS analyses of extracts of chlorinated water showed the compounds to be present in the extracts. Mutagenicity assays showed red-MX to generate less than 2% of the activity MX generates and ox-MX to be nonmutagenic (strain TA100-S-9). This finding clearly shows that not only the cis arrangement of $CHCl_2$ and Cl groups at the double bond but also the aldehyde group is a critical factor for governing the mutagenicity of MX.

Horth et al. (12) suggested that brominated analogues of MX may be formed during chlorination of water containing bromide. They tested the hypothesis by studying the products of chlorination of the phenolic amino acid tyrosine in the presence of bromide. (Tyrosine has previously been shown to form MX and EMX upon chlorination.) The results of SIM-mode GC/MS analyses indicated that brominated MX-analogues were produced. Three of possible five brominated MX isomeres were synthesized and tested for mutagenicity (Fig. 3). Two of the compounds (BMX-2 and BMX-3) were at least equally active as MX, while the third compound (BMX-1) generated considerably less mutagenicity, although still a potent mutagen. No report on the identification of these compounds in actual samples of chlorinated drinking water has been published yet.

It might be the case that the unidentified mutagens are compounds not structurally related to MX. A possible approach to structural determination of these mutagens is the use of the fractionation procedure previously used for the identification of MX and EMX. Success in this work demands that the residual activity is caused by one or a few highly active compounds. In previously performed fractionation work by Meier et al. (29) and by Maruoka (27), mutagenic fractions were obtained which probably did not contain MX or EMX. However, structural identification of the active constituents has not been reported yet. Kronberg and Christman (25) found recently that MX in highly active extracts of aqueous solutions of fulvic acids chlorinated at pH 2, accounting for only 24% of the total mutagenicity. In view of the high yield of mutagenicity attributable to mutagens other than MX, this water might serve as a suitable material for the development of methods for chemical and chromatographic fractionation of yet unidentified mutagens.

REFERENCES

1. Backlund, P., L. Kronberg, G. Pensar, and L. Tikkanen (1985) Mutagenic activity in humic water and alum flocculated humic water treated with alternative disinfectants. Sci. Total Env. 47:257-264.
2. Backlund, P., E. Wondergem, K. Voogd, and A. de Jong (1989) Mutagenic activity and presence of the strong mutagen 3-chloro-4-(dichloromethyl)-5-hydroxy-2(5H)-furanone (MX) in chlorinated raw- and drinking waters in The Netherlands. Sci. Total Env. 84:273-282.
3. Barnes, D., and F. Wilson (1981) Chemistry and Unit Operations in Water Treatment, Applied Science, Essex, England, Chap. 5.
4. Bull, R.J. (1982) Health effects of drinking water disinfectants and disinfectant by-products. Env. Sci. Tech. 16:554A-559A.
5. Cheh, A.M., J. Skochdopole, P. Koski, and L. Cole (1980) Nonvolatile mutagens in drinking water: Production by chlorination and destruction by sulfite. Science 207:90-92.
6. de Leer, E.W.B. (1987) Aqueous chlorination products: The origin of organochlorine compounds in drinking and surface water. PhD Dissertation, Delft University of Technology, The Netherlands.
7. Foster, R. (1984) Mutagenicity testing of drinking water using freeze-dried extracts. Soc. Appl. Bacteriol. Tech. Ser. 19:375-391.
8. Glatz, B.A., C.D. Chriswell, M.D. Arguello, H. Svec, J.S. Fritz, S.A. Grimm, and M.A. Thomson (1978) Examination of drinking water for mutagenic activity. J. Am. Water Works Assoc. 78:465-468.
9. Hemming, J., B. Holmbom, M. Reunanen, and L. Kronberg (1986) Determination of the strong mutagen 3-chloro-4-(dichloromethyl)-5-hydroxy-2(5H)-furanone in chlorinated drinking water. Chemosphere 15:549-556.
10. Holmbom, B., R.H. Voss, R.D. Mortimer, and A. Wong (1984) Fractionation, isolation and characterization of Ames mutagenic compounds in kraft chlorination effluents. Env. Sci. Tech. 18:333-337.
11. Horth, H.B. Crathorne, R.D. Gwilliam, C.P. Palmer, J.A. Stanley, and M.J. Thomas (1987) Technique for the fractionation and identification of mutagens produced by water treatment chlorination. In Organic Pollutants in Water, I.H. Suffet, and M. Malaiyandi, eds. Advances in Chemistry Series, No. 214, American Cancer Society, Washington, pp. 659-674..
12. Horth, H., M. Fielding, T. Gibson, H.A. James, and H. Ross (1989) Identification of Mutagens in Drinking Water (EC 9105 SLD). PRD 2038-M. Water Research Centre, Marlow, Bucks, United Kingdom.
13. Ishiguro, Y., J. Santodonato, and M.W. Neal (1988) Mutagenic potency of chlorofuranones and related compounds in Salmonella. Env. and Molec. Muta. 11:225-234.

14. Johnson, J.D., R.F. Christman, D.L. Norwood, and D.S. Millington
     (1982) Reaction products of aquatic humic substances with chlorine.
     Env. Health Perspect. 46:63-71.
15. Jolley, R.L. (1981) Concentrating organics in water for biological
     testing Env. Sci. Tech. 15:874-880.
16. Jolley, R.L., and J.H. Carpenter (1983) A review of the chemistry
     and environmental fate of reactive oxidant species in chlorinated wa-
     ter.   In Water Chlorination: Environmental Impact and Health Ef-
     fects, R.L. Jolley, W.A. Brungs, J.A. Cotruvo, R.B. Cumming,
     J.S. Mattice, and V.A. Jacobs, eds. Ann Arbor Science, Ann Ar-
     bor, Michigan, Vol. 4, pp. 3-47.
17. Kool, H.J., C.F. van Kreijl, and M. Verlaan-de Vries (1987) Con-
     centration, fractionation and characterisation of organic mutagens in
     drinking water.   In Organic Pollutants in Water, I.H. Suffet, and M.
     Malaiyandi, eds. Advances in Chemistry Series, No. 214, American
     Chemical Society, Washington, DC., pp. 605-625.
18. Kopfler, F.C., H.P. Ringhand, W.E. Coleman, and J.R. Meier
     (1985) Reactions of chlorine in drinking water, with humic acids and
     in vivo.   In Water Chlorination: Environmental Impact and Health Ef-
     fects, R.L. Jolley, R.J. Bull, W.P. Davis, S. Katz, M.H. Roberts,
     and V.A. Jacobs, eds. Lewis Publishers, Chelsea, Michigan, Vol.
     5, pp. 161-173.
19. Kronberg, L., B. Holmbom, and L. Tikkanen (1985) Mutagenic activ-
     ity in drinking water and humic water after chlorine treatment.
     Vatten 41:106-109.
20. Kronberg, L., B. Holmbom, and L. Tikkanen (1986) Properties of
     mutagenic compounds formed during chlorination of humic water.   In
     Proceedings of the Fourth European Symposium on Organic Micropol-
     lutants in the Aquatic Environment, A. Bjorseth, G.Angeletti, and
     D. Reidel, eds. Dordrecht, Holland, pp. 449-454.
21. Kronberg, L., B. Holmbom, M. Reunanen, and L. Tikkanen (1988)
     Ames mutagenicity and concentration of the strong mutagen 3-chloro-
     4-(dichloro-methyl)-5-hydroxy-2(5H)-furanone and of its geometric
     isomer (E)-2-chloro-3-(dichloromethyl)-4-oxobutenoic acid in chlo-
     rine-treated humic and drinking water extracts. Env. Sci. Tech.
     22:1097-1103.
22. Kronberg, L., and T. Vartiainen (1988) Ames mutagenicity and con-
     centration of the strong mutagen 3-chloro-4-(dichloromethyl)-5-hydro-
     xy-2(5H)-furanone and of its geometric isomer (E)-2-chloro-3-(di-
     (dichloromethyl)-4-oxobutenoic acid in chlorine-treated tap waters.
     Mutat. Res. 206:177-182.
23. Kronberg, L., B. Holmbom, and L. Tikkanen (1989) Identification of
     the strong mutagen 3-chloro-4-(dichloromethyl)-5-hydroxy-2(5H)-
     furanone and of its geometric isomer (E)-2-chloro-3-(dichloromethyl)-
     4-oxobutenoic acid in mutagenic fractions of chlorine-treated humic
     water and in drinking waters. In Water Chlorination: Environmental
     Impact and Health Effects, R.L. Jolley et al., eds. Lewis Publish-
     ers, Chelsea, Michigan, Vol. 6, pp. 137-146 (in press).
24. Kronberg, L., and R.F. Christman (1989) Chemistry of mutagenic
     by-products of water chlorination. Sci. Total Env. 81/82:219-230.
25. Kronberg, L., and R.F. Christman (1989) (unpubl. data).
26. Loper, J.C. (1980) Mutagenic effects of organic compounds in drink-
     ing water. Mutat. Res. 76:241-268.
27. Maruoka, S. (1986) Analysis of mutagenic by-products produced by
     chlorination of humic substances by thin layer chromatography and
     high-performance liquid chromatography. Sci. Total Env. 54:195-
     -205.

28. Meier, J.R., H.P. Ringhand, W.E. Coleman, K.M. Schenck, J.W. Munch, R.P. Streicher, W.H. Kaylor, and F.C. Kopfler (1986) Mutagenic by-products of chlorination of humic acid. Env. Health Perspect. 69:101-107.
29. Meier, J.R., R.B. Knohl, W.E. Coleman, H.P. Ringhand, J.W. Munch, W.H. Kaylor, R.P. Streicher, and F.C. Kopfler (1987) Studies on the potent bacterial mutagen 3-chloro-4-(dichloromethyl)-5-hydroxy-2(5H)furanone: Aqueous stability, XAD recovery and analytical determination in drinking water and in chlorinated humic acid solutions. Mutat. Res. 189:363-373.
30. Meier, J.R. (1988) Genotoxic activity of organic chemicals in drinking water. Mutat. Res. 196(3):211-245.
31. Monarca, S., J.K. Hongslo, A. Kringstad, and G.E. Carlberg (1985) Microscale fluctuation assay coupled with Sep-Pak concentration as a rapid and sensitive method for screening mutagens in drinking water. Water Res. 19:1209-1216.
32. Morris, J.C. (1978) The chemistry of aqueous chlorine in relation to water chlorination. In Water Chlorination: Environmental Impact and Health Effects, R.L. Jolley, ed. Ann Arbor Science, Ann Arbor, Michigan, Vol. 1, pp. 21-35.
33. Oliver, B.G. (1978) Chlorinated non-volatile organics produced by the reaction of chlorine with humic materials. Canadian Res. 11:21-22.
34. Padmapriya, A.A., G. Just, and N.G. Lewis (1985) Synthesis of 3-chloro-4-(dichloromethyl)-5-hydroxy-2(5H)-furanone. Can. J. Chem. 63:828-832.
35. Ringhand, P.H., J.R. Meier, F.C. Kopfler, K.M. Schenck, W.H. Kaylor, and D.E. Mitchell (1987) Importance of sample pH on the recovery of mutagenicity from drinking water by XAD resins. Env. Sci. Tech. 21:382-387.
36. Rook, J.J. (1974) Formation of haloforms during chlorination of natural waters. Water Treat. Exam. 23:234-243.
37. Simmon, V.F., and R.G. Tardiff (1976) Mutagenic activity of drinking water concentrates. Mutat. Res. 38:389-390.
38. Streicher, R.B. (1987) Studies of the products resulting from the chlorination of drinking water. PhD Dissertation, The University of Cincinnati, Ohio.
39. Sweeney, A.G., and A.M. Cheh (1985) Production of mutagenic artifacts by the action of residual chlorine on XAD-4 resin. J. Chromatogr. 325:95-102.
40. van Rossum, P.G. (1985) Progress in the isolation and characterization of non-volatile mutagens in a drinking water. Sci. Total. Env. 47:361-370.
41. Vartiainen, T., and A. Liimatainen (1986) High levels of mutagenic activity in chlorinated drinking water in Finland. Mutat. Res. 167:29-34.
42. Wigilius, B., H. Borén, G.E. Carlberg, A. Grimwall, and M. Möller (1985) A comparison of methods for concentrating mutagens in drinking water: Recovery aspects and their implications for the chemical character of major unidentified mutagens. Sci. Total Env. 47:265-272.
43. Wilcox, P., F. van Hoof, and M. van der Gaag (1986) Isolation and concentration of mutagens from drinking water. In Proceedings of the XVIth Annual Meeting of the European Environmental Mutagen Society, A. Leonard and M. Kirsch-Volders, eds., Brussels, Belgium, pp. 92-103.

GENOTOXIC AND CARCINOGENIC PROPERTIES OF CHLORINATED

FURANONES: IMPORTANT BY-PRODUCTS OF WATER CHLORINATION

J.R. Meier,[1] A.B. DeAngelo,[1] F.B. Daniel,[1]
K.M. Schenck,[1] J.U. Doerger,[1] L.W. Chang,[1]
F.C. Kopfler,[2] M. Robinson,[2] and H.P. Ringhand[2]

[1]Genetic Toxicology Division

[2]Environmental Toxicology Division
Health Effects Research Laboratory
U.S. Environmental Protection Agency
Cincinnati, Ohio   45268

INTRODUCTION

Mutagenic activity is frequently detectable in organic concentrates of drinking water derived from surface waters (11). Because most mutagens tested to date are animal carcinogens and because mutagens may induce heritable alterations in germ cells, there is concern as to whether or not the presence of mutagenic chemicals in drinking water represent an acceptable human health risk. Identification of the mutagenic components of drinking water is needed before an accurate assessment of the health risks can be made. Methods have been developed for the isolation and identification of mutagens in drinking water (24), but the task of ascribing mutagenicity levels to specific chemical contaminants has proved difficult.

Studies comparing the mutagenic activity of raw and finished drinking water samples have shown that the mutagenicity in drinking water is often attributable to by-products from chlorine disinfection (13). The reaction of chlorine with humic materials also results in the formation of mutagens, and provides a reasonable model for the study of mutagens formed during water chlorination (14,15). Recently, the highly potent bacterial mutagen, 3-chloro-4-dichloromethyl-5-hydroxy-2(5H)-furanone (MX), has been identified and quantified in chlorinated humic materials (8,17) and in drinking water in the United States, Finland, and Great Britain (8-10,17). Of a total of 37 drinking water samples examined thus far, MX has been estimated to be present at concentrations ranging from 2 to 67 ng/l, and to account for between 5 and 60% of the mutagenicity present. The discovery of MX is particularly significant in light of the

Genetic Toxicology of Complex Mixtures
Edited by M. D. Waters *et al.*
Plenum Press, New York, 1990

fact that all previously identified mutagens as a whole accounted for less than 10% of the mutagenicity of chlorinated humic solutions (15) or drinking water concentrates (6). This chapter summarizes our efforts to date aimed at the characterization of the genotoxic and carcinogenic properties of MX.

MATERIALS AND METHODS

The Salmonella mutagenicity assay was carried out according to the method of Maron and Ames (12). S. typhimurium strains TA1535, TA92, TA2410, and TA100 were a gift of Dr. Bruce Ames (Berkeley, CA). The SOS/Umu test was conducted according to the method of Whong et al. (25) with minor modification. The chromosomal aberration assay in Chinese hamster ovary (CHO) cells and mouse bone marrow micronucleus assay was done as described elsewhere (16).

Induction of nuclear anomalies in the small intestine of mice was evaluated using the method of Blakey et al. (1) as modified by Daniel et al. (4). Briefly, MX was administered in a single dose by oral gavage to groups of ten male B6C3F1 mice at levels of 0, 60, 80, and 100 mg/kg. Twenty-four hours later the animals were sacrificed, the duodenums removed, sectioned, and stained, and nuclear anomalies (i.e., micronuclei, pyknotic nuclei, and karyorrhectic nuclei) were scored in the crypt cells (10 crypts/animal).

DNA strand breaks were analyzed in the human lymphoblastic cell line CCRF-CEM American Type Culture Collection (ATCC, Rockville, MD). Cells were grown in RPMI-1640 medium with 10% fetal bovine serum supplemented with 20 mM Hepes, 20 mM L-glutamine, and 50 µg/ml Gentamicin. The cells were suspended in serum-free medium to a concentration of 2×10^6 cells/ml, and treated with test chemicals for 2 hr. The cells were centrifuged and the pellets were resuspended in 1 ml of phosphate-buffered saline containing 0.02 M EDTA. DNA strand breaks were analyzed via the DNA alkaline unwinding procedure as previously described by Daniel et al. (3).

Methods for examining the disposition of ^{14}C-MX in rats are described in detail elsewhere (21). Briefly, 3-^{14}C-MX was administered once in water by oral gavage to six male, F344 rats. At 8, 24, 32, and 48 hr after dosing, urine and feces were collected and at sacrifice (48 hr) tissue and blood samples were taken. Expired air was also collected over the 48-hr period using ethanolamine and activated carbon traps. Samples were prepared and counted for ^{14}C-radioactivity using a liquid scintillation spectrometer.

The mouse skin tumor initiation-promotion assay was conducted using SENCAR mice. A single dose of MX in a volume of 0.2 ml was administered topically or orally at levels of 5, 16, 28, and 50 mg/kg with either acetone (topically) or distilled water (orally) as a vehicle. Two weeks after treatment with MX, a tumor-promotion schedule involving applications of 1.0 µg of 12-0 tetradecanoyl-phorbol-13-acetate (TPA) in 0.2 ml acetone was begun. TPA was given three times weekly for 30 weeks. Tumors (papillomas) were charted by location and size from weekly observations and were included in the cumulative tumor count only if they were observed for at least three consecutive weeks.

RESULTS AND DISCUSSION

In Vitro Studies

Previous work had shown that MX is among the most potent mutagens in the Ames assay, producing a response of ∿13,000 TA100 revertants/nmole. Figure 1 shows the effect of different genetic backgrounds in Salmonella on the mutagenic potency of MX. In the absence of the plasmid pKM101, the response is reduced approximately 100-fold (compare TA1535 vs TA100), indicating the apparent importance of error-prone repair processing of DNA damage in the mutagenic response. However, even in the presence of pKM101, mutagenicity is substantially reduced (∿40-fold) when an intact excision repair system is present (compare TA2410 vs TA92). The presence of the deep rough (rfa) mutation in the outer membrane lipopolysaccharide (LPS) also increases the mutagenic response approximately three-fold (compare TA100 vs TA2410), indicating that a normal LPS is a partial barrier to penetration of MX into the cell. These results illustrate the importance of factors that have been genetically engineered into the Ames tester strains in the mutagenic response induced by MX.

The importance of SOS processing for induction of mutations in Salmonella is borne out by the results of the SOS/Umu assay. This assay measures DNA damaging activity of chemicals by indirect measurement of induction of umuCD gene expression. The umuCD genes are analogous in function and partially homologous in structure to the mucAB genes present in the pKM101 plasmid (20). Both genes code for proteins involved

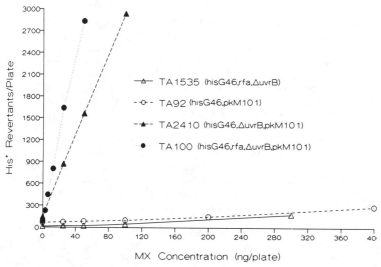

Fig. 1.   Effect of DNA repair and LPS mutations on mutagenesis by MX in Salmonella.

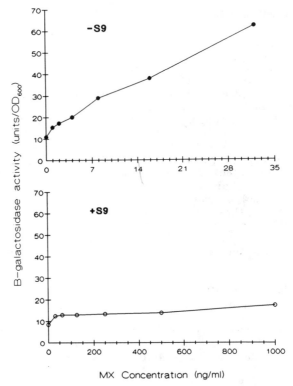

Fig. 2.   Induction of DNA damage by MX in Salmonella as measured in
           the SOS/Umu test.

in the error-prone repair of DNA damage.   When tested for activity
without S-9 activation in the SOS/umuC assay, MX induced a clear
positive response (Fig. 2).   The β-galactosidase activity increased
linearly with dose over a range of 1 to 32 μg/ml).   When tested with
metabolic activation, a marginal increase in response was observed, and
only at much higher concentrations (30-1,000 μg/ml). The substantially
lower activity when S-9 is added is consistent with results previously
seen in the Salmonella mutagenicity assay (16).

    MX has also been examined for genotoxic activity in cultured
mammalian cells.   In CHO cells, MX induced significant increases in structural chromosomal aberrations at a dose of 4 μg/ml without metabolic activation and at 75 μg/ml in the presence of a metabolic activation system
(Tab. 1).   The aberrations were mainly chromatid deletions, but a few
chromatid exchanges were also observed.

    The DNA strand-breaking potential of MX for human CCRF-CEM cells
has been evaluated using an alkaline unwinding procedure.   In this assay, the fraction of DNA that remains double-stranded after separation of
DNA strands in alkaline solution is inversely related to the number of
DNA strand breaks induced by certain chemicals (3).   The results in Fig.

Tab. 1.   Induction of structural chromosome aberrations by MX in CHO
          cells [adapted from Meier et al. (13)]

Treatment	Mitotic Index (%)	% Cells With Aberrations	No. Aberrations Per Cell
Without Metabolic Activation			
Deionized H_2O	4.6	0	0
MMS (110 μg/ml)	1.4*	92*	3.99*
MX: 1 μg/ml	7.0	0	0
2 μg/ml	7.8	3	0.03
4 μg/ml	2.6*	22*	0.31*
With Metabolic Activation			
Deionized H_2O	7.9	1	0.01
CP (75 μg/ml)	1.0	19*	0.30*
18.8 μg/ml	8.4	0	0
37.5 μg/ml	10.6	0	0
75.0 μg/ml	4.2*	10*	0.30*

[a]Adapted from Meier et al (13).

*Significantly different (p <0.050) from the appropriate solvent control.

MMS - methyl methanesulfonate; CP - cyclophosphamide.

3 show that with increasing MX concentrations, a dose-related decrease in
the fraction of double-stranded DNA was observed.   On a molar basis,
the response with MX was found to be comparable (one-half to one-third)
that observed for dimethylsulfate, a strong direct-acting SN2-type
alkylating agent.   These results clearly show that MX possesses genotoxic
activity in vitro both for bacterial and mammalian cells.

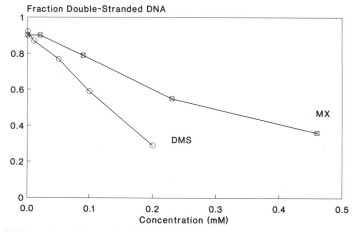

Fig. 3.   DNA strand breaks induced by MX and dimethylsulfate (DMS) in
          human lymphoid cells.

Tab. 2.   Evaluation of MX for activity in the mouse bone marrow micro-
          nucleus assay [adapted from Meier et al. (13)]

| | % Micronucleated PCE | | | |
| Treatment | 48 Hr Sacrifice | | 72 Hr Sacrifice | |
	Male	Female	Male	Female
Deionized H_2O	0.12	0.44	0.28	0.08
TMP (1250 mg/kg/d)	1.67*	2.74*	-	-
MX: 22.5 mg/kg/d	0.20	0.35	0.24	0.12
45 mg/kg/d	0.32	0.32	0.12	0.16
90 mg/kg/d	0.13	0.24	0.27	0.20

[a]Adapted from Meier et al (13).

*p <0.01.

TMP = trimethylphosphate

In Vivo Studies

The mouse bone marrow micronucleus test is a widely used in vivo
test for the detection of chromosomal damage or damage of the spindle ap-
paratus (22). The results of testing MX for micronuclei induction are
presented in Tab. 2. The doses were chosen based on an estimated oral
LD_{50} value in mice of 128 mg/kg (16). Even after oral administration of
two doses, spaced 24 hr apart, of 90 mg/kg (approximately 70% of the
LD_{50}), no evidence of increases in micronucleated polychromatic
erythrocytes were observed at either 48 hr or 72 hr after the initial
dose. Thus, MX does not appear to possess clastogenic activity in vivo
for mouse bone marrow despite its relatively high clastogenic activity for
mammalian cells in vitro.

Several possible explanations exist for the failure of MX to induce
micronuclei in bone marrow: (1) it may be poorly absorbed following oral
administration; (ii) it may react immediately with cells of the
gastrointestinal (GI) tract; and (iii) it may be rapidly metabolized to an
inactive species in the GI tract, liver, or systemic circulation. Several
studies have been conducted to examine these possibilities.

The disposition of ^{14}C-labeled MX was followed after oral administration of the compound to rats. The results of this study are summarized in Fig. 4. Further details of this work are published elsewhere
(21). The data indicate that a substantial fraction of MX or its
metabolites (>40%) was absorbed into the systemic circulation and either
distributed in tissues or eliminated in the urine. Whether the radioactivity in the feces (>44%) results from incomplete absorption or from biliary
excretion remains to be resolved. Only a small percentage (4.7) of the

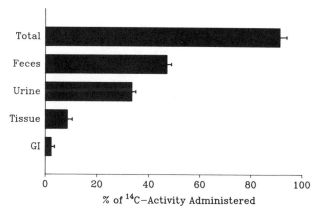

% of ^{14}C−Activity Administered

Fig. 4.   Tissue distribution and cumulative excretion of total radioactivity after oral administration of ^{14}C-MX to rats.

dose was found in the tissues at 48 hr postadministration, and the
^{14}C-activity was not selectively concentrated by any particular tissue.
None of the ^{14}C-activity was eliminated via the lungs as either CO_2 or as
volatile organic compounds.

The ability of MX to induce nuclear anomalies in cells of the GI tract
was also examined (4). Nuclear anomalies (including both micronuclei and
apoptotic bodies) are produced by agents which induce tumors of the GI
tract in animals, and an assay using this endpoint has been proposed as
a short-term in vivo test for detecting small intestinal and colon

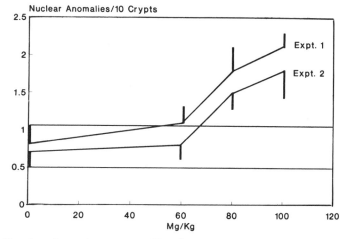

Fig. 5.   Duodenal nuclear anomalies induced by MX in male B6C3F1 mice.

carcinogens (1,7). Figure 5 shows the results of two independent experiments in which crypt cells of the duodenum of mice were evaluated for
nuclear anomalies 24 hr after a single oral gavage administration of MX.
Dose-related increases in nuclear anomalies were observed; these were
significantly above background at the two highest doses. It should be
noted that the significant responses were observed only at close to toxic
doses (approximately 60 and 80% of the LD_{50}). Furthermore, the
response was weak in comparison with that observed for the intestinal
carcinogen, methyl nitrosourea (MNU), which was administered at a ten-
fold lower dose. The response with MX was 1.8 nuclear anomalies/10
crypts (approx. 100 cells) as compared to 19.8 nuclear anomalies/10
crypts for MNU.

Preliminary evaluation of the carcinogenic activity of MX has been
done by examining skin tumor induction in SENCAR mice using an initiation-promotion protocol. The results are presented in Tab. 3. No tumors were observed in mice which received MX alone (without TPA promotion), given either topically or orally. In mice treated topically with
MX, followed by TPA promotion, the incidence and number of skin tumors
were not significantly higher than those in control animals. A highly significant tumor response was seen in animals treated with the positive control compound, benzo(a)pyrene, as expected. Following oral administration, both the tumor incidence and number of tumors/mouse were significantly elevated at the 16, 28, and 50 mg/kg doses of MX based on

Tab. 3. Initiation of skin tumors in SENCAR mice at 24 weeks after administration of MX followed by TPA promotion

Treatment	Topical		Oral	
	Tumors Per Mouse	No. Mice With Tumors	Tumors Per Mouse	No. Mice With Tumors
Vehicle[a]+ TPA	0.43	8/39	0.03[b]	1/39
1 mg/kg BaP + TPA	8.95	20/20	----	----
50 mg/kg MX (No TPA)	0.00	0/20	0.00	0/20
5 mg/kg MX + TPA	0.23	6/40	0.15[c]	5/40
16 mg/kg MX + TPA	0.28	4/38	0.18[c]	6/39[d]
28 mg/kg MX + TPA	0.08	2/40	0.38[c]	7/40[d]
50 mg/kg MX + TPA	0.35	10/39	0.48[c]	7/39[d]

[a] Vehicle was acetone topically or distilled water orally.

[b] Overall analysis significant at p = 0.014 by Logrank trend test.

[c] Significantly greater than vehicle control at p < 0.05 by Logrank trend
test.

[d] Significantly greater than vehicle control at p < 0.05 by Likelihood
Rationtest.

pairwise comparisons to the control group. An overall trend analysis also revealed significantly higher number of tumors/mouse in the treated vs control animals. It may seem unusual that MX should induce tumors when given orally but not topically, especially since the SENCAR mouse assay has been designed primarily for examining tumor induction following topical application. However, previously both ethyl carbamate and acrylamide also have been found to induce substantially greater numbers of skin tumors when given topically than when given orally (2). Perhaps of greater concern is the wide difference in vehicle control values between the topical and oral exposure. The only difference between these two groups was that acetone was the vehicle topically, whereas distilled water was given orally. The historical control tumor incidence values for our lab are 8% for either route. Thus, in the present study, the tumor values for the topical and oral control groups are somewhat higher and lower, respectively, than normal. In order to minimize the influence of low control values on the significance of the results, data at 28 weeks after dosing were also analyzed. By this time, the control values had risen to 7.5% tumor incidence and approximately 0.1 tumor/animal. The values for the treated groups had also increased (data not shown). Reanalysis, using the 28-wk data, revealed that the number of tumors/mouse were still significantly higher ($p < 0.05$) than control at the 16, 28, and 50 mg/kg MX dose levels, and the overall trend analysis was significant at $p < 0.028$. However, analysis of the tumor incidence data showed that only the 16 mg/kg dose was significantly different from control at $p < 0.05$; the p-values for the 28 and 50 mg/kg were both 0.099.

The results from the toxicological studies which have been conducted thus far on MX may be summarized as follows:

(i)   MX exhibits potent direct-acting genotoxic activity for both bacterial and mammalian cells in vitro, and forms a DNA adduct(s) which remains uncharacterized (18). The DNA repair potential of the bacterial cells has a major influence on the mutagenic potency of the compound.

(ii)   Structure-activity studies, reported elsewhere (23), indicate that the dichloromethyl substituent is a key structural element in the high potency of MX as a bacterial mutagen.

(iii)   MX appears to be inactive as a clastogen in mouse bone marrow. Although the tissue distribution studies indicate that a substantial fraction of orally administered MX is absorbed and systemically distributed, these studies have not distinguished between whether MX is present as the parent compound or an unreactive metabolite. In vitro studies have shown that the genotoxic potential of MX is almost completely abolished by liver cytosol, and this inactivation can be partially explained by conjugation with glutathione via glutathione-S-transferases (19). Thus, detoxification by the liver may explain the lack of genotoxicity for the bone marrow.

(iv)   MX is marginally active for inducing nuclear anomalies in the small intestine of mice, but only at doses near the LD_{50} (4). DNA adduct formation has also been demonstrated via ^{32}P-post-labeling for stomach cells of MX-treated rats following a 60 mg/kg oral dose (5). Taken together, these findings suggest that the cells of the GI tract may be targets for tumor induction by MX. This issue remains to be explored.

(v) MX appears to possess skin tumor initiating activity when administered orally. However, this result needs confirmation, especially in view of problems with control values.

ACKNOWLEDGEMENT

We wish to thank Ms. Shin-Ru Wang and Dr. Judy Stober for statistical analysis of the skin tumor data.

This document has been reviewed in accordance with U.S. Environmental Protection Agency policy and approved for publication. Approval does not signify that the contents necessarily reflect the views or policies of the Agency, nor does mention of trade names or commercial products constitute endorsement or recommendation for use.

REFERENCES

1. Blakey, D.H., A.M. Duncan, M.J. Wargovich, M.T. Goldberg, W.R. Bruce, and J.A. Heddle (1985) Detection of nuclear anomalies in the colonic epithelium of the mouse. Cancer Res. 45:242-249.
2. Bull, R.J., M. Robinson, R.D. Laurie, E. Greisiger, J.R. Meier, and J. Stober (1984) Carcinogenic effects of acrylamide in Sencar and A/J mice. Cancer Res. 44:107-111.
3. Daniel, F.B., D.L. Haas, and S.M. Pyle (1985) Quantitation of chemically induced DNA strand breaks via an alkaline unwinding assay. Analyt. Biochem. 144:390-402.
4. Daniel, F.B., G.R. Olson, and J.A. Stober (1989) The induction of G.I. tract nuclear anomalies by 3-chloro-4(dichloromethyl)-5-hydroxy-2(5H)-furanone, a chlorine disinfection by-product, in the male B6C3F1 mouse (ms. in prep.).
5. Daniel, F.B., E.L.C. Lin, H.P. Ringhand, R.G. Miller, A.B. DeAngelo, and J.R. Meier (1989) 3-Chloro-4(dichloromethyl)-5-hydroxy-2(5H)-furanone, a potent mutagen isolated from chlorine-disinfected drinking water, forms DNA adducts in vivo (ms. in prep.).
6. Fielding, M., and H. Horth (1986) Formation of mutagens and chemicals during water treatment chlorination. Water Supply 4:103-126.
7. Goldberg, M.T., and P. Chidiac (1986) An in vivo assay for small intestine genotoxicity. Mutat. Res. 164:209-215.
8. Hemming, J., B. Holmbom, M. Reunanen, and L. Kronberg (1986) Determination of the strong mutagen, 3-chloro-4-(dichloromethyl)-5-hydroxy-2(5H)-furanone in chlorinated drinking and humic waters. Chemosphere 15:549-556.
9. Horth, H., M. Fielding, T. Gibson, H.A. James, and H. Ross (1989) Identification of mutagens in drinking water. Water Research Centre Technical Report. PRD 2038-M, Medmenham, England.
10. Kronberg, L., and T. Vartianen (1988) Ames mutagenicity and concentration of the strong mutagen 3-chloro-4-(dichloromethyl)-5-hydroxy-2(5H)-furanone and of its geometric isomer (E)-2-chloro-3-(dichloromethyl)-4-oxo-butenoic acid in chlorine-treated tap waters. Mutat. Res. 206:177-182.

11. Loper, J.C. (1980) Mutagenic effects of organic compounds in drinking water. Mutat. Res. 76:241-268.

12. Maron, D.M., and B.N. Ames (1983) Revised methods for the Salmonella mutagenicity test. Mutat. Res. 113:173-215.

13. Meier, J.R. (1988) Genotoxic activity of organic chemicals in drinking water. Mutat. Res. 196:211-245.

14. Meier, J.R., R.D. Lingg, and R.J. Bull (1983) Formation of mutagens following chlorination of humic acid: A model for mutagen formation during drinking water treatment. Mutat. Res. 118:25-41.

15. Meier, J.R., H.P. Ringhand, W.E. Coleman, J.W. Munch, R.P. Streicher, W.H. Kaylor, and K.M. Schenck (1985) Identification of mutagenic compounds formed during chlorination of humic acid. Mutat. Res. 157:111-122.

16. Meier, J.R., W.F. Blazak, and R.B. Knohl (1987) Mutagenic and clastogenic properties of 3-chloro-4-(dichloromethyl)-5-hydroxy-2(5H)-furanone, a potent bacterial mutagen in drinking water. Env.Muta. 10:411-424.

17. Meier, J.R., R.B. Knohl, W.E. Coleman, H.P. Ringhand, J.W. Munch, W.H. Kaylor, R.P. Sreicher, and F.C. Kopfler (1987) Studies on the potent bacterial mutagen, 3-chloro-4-(dichloromethyl)-5-hydroxy-2(5H)-furanone: Aqueous stability, XAD recovery and analytical determination in drinking water and in chlorinated humic acid solutions. Mutat. Res. 189:363-373.

18. Meier, J.R., A.B. DeAngelo, F.B. Daniel, K.M. Schenck, M.F. Skelly, and S.L. Huang (1989) DNA adduct formation in bacterial and mammalian cells by 3-chloro-4-(dichloromethyl)-5-hydroxy-2-(5H)-furanone (MX). Env. Molec. Muta. 14(Suppl. 15):128.

19. Meier, J.R., R.B. Knohl, B.A. Merrick, and C.L. Smallwood (1989) Importance of glutathione in the in vitro detoxification of 3-chloro-4-dichloromethyl-5-hydroxy-2(5H)-furanone, an important mutagenic by-product of water chlorination. In Water Chlorination: Chemistry, Environmental Impact and Health Effects, Vol. 6, R.L. Jolley et al., eds. Lewis Publishers, Ann Arbor, MI (in press).

20. Perry, K.L., S.J. Elledge, B.B. Mitchell, L. Marsh, and G.C. Walker (1985) umuDC and mucAB operons whose products are required for UV light and chemical-induced mutagenesis: UmuD, MucA, and LexA proteins share homology. Proc. Natl. Acad. Sci., USA 82:4331-4335.

21. Ringhand, H.P., W.H. Kaylor, R.G. Miller, and F.C. Kopfler (1989) Synthesis of $3-^{14}C-3-$chloro-4-(dichloromethyl)-5-hydroxy-2(5H)-furanone and its use in a tissue distribution study in the rat. Chemosphere 18:2229-2236.

22. Schmid, W. (1976) The micronucleus test for cytogenetic analysis. In Chemical Mutagens, Principles and Methods for Their Detection, Vol. 4, A. Hollaender, ed. Plenum Press, New York, pp. 31-53.

23. Streicher, R.P., H. Zimmer, J.R. Meier, R.B. Knohl, F.C. Kopfler, W.E. Coleman, J.W. Munch, and K.M. Schenck (1989) Structure-mutagenic activity relationships in chlorinated α, β-unsaturated carbonyl compounds (sub. for publ.).

24. Tabor, M.W., and J.C. Loper (1985) Analytical isolation, separation and identification of mutagens from nonvolatile organics of drinking water. J. Env. Analyt. Chem. 19:281-318.

25. Whong, W.-Z., Y.-E. Wen, J. Steward, and T. Ong (1986) Validation of the SOS/Umu test with mutagenic complex mixtures. Mutat. Res. 175:139-144.

ASSESSMENT AND IDENTIFICATION OF

GENOTOXIC COMPOUNDS IN WATER

J.K. Fawell and H. Horth

WRc Medmenham
Medmenham, Nr. Marlow
Bucks, SL7 2HD United Kingdom

INTRODUCTION

At the end of the 1970s, it was shown that only about 15% of the organic matter in drinking water was amenable to identification by gas chromatography-mass spectrometry (GC-MS) (9,11). However, there were indications from some epidemiological studies that an association may exist between chlorinated drinking water derived from surface sources and some cancers of the gastrointestinal (GI) tract and bladder (1,3,5).

In order to investigate whether components of the unidentified fraction were implicated in this association, a number of groups turned to mutagenicity assays in bacteria to determine whether genotoxic and, therefore, potentially carcinogenic compounds were present in this fraction.

The subject has been reviewed in some detail by Kool et al. (20) and Meier (23). It is, therefore, the intention of this chapter to discuss the approaches that have been used to determine whether such compounds constitute a hazard to man, and to highlight the difficulties and uncertainties in testing complex mixtures containing many unidentified components.

BACTERIAL ASSAYS

Tests with Unconcentrated Samples

Initially it was considered that the use of a very sensitive assay, the liquid based fluctuation test, would enable simple studies to be carried out with unconcentrated water samples. Although positive results were obtained with Salmonella typhimurium strains TA98 and TA100 in the absence of rat liver S-9 mix, the response obtained was very weak. In addition, closer examination of the assays showed that nutrients present in water enabled the bacteria to utilize the available histidine more effi-

Genetic Toxicology of Complex Mixtures
Edited by M. D. Waters *et al.*
Plenum Press, New York, 1990

197

ciently. This increased the number of natural mutations to histidine independence and gave possible false positive results. The amount of testing required to confirm a positive was out of proportion to the value of such a marginal result (7,13).

However, the first problem in testing complex mixtures had emerged. There are other components in the mixture than the compounds of interest which may show biological effects and influence the interpretation of results.

Tests with Concentrated Samples

In order to combat the problem of such a weak response with unconcentrated samples, concentration procedures were used to increase the dose level of the genotoxic material in the test. Two techniques were chosen: freeze drying, which would concentrate all but the most volatile components, and adsorption to macroreticular (XAD) resin with subsequent solvent elution. The uncertainty with the XAD-resin technique was that it only concentrated a more selective range of organics and it was, therefore, not known if the compounds of interest would be included.

Clear positive results were obtained with both S. Typhimurium TA100 and TA98 in the absence of rat liver S-9 mix with both freeze-dried and XAD extracts. The level of activity was similar with both extracts, which gave credence to the view that, fortuitously, XAD, despite its selectivity, concentrated the genotoxic components of the mixture. The use of freeze-drying as a concentration technique was slow and the amounts of extract produced were also small.

A variety of XAD resins and solvents were used by a number of different workers. The results obtained by the different groups were, however, very similar (29). Raw surface water was usually nonmutagenic although a low level of activity in TA98 was sometimes observed. Activity in TA100 and an increase in activity in TA98 were invariably seen after chlorination of surface waters.

Studies of reuse found little difference between the extent of reuse and mutagenic activity. Moreover, it was also demonstrated that upland waters in the United Kingdom which receive no effluent, and chlorinated humic acid solutions showed a similar pattern of mutagenicity to drinking water derived from lowland rivers. This indicated that the mutagen(s) was formed as a consequence of this reaction between naturally occurring components of the water and chlorine applied in the disinfection process (2,30).

Concentration on XAD was carried out by adsorbing the organics at sample pH (close to pH 7.0). Subsequently, it was demonstrated that acidification of the effluent from the column to pH 2.0 and passage through a second column gave additional extracts of a similar or greater mutagenic response on testing in bacterial assays (28). This finding showed that the mutagenicity in chlorinated drinking water was due to a number of mutagens which appeared to be chemically different.

Significance of Positive Findings

Finding positive results in bacterial assays indicated the presence of compounds which were capable of inducing gene mutations. These findings did not demonstrate a hazard to consumers, since they had not been shown to be active in mammalian cells. In addition, the necessity for using concentrated extracts meant that the mutagens could be present at concentrations which would be of no significance to the health of consumers.

Three approaches were available to determine whether a hazard did exist. The first was to demonstrate whether the extracts were active in a battery of tests in higher cells. The second approach was to identify the compounds responsible for the mutagenicity. This approach had clear advantages in assessing hazard because testing could be carried out on pure compounds thus avoiding problems of testing unknown mixtures. It particularly avoided the problem of the toxicity of compounds in the mixture other than those of interest which could result in the dose levels used in testing being lower than desirable. The third approach was the use of long-term, animal bioassays, but this approach was rejected on the basis of enormous quantities of sample required, and uncertainties over dose levels as well as prohibitive costs.

TESTS WITH IN VITRO HIGHER CELL SYSTEMS

There are a number of well-established in vitro tests which can be used to demonstrate genotoxicity in eukaryotic cells. These test systems detect the ability to induce both point mutations as detected in bacteria and to produce chromosome damage.

Testing frequently follows a tiered procedure, and that recommended by the U.K. Department of Health is fairly typical (6). The full battery of recommended assay systems includes tests in prokaryotic cells for point mutations, the tests in bacterial cells for which positive results had already been obtained. Tests in eukaryotic cells include tests for the induction of cytogenetic changes in cells in tissue culture, point mutations in cells in tissue culture or recessive lethal mutations in Drosophila, and in vivo tests for chromosome damage or induction of dominant lethals.

Chromosome Damage

There are several in vitro assays for assessing chromosome damage, including cytogenetics in Chinese hamster ovary cells (CHO cells) and human lymphocytes. Initial studies with CHO cells with XAD extracts of laboratory chlorinated water adsorbed at pH 7.0 indicated low cytogenetic activity. However, when the cells were treated with the extracts in serum-free medium for three hours a much higher level of activity was observed. These findings were confirmed with extracts of final drinking waters from both lowland river sources and upland sources. The importance of chlorine was confirmed by testing samples of raw water taken from the intake to the water treatment works and of raw water chlorinated in the laboratory. There was very low activity in the raw water and consistently high activity in chlorinated raw water and final treated water (31-33). Such findings indicated that the mutagens present in chlorinated drinking water were capable of damaging chromosomes in mammalian cells; however, the presence of protein in the media in the first three hours of exposure substantially reduced activity. This was confirmed by adding a series of different concentrations of exogenous protein which gave a clear inverse dose response, the activity declining with increasing protein concentration (Tab. 1).

Tab. 1.   Percentages of CHO cells with aberrations following treatment
          with water sample X036 in the presence of different levels of
          serum

Treatment	% serum in treatment medium	Number of metaphases analysed	% cells with aberrations[*]		
			Chromatid	Chromosome	Total
X036 1.75 l/ml	0%	100	82 (79)	38 (38)	86 (83)
"	2%	100	58 (49)	23 (22)	66 (56)
"	5%	100	37 (23)	10 (10)	44 (30)
"	10%	100	19 (17)	6 (6)	24 (23)
"	15%	100	22 (17)	4 (3)	24 (19)
MNNG 1 µg/ml	0%	100	91 (90)	65 (63)	96 (95)
"	15%	100	60 (53)	37 (35)	70 (66)
Untreated control	0%	100	9 (2)	2 (2)	10 (3)
	15%	100	11 (2)	2 (1)	13 (4)

* Figures in brackets exclude cells with gaps only

Tab. 2.   Percentage of CHO cells with aberrations following treatment
          with XAD extracts of humic acid solutions

Treatment	Number of metaphases analysed	% cells with aberrations[*]		
		Chromatid	Chromosome	Total
Unchlorinated				
4 l/ml	100	11(3)	5 (1)	14 (3)
3 l/ml	100	15 (2)	5 (3)	20 (4)
2 l/ml	100	6 (2)	2 (1)	8 (3)
1 l/ml	100	20 (3)	5 (4)	24 (7)
Chlorinated				
2.0 l/ml	100	71 (63)	37 (30)	77 (67)
1.75 l/ml	100	34 (30)	20 (14)	44 (34)
1.5 l/ml	100	56 (54)	46 (35)	66 (58)
1.25 l/ml	100	44 (32)	26 (24)	52 (49)
1.0 l/ml	100	41 (31)	24 (20)	55 (45)
0.5 l/ml	100	11 (2)	7 (4)	13 (4)
0.25 l/ml	100	17 (0)	2 (2)	16 (2)
MNNG 1 µg/l	100	32 (22)	30 (18)	52 (37)
Control	100	8 (5)	0 (0)	8 (5)

* Figures in brackets exclude cells with gaps only

That mutagens could be produced as a result of the reaction between chlorine and humic acids was also demonstrated in CHO cells (Tab. 2). To confirm the findings in CHO cells, studies were carried out to determine the clastogenic activity in human lymphocyte cultures. In whole blood cultures there was no activity, but when experiments were carried out with separated lymphocytes, activity was demonstrated, although this was not as great as in CHO cells. The lack of activity in whole blood cultures was perhaps not surprising, considering the previous findings in CHO cells in the presence of serum. This once again demonstrated the amelioration of activity by exogenous protein and pointed to the mutagens being highly reactive compounds.

Only very limited studies have been carried out on the clastogenicity of XAD extracts of drinking water adsorbed at pH 2.0. Although clastogenicity has been observed, these extracts are considerably more cytotoxic than those adsorbed at pH 7.0 which causes problems in carrying out satisfactory in vitro cytogenetic assays, and the dose response appears to be less clear than with samples adsorbed at pH 7.0.

Point Mutations in Mammalian Cell Lines

Although one group of workers reported finding induction of ouabain-resistant mutations in V79 cells with reverse osmosis concentrates of a recycled water, this was only active in the presence of S-9 activation (12). They did not examine chlorinated drinking water.

Other workers have shown that drinking water extracts produce positive results in the mouse lymphoma mutagenesis assay (22,27). WRc have studied drinking water samples for point mutational activity in the mouse lymphoma assay, in Chinese hamster V79/4 cells and CHO cells (32). In the three assays employed, concentrated drinking water extracts (pH 7.0) showed, at best, only marginal activity. The extracts, however, induced a high frequency of structural chromosome aberrations in the same cells treated under identical conditions.

Drosophila Studies

The first studies in Drosophila to determine whether sex-linked recessive lethal mutations were induced by concentrated XAD extracts of drinking water adsorbed at pH 7.0 proved negative (32,33). In a second Drosophila study, exposure was optimized so that some toxicity was apparent, but no indication of an increase in mutations was observed.

IN VIVO STUDIES

Mice dosed orally by gavage with pH 7.0 extracts of the equivalent of an average human water consumption over one, three, and ten years were examined for cytogenetic damage in bone marrow. This study was negative (31). Negative findings for the induction of micronuclei in mouse bone marrow with concentrates of drinking water treated with chlorine and other alternative disinfectants have also been reported by Meier et al. (26). Meier and Bull (24) also reported that chlorinated humic acid at very high total organic carbon (TOC) concentrations failed to induce either bone marrow micronuclei or spermhead abnormalities in mice.

Further Studies In Vivo

The data on the chlorination-derived mutagens had shown that the activity detected in bacteria, particularly TA100, was also capable of inducing chromosome damage in mammalian cells in vitro. Activity at both levels was substantially reduced by the presence of serum proteins or S-9 fraction indicating significant protein binding and possibly enzymic detoxification.

The lack of activity in vivo observed in the work heretofore described would appear to be very reassuring with regard to the possibility of health effects in man, but there are several reasons why this view could not be wholly substantiated. First, no in vivo studies with mutagens adsorbed at pH 2.0 had been carried out. Second, the extent of protein binding and inactivation of the mutagens by rat liver S-9 fraction would lead to the prediction that the mutagens would not reach bone marrow, the target organ, in an active form in the in vivo studies. In this respect the studies in Drosphila may provide some additional reassurance. It would therefore be reasonable to expect that mutagens produced by chlorination would not pose a hazard of systemic effects in man.

This, however, does not account for the cells of the gastrointestinal tract which would be those with which the mutagens would first come into contact. In addition, there are no data to support irreversible protein binding, and if the mutagens were absorbed and bound to protein, the form in which they were excreted is unknown. If excreted in an active form in the urine, this would provide a biologically plausible explanation for the findings in the case control study of bladder cancer by Cantor et al. (3).

Perhaps the most important additional reason for caution in interpretation lies with the problem of assessing dose levels in the in vivo studies in which the problem was overcome by using arbitrary doses. These assays are based on a high dose of the compound under test and there was no way of assessing the actual dose of mutagens in the mixtures. This major difficulty will be discussed in more detail below.

To make further progress in determining whether the mutagens from chlorination pose a hazard to man, it was necessary to examine the extracts in a short-term test which would give suitable data in the target organs of interest, the GI tract and the bladder. Only three possible assays were available, these being the induction of nuclear anomalies in colon (14,15), which could also be applied to other parts of the GI tract; assessment of sister chromatid exchanges in intestinal cells (4); and measurement of unscheduled DNA synthesis (UDS) in the stomach (10). Technically, the best established of these was the induction of nuclear anomalies in colon. Consideration was also given to other techniques such as treatment of GI cells in vivo followed by stripping and measurement of UDS in vitro. However, such techniques would require considerable developmental work with no guarantee that they would be more successful or sensitive than existing assays.

The nuclear anomaly assay (NAA) was chosen with the option of using an assay for induction of sister chromatid exchanges (SCEs), in the event of clear positive results, to confirm that the nuclear anomalies arose from damage to the DNA rather than a shower of apoptosis induced by cytotoxicity. However, in order to overcome previous concerns regarding dose levels, the XAD adsorption concentration procedure was modified to increase the concentration of the extracts (Fig. 1). In this way the dose level could be increased to 100 litres per mouse.

Difficulties were encountered with this concentration procedure, and finally because of precipitation of solid matter a compromise was reached in which concentration only went as far as a stage when a small amount of precipitate began to appear. On testing in TA100, the concentrate was extremely mutagenic but the precipitate also showed activity when redissolved. Therefore, some loss of activity was starting to occur. Clearly, there were risks of changing the mutagens by approaching this level of

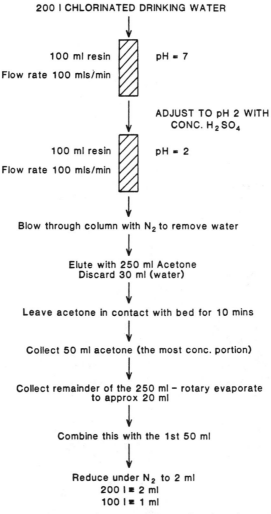

Fig. 1.   Concentration procedure for in vivo studies.

Tab. 3.   Summary of findings from preliminary nuclear anomaly assay
          with pH 7.0 extract (aberrations per 100 cells/nuclei)

	Untreated	Blank concentrate	Water concentrate	1,2-dimethyl-hydrazine
MITOSIS				
Liver	1.5	1.8	4.5	0.8
Stomach	4.0	3.5	4.8	4.8
Duodenum	5.5	9.5	4.5	5.0
Jejunum	5.9	4.5	7.5	3.1
Ileum	7.8	7.5	7.3	5.5
Colon	1.5	2.3	2.9	1.4
Rectum	2.5	4.5	6.0	2.3
TOTAL ABERRATIONS[*]				
Liver	0.8	0.3	1.0	23.8
Stomach	0.8	0.3	0	2.5
Duodenum	1.3	0.8	0.3	8.8
Jejunum	1.0	0.5	0.8	8.3
Ileum	0.8	0.1	0.1	5.8
Colon	0.9	0.1	0.3	10.5
Rectum	0	0.3	0.3	8.8
MNP[**]				
Bone marrow	0.8	0.8	1.0	0

* Apoptosis + keryorrhexis + pyknosis
** MNP - micronucleated polychromatic erythrocytes

concentration.   However, it was considered important to achieve a dose
level which would be close to that causing toxicity to the mice since a
high dose was required to engender real confidence in a negative result.

A series of preliminary experiments were planned in order to investigate the use of the assay with extracts containing a complex mixture.   In
the initial preliminary study, two male and two female CD-1 mice were
dosed by gavage three times at 2-hr intervals with either 0.33 ml of extract of chlorinated drinking water adsorbed at pH 7.0 or a blank extract
of laboratory chlorinated, carbon filtered, deionized water.   The animals
were starved for 5 hr before dosing. The positive control was a single
oral dose of 1,2-dimethylhydrazine (DMH.2HCl).   The animals were killed
24 hr after the middle dose.   No signs of toxicity were observed and
there were no signs of an increase in nuclear anomalies in the tissues examined (Tab. 3).

In order to increase the dose level in a subsequent preliminary
study which included pH 7.0 and pH 2.0 extracts and their respective
blanks, the dose volume was increased to 3 x 1 ml of extract per animal,
equivalent to 300 l/animal.   In this case there was severe toxicity and a
number of deaths occurred in both the drinking water extract and blank
groups with histopathological changes in the GI tract of the pH 2.0 ex-

tract animals, consistent with a direct toxic effect on the mucosa. However, in the surviving animals there was no sign of a positive response in the pH 7.0 extract group and only a questionable effect in the stomach of the one surviving animal receiving the pH 2.0 extract.

A subsequent study was carried out in which a preliminary toxicity study was performed with both pH 7.0 and pH 2.0 extracts in order to give a maximum dose which was not lethal (180 l and 90 l/mouse, respectively). In this case no blanks were prepared for two reasons. First, the effort involved in producing large quantities of extract is massive, taking several weeks, but preparation of blanks is even more difficult. Second, if the extracts are negative then there is no need for the blanks.

The results of this final assay showed that there were no effects in bone marrow with either pH 7.0 or pH 2.0 extracts. Preliminary scanning of the slides has shown that there is no sign of clear evidence for nuclear anomalies in GI tract or bladder.

The negative results obtained with the in vivo studies are, even within the limitations of the technique, reassuring. The dose levels used in the NAA are substantially above anything used before in testing extracts. The dose per mouse was approximately 8,333 l/kg or the equivalent of the water-intake of a 60 kg person over nearly 700 years.

IDENTIFICATION OF MUTAGENS IN DRINKING WATER

This proved to be a difficult task due to the complexity of the organic material and the low concentrations of the mutagens in drinking water. The approach taken at WRc to the identification of mutagens in drinking water has been described by Horth (18) and Horth et al. (19). Since it was clear that much of the mutagenicity detected in U.K. drinking water was introduced as a result of disinfection with chlorine, the work has focused on studying chlorination by-products. In addition to comparing treated water sampled before and after final chlorination and fractionating concentrated extracts to isolate the mutagens, precursors of chlorination-derived mutagenicity were investigated and suitable model compounds studied in laboratory chlorination experiments.

Several mutagenic compounds were identified in "neutral" extracts of drinking water (concentrated by XAD adsorption at sample pH). These compounds were typically found in drinking water at low $\mu g\ l^{-1}$ concentrations but together accounted for less than 10% of the mutagenicity detected in "neutral" extracts (8). Similar findings were reported by Meier et al. (25) who calculated that the contribution of several mutagens identified in "acidic" extracts of chlorinated humic acid solutions was about 6%.

The most significant progress in the identification of drinking water mutagens was achieved with the identification of 3-chloro-4-(dichloromethyl)-5-hydroxy-2(5H)-furanone (MX) in extracts of chlorinated water with a high humic content and in Finnish drinking water (16). MX is a highly potent mutagen first identified by Holmbom et al. (17) in chlorinated pulp effluent. In a survey of Finnish drinking waters, MX was detected (in extracts prepared by XAD adsorption at pH 2) at concentrations ranging from 5 to 67 ng l^{-1} (21). It was calculated that MX in these waters

Tab. 4. Mutagenicity, MX and approximate contribution of MX to mutagenicity of XAD extracts of treated water

Sample no.	Source	Extract	Slope[1] (revts.l^{-1} ml)	MX (ng l^{-1})[2]	Contribution MX (%)
1	Blank[3]	A[4]	0.46(ns)	nd	–
		B	1.67(ns)	nd	–
		C	0.65(ns)	nd	–
2	Lowland	A	25.8	3	30
		B	14.6	5	80
		C	38.5	4	30
3	Lowland	A	14.3	nd	–
		B	12.7	6	160
		C	33.9	5	40
4	Lowland	C	18.6	<3	<50
5	Lowland	C	35.9	5	30
6	Upland	A	17.7	<3	<50
		B	23.8	7	90
		C	50.2	7	40
7	Upland	A	5.7	nd	–
		B	14.7	3	70
		C	18.0	2	40
8	Upland	A	60.3	+	
		B	55.1	+	
		C	112.8	41	34
9	Upland	C	33.9	8	60
10	Upland	C	82.4	23	60

(1) Slope values were calculated from dose response graphs of estimated number of revertants per well against litre-equivalents of extract (litre per ml incubation medium)
(2) per litre-equivalent extract
(3) deionised/active carbon treated water, chlorinated
(4) A = extract prepared by XAD adsorption at sample pH ('neutral extract'),
     B = effluent of A concentrated by XAD adsorption at pH 2 ('acid extract'),
     C = concentrated by XAD adsorption at pH 2 directly ('neutral and acid combined extract')
ns = not significantly different from negative controls (p>0.05)
nd = not detected
+ = detected, not quantifiable because of interference

accounted for between 15 and 57% of the mutagenicity depending on the water (21).

MX was subsequently identified also as a product of the chlorination of the amino acid tyrosine in our experiments with chlorinated model compounds and in U.K. drinking water (18,19). Table 4 shows the results of a limited survey of U.K. drinking waters from different treatment plants using either upland or lowland river water as their source. Lowland river waters in the U.K. typically receive industrial and domestic sewage treatment effluents, whereas upland waters are not contaminated by such effluents.

Several of the water samples were first extracted by XAD-2 adsorption at sample pH (extract A) followed by acidification of the effluent to pH 2 and adsorption on another XAD column (extract B). A separate portion of the water sample was acidified directly to pH 2 and passed through an XAD column (extract C). XAD-2 columns were eluted with diethyl ether. A portion of each concentrated extract was exchanged into ethylacetate and tested for mutagenicity against strain TA100 in the fluctuation test (18,19). Another portion of each extract was methylated and anlayzed by GC-MS with multiple ion detection (GC-MS-MID). The analytical method for MX was essentially the same as that of Hemming et al. (16), with minor modifications (19). The recovery of MX (in extracts C) was $80 \pm 12\%$. The contributions of MX to the mutagenicity of extracts was calculated from the mutagenic response of the extracts (slope value), the amount of MX in the extracts, and the mutagenic response of MX tested in the same assays as the extracts.

Generally the mutagenic activity (slope value) of extract A (concentrated at sample pH) and extract B (effluent of A extracted at pH 2) added up was similar to the mutagenic activity of extract C (extracted at pH 2 directly). This suggested that the concentration on XAD-2 at pH 2 recovered the total activity extractable at neutral and pH 2, i.e., including the "neutral" and "acidic" fraction of the mutagenicity.

MX was detected in all treated waters (extracts B and C, Tab. 4); most were below 10 ng 1^{-1}. The higher levels of MX (23 and 41 µg 1^{-1}) were found in upland waters, but as the number of samples was limited, it is not possible to draw conclusions with respect to differences between upland and lowland waters.

It is interesting to note that MX was detected in three out of five extracts concentrated at sample pH, although close to the limit of detection. These results indicate that small amounts of MX may be recovered at sample pH and therefore account for a proportion of the mutagenic activity detected in such extracts.

The contribution of MX to the total XAD-extractable mutagenicity (extract C) ranged from 30 to 60%, or greater than 70% of the acid fraction (extract B). In one extract (extract B, sample 2) the mutagenicity was lower than expected from the MX concentration (160% contribution of MX). This could be due to toxic or antagonistic effects suppressing the mutagenic response, and suggests that it may not be valid to calculate the contribution of MX to mutagenicity based on the assumption of an additive response. To investigate this, several extracts were spiked with MX and tested for their mutagenic response with and without the added

MX. In eight cases, the response of the spiked extracts was between 88 and 123% of the expected (additive) response, a normal range of variation in mutagenicity test results. In two cases, the spiked extracts gave 60 and 62% of the anticipated response, and in two cases, 130 and 138%. The latter results could indicate suppression or enhancement of the mutagenic response due to toxicity/antagonism or synergism, respectively. On the whole the results indicated that, while it is not unreasonable to calculate the contribution made by MX to the observed mutagenicity, such values can only be regarded as rough estimates.

In total, between 35 and 65% (depending on the water) of the XAD extractable mutagenicity of U.K. drinking waters have been explained. Of this, 30 to 60 % is due to a few ng 1^{-1} of the single compound MX (recovered mainly in the acid fraction), whereas several mutagens present at low μg 1^{-1} concentrations (recovered in the neutral fraction) account for less than 5% of the total mutagenicity. Between 35 and 65% of the activity has not been explained by this research.

Brominated organic compounds are frequently detected in drinking water, including several mutagens such as bromoform, chlorodibromomethane, bromopropane, etc. Such compounds are formed as a result of bromide, which is often present in raw water. For this reason it was considered likely that brominated analogues of MX may be produced and that these were also likely to be potent mutagens.

Three brominated analogues of MX were synthesized:

3-bromo-4-(dichloromethyl)-5-hydroxy-2(5H)-furanone (BMX-1),
3-chloro-4-(dibromomethyl)-5-hydroxy-2(5H)-furanone (BMX-2), and
3-bromo-4-(dibromomethyl)-5-hydroxy-2(5H)-furanone (BMX-3).

All three proved to be potent mutagens with the same overall characteristics as MX (see Tab. 5), indicating that very low concentrations in water (around 1 ng 1^{-1} of each) could account for considerable levels of mutagenic activity. (There are five possible BMX isomers, although only three were synthesized.)

The brominated analogues of MX were methylated using the same technique as for MX and analyzed by GC-MS-MID. Tyrosine, which proved to be a suitable model with which to study the production of chlorination-derived mutagens including MX (18), was chlorinated after addition of bromide to the aqueous solution. The products were concentrated by XAD-2 adsorption at pH2, methylated and analyzed by GC-MS-MID. Little or no MX (which is produced in the absence of bromide) was detected, whereas BMX-1, BMX-2, and BMX-3 were all detected (19).

This preliminary experiment showed that highly mutagenic, brominated analogues of MX can be produced during aqueous chlorination when bromide is present. Using the tyrosine model allowed the production of considerably higher levels of BMX which were easier to detect than those likely to occur in drinking water. The analytical technique needs to be developed to determine whether brominated analogues of MX are present in drinking water derived from bromide-containing raw waters. If these compounds were found, they could account for a significant proportion of the unexplained mutagenicity in some drinking waters.

Tab. 5. Mutagenic activity of MX and brominated analogues BMX-1, -2, and -3 to strains TA100 and TA98 with and without S-9 metabolic activation

Test compound	Slope value (and standard deviation)*			
	TA100-S9	TA98-S9	TA100+S9	TA98+S9
MX	1.95 (0.33)	0.12 (0.02)	0.18 (0.06)	ns
BMX-1	0.13 (0.02)	ns	0.16 (0.06)	ns
BMX-2	2.54 (0.38)	0.23 (0.03)	0.22 (0.05)	ns
BMX-3	2.46 (0.20)	0.07 (0.01)	0.17 (0.04)	ns

* Slope values were calculated from dose response graphs of estimated number of revertants per well against ng test compound per ml incubation medium

ns = not significantly different from negative control (p >0.05)

MX STUDIES IN VIVO

MX has been tested in vitro and has been shown to be very highly mutagenic in bacterial systems. It causes in vitro cytogenetic damage but does not induce positive effects in mouse bone marrow in vivo (23). We have tested MX in the mouse nuclear anomaly assay at a dose approaching the LD50. Groups of 5 male and 5 female CD-1 mice were given an oral dose of 144 mg/kg body weight of MX (Tab. 6). There were no changes observed in bone marrow or in bladder, although cyclophosphamide was positive in both organs. In addition, there were no changes observed in the GI tract although there appeared to be some questionable activity in the stomach in the preliminary assay and small intestine in the full assay. There were, however, clear signs of tissue damage in the stomach and, to a lesser extent, the small intestine.

DISCUSSION

In assessing the potential for mutagens produced by the chlorination of drinking water to pose a hazard to consumers, we have tried to follow a logical progression of testing for genetic toxicity. We recognize that testing complex mixtures is not an ideal approach since there are so many difficulties and uncertainties involved; therefore, identification of mutagens is an important part of the program.

Among the difficulties encountered was that of delivering a sufficiently high dose of the mutagens for the type of in vivo short-term assay. Even when a sufficiently high dose to induce toxicity can be achieved, it remains uncertain whether that toxicity is due to the mutagens, other components of the mixture, or a combination of the two. In the case of short-term assays such as assessing chromosome damage in the bone marrow of mouse or NAA, the dose must be close to that causing acute toxicity for a negative result to have any value. This presents additional problems when testing complex mixtures such as concentrated extracts of drinking water. It is often difficult to achieve a sufficiently

Tab. 6.   Incidence of nuclear anomalies after treatment with MX

Tissue	Treatment			
	Untreated	MX	DMH.2HCl	Cyclophosphamide
Bone marrow	12	9	80$^+$	591^{++}
Liver	52	54	427$^+$	142?
Bladder TE	6	10	12	(18)?
Bladder LP	3	11	44?	(294)$^{++}$
Stomach NG	3	11	10	53$^+$
Duodenum	49	59	1535^{++}	3183^{++}
Jejunum/Ileum	59	90?	1375^{++}	3874^{++}
Colon	140	147	1608^{++}	2487^{++}
Rectum	70	56	1508^{++}	1187^{++}

*   Bone marrow : micronuclei per 10 000 polychromatic erythrocytes
    Stomach   : aberrations per 20 ts sections
    Otherwise : aberrations per 10 ls sections

TE   transitional epithelium
LP   lamina propria
NG   non-glandular forestomach

Interpretation of results : ? questionable response
                             + moderate response
                            ++ strong response

high concentration of the extract to induce toxicity with a practical dose volume.   There is variation from one sample to another, even from the same site, and there is uncertainty about the period for which these samples for use in in vivo assays are stable in storage.   In addition, there are major logistical problems in producing sufficient extract, which indeed is the major part of the effort needed for each in vivo experiment.

It is, therefore, desirable to pursue hazard assessment on identified components of the mixture.   The criticism that interactions may be important is countered by the artificially high concentrations used in these assays and the fact that the extraction techniques are highly selective.   Reconstituted mixtures can of course be tested at a later stage if this is considered desirable.   However, the aim must be to simplify the problem in order to provide answers in which we have adequate confidence.   A negative result with a complex mixture is always open to some doubt.

Clearly, the findings support the contention that MX is unlikely to be hazardous to consumers as far as the GI tract is concerned.   Although no increase in nuclear anomalies was seen in the bladder and this is reassuring, it would be of value to obtain further data to substantiate this position.

In spite of this reservation, the low levels of MX generally found in drinking water in combination with the amelioration of MX mutagenicity by protein would indicate that MX is unlikely to pose a major hazard to consumers.

The present position of the assessment of potential hazard from drinking water mutagens deriving from chlorination is that some reassurance can be obtained from the results to date within the limits of the techniques available.

On the positive side, mutagens adsorbed at sample pH (near neutral) have shown consistently negative results in vivo even at toxic dose levels. The more limited studies with mutagens adsorbed at pH 2.0 also indicate a negative result in vivo. The data on protein binding and/or enzymic detoxification support the contention that the body is well able to deal with these compounds and the epidemiological studies indicate that any risk of GI cancer is small enough to be uncertain.

The identification of MX in pH 2.0 extract has been a major step forward. The data on MX so far indicate that, although it is an extremely potent bacterial mutagen, it is at worst only very weakly active in vivo.

On the negative side, the dose level of MX in the pH 2.0 extracts, assuming a concentration of 30 ng/l in the original water and 100% recovery, would be 3 µg/mouse as opposed to a dose of 1.7 mg/mouse in the NAA on pure MX. The comparative dose of other mutagens may well be similar.

There are also a number of outstanding problems. For example, little of the mutagenicity adsorbed at neutral pH has been satisfactorily explained in terms of compounds identified. The most convincing epidemiological finding is the correlation between chlorinated drinking water consumption and bladder cancer in nonsmokers, although even this remains to be confirmed as a causal link. The negative findings in bladder with NAA may provide some reassurance but this is unlikely to be a particularly sensitive assay in a tissue with a relatively low cell turnover. Therefore, further investigation is required to give adequate reassurance.

There are other potent mutagens which are likely to be present in the pH 2.0 extracts for which there are no data on activity in vivo.

The present analysis with regard to the risk/benefit equation on the use of chlorine as a disinfectant in drinking water is that the benefits of chlorination in controlling waterborne disease are well established, while the risks must be considered not proven. The value of chlorination, particularly in developing countries where in rural areas even basic water treatment is a major problem, should not be underestimated. It is, therefore, important not only to make a thorough assessment of the risks from mutagens formed by chlorine before rejecting chlorination as a water treatment process, but also to make a thorough assessment of the benefit.

The testing of complex mixtures derived from drinking water has played an important part in both the assessment and identification of the genotoxic compounds responsible. However, the limitations and uncertainties of such testing dictate that further work on identified components of the mixture is necessary before a final assessment can be made.

ACKNOWLEDGEMENTS

This work was funded by the Department of the Environment whose permission to publish has been given. Nuclear Anomaly Assays were carried out by Huntingdon Research Centre.

REFERENCES

1. Beresford, S.A.A., L.M. Carpenter, and P. Powell (1984) Epidemiological studies of water re-use and type of water supply. Water Research Centre, Technical Report TR 216, Medmenham.
2. Bull, R.J., N. Robinson, J.R. Meier, and J. Stoker (1982) The use of biological assay systems to assess the relative carcinogenic hazards of disinfection by-products. Env. Health Persp. 46:215-227.
3. Cantor, K.P., R. Hoover, P. Hartge, T.J. Mason, D.T. Silverman, R. Altman, D.F. Austin, M.A. Child, C.R. Key, L.D. Marrett, M.H. Myers, A.S. Narayana, L.I. Levin, J.W. Sullivan, G.M. Swanson, D.B. Thomas, and D.W. West (1987) Bladder cancer, drinking water source, and tap water consumption: A case-control study. JNCI 79:1269-1279.
4. Couch, D.B., J.D. Gingerich, E. Stuart, and J.A. Heddle (1986) Induction of sister chromatid exchanges in marine chlorine tissue. Env. Mut. 8:579-588.
5. Crump, K.S., and H.A. Guess (1982) Drinking water and cancer: A review of recent epidemiological findings and an assessment of risks. Annual Rev. of Pub. Health 3:229-357.
6. Department of Health and Social Security (1981) Guidelines for the Testing of Chemicals for Mutagenicity. Produced by Committee on Mutagenicity of Chemicals in Food, Consumer Products and the Environment. HMSO, London.
7. Forster, R., M.H.L. Green, R.D. Gwilliam, A. Priestley, and B.A. Bridges (1983) Use of the fluctuation test to detect mutagenic activity in unconcentrated samples of drinking waters in the United Kingdom. In Water Chlorination, Environmental Impact and Health Effects, Vol. 4, R.L. Jolley et al., eds., Ann Arbor Science, Ann Arbor, pp. 1189-1197.
8. Fielding, M., and H. Horth (1986) Formation of mutagens and chemicals during water treatment chlorination. Water Supply 4:103-126.
9. Fielding, M.F., and R.F. Packham (1977) Organic compounds in drinking water and public health. J. Inst. of Water Eng. and Sci. 31:353-375.
10. Furihata, C., Y. Yamawaki, S. Jim, H. Moriya, K. Kodama, T. Matsoshima, T. Ishikawa, S. Takayama, and M. Nakadate (1984) Induction of unscheduled DNA synthesis in rat stomach minosa by glandular stomach carcinogens. JNCI 72:1327-1334.
11. Garrison, A.W. (1976) International Reference Centre for Water Supply, Technical Paper No. 9, The Hague.
12. Gruener, N., and M.P. Lockwood (1979) Mutagenicity and transformation by recycled water. J. Tox. and Env. Health 5:663-670.
13. Harrington, T.R., E.R. Nestmann, and D.J. Kowbel (1983) Suitability of the modified fluctuation assay for evaluating the mutagenicity of unconcentrated drinking water. Mutat. Res. 120:97-103.

14. Heddle, J.A., D.H. Blakey, A.M.V. Duncan, M.T. Goldberg, N. Newmark, J.M. Wargovich, and W.R. Bruce (1982) Micronuclei and related nuclear anomalies as a short-term assay for colon carcinogens. In Banbury Report No. 13-Indicators of Genotoxic Exposure. Cold Spring Harbor Lab. Publication, pp. 367-375.

15. Heddle, J.A., M. Hite, B. Kirkhart, K. Mavourin, J.T. MacGregor, G.W. Newell, and M.F. Salamore (1983) The induction of micronuclei as a measure of genotoxicity. A report of the U.S. Environmental Protection Agency Gene-Tox. Program. Mutat. Res. 123:61-118.

16. Hemming, J., B. Holmbom, M. Reunanen, and L. Kronberg (1986) Determination of the strong mutagen 3-chloro-4-(dichloromethyl)-5-hydroxy-2(5H)-furanone in chlorinated drinking and humic waters. Chemosphere 15:549-555.

17. Holmbom, B., R.H. Voss, R.D. Mortimer, and A. Wong (1978) Fractionation, isolation, and characterisation of Ames mutagenic compounds in Kraft chlorination effluents. Env. Sci. and Tech. 18:333-337.

18. Horth, H. (1989) Identification of mutagens in drinking water. Aqua 38:80-100.

19. Horth, H., M. Fielding, T. Gibson, H.A. James, and H. Ross (1989) Identification of Mutagens in Drinking Water. WRc Report PRD 2038-M, WRc Medmenham, Marlow, Bucks, SL7 2HD, United Kingdom.

20. Kool, H.J., C.F. van Kreijl, and B.C.J. Zoetman (1983) Toxicology assessment of organic compounds in drinking water. CRC Crit. Rev. in Env. Con. 12:307-357.

21. Kronberg, L. (1987) Mutagenic Compounds in Chlorinated Humic and Drinking Water. Dissertation, Abe Akademi, Abo/Turku, Finland.

22. Lee, P.S., C.J. Rudd, and J.R. Meier (1986) Use of the mouse lymphoma mutagenesis assay to compare the activities of concentrates of treated drinking water. Env. Muta. 8(Suppl. 6):116.

23. Meier, J.R. (1988) Genotoxic activity of organic chemicals in drinking water. Mutat. Res. 196:211-245.

24. Meier, J.R., and R.J. Bull (1985) Mutagenic properties of drinking water disinfectants and by-products. In Water Chlorination, Environmental Impact and Health Effects, Vol. 5, R.L. Jolley et al., eds.. Lewis, Chelsea, pp. 207-236.

25. Meier, J.R., H.P. Ringhand, W.E. Coleman, J.W. Munch, R.P. Streicher, W.H. Kylor, and K.M. Schenk (1985) Identification of mutagenic compounds formed during chlorination of humic acids. Mutat. Res. 157:111-122.

26. Meier, J.R., C.J. Rudd, W.F. Blazak, E.S. Riccio, and R.G. Miller (1986) Comparison of the mutagenic activities of water samples disinfected with ozone, chlorine dioxide, monochloramine or chlorine. Env. Muta. 8:55.

27. Tye, R.J. (1983) Comparison of ames and mouse lymohoma LS 1787 test results on extracts derived from water samples. Env. Muta. 5:953.

28. Van der Gaag, M.A., A. Noordsij, and J.P. Orange (1982) Presence of mutagens in Dutch surface waters and effects of water treatment processes for drinking water preparation. In Mutagens in Our Environment, Alan R. Liss, New York, pp. 277-286.

29. Wilcox, P., F. van Hoof, and M. van der Gaag (1986) Isolation and characterisation of mutagens from drinking water. In Proceedings of XVIth Annual Meeting of the European Environmental Mutagen Society, A. Léonard and M. Kirsch-Volders, eds., Brussels, pp. 92-103.

30. Wilcox, P., and H. Horth (1984) Microbial mutagenicity testing of water samples. In Freshwater Biological Monitoring, D. Pascoe and R.W. Edwards, eds. Pergamon, London, pp. 131-141.

31. Wilcox, P., and S. Williamson (1986) Mutagenic activity of concentrated drinking water samples. Env. Health Persp. 69:141-149.

32. Wilcox, P., S. Williamson, and R. Tye (1987) Mutagenic activity of concentrated water extracts in cultured mammalian cells and Drosophila melanogaster. In Water Chlorination, Environmental Impact and Health Effects, R.L. Jolley et al., eds., Vol. 6 (in press).

33. Wilcox, P., S. Williamson, D.C. Lodge, and J. Bootman (1988) Concentrated drinking water extracts which cause bacterial mutation and chromosome damage in CHO cells do not induce sex-linked recessive lethal mutations in Drosophila. Mutagenesis 3:381-387.

METHODS TO MEASURE GENOTOXINS IN WASTEWATER: EVALUATION

WITH IN VIVO AND IN VITRO TESTS*

Marten A. van der Gaag,[1] Laury Gauthier,[2] Arie
Noordsij,[3] Yves Levi,[4] and M. Nicoline Wrisberg[5]

[1]Institute for Inland Water Management and
Waste Water Treatment
Lelystad, The Netherlands
[2]Université Paul Sabatier
Toulouse, France
[3]The Netherlands Waterworks Testing and
Research Institute
Nieuwegein, The Netherlands
[4]Compagnie Générale des Eaux
Anjou-Recherche Maisons-Lafitte, France
[5]Technical University of Denmark
Lyngby, Denmark

GENOTOXINS IN WASTEWATER

Poor Detection of Potential Black List Substances

Genotoxic carcinogens are black list substances, whose discharge via wastewater should be reduced by best technical means. Detection of carcinogens as such in wastewater is at present nearly impossible. Identification combined with a literature search on the toxic properties is therefore still a favored approach, for want of something better. A major part of the discharged effluent will escape scrutiny, however, either because identification is not possible, or because toxicological information about identified compounds is not available.

Attempts to get a better definition of the genotoxic load of effluents were most often based on short-term bacterial assays such as the Ames test, performed either directly in the wastewater or after preconcentration of the organics (5,6,22). Tests with tissue culture systems have also been used to assess the genotoxicity of extracts (18). Bacterial assays

---

* This paper is dedicated to the memory of Professor André Jaylet (deceased May 29, 1989) who should have been among us in Washington.

performed directly in water often yield negative or ambiguous results (31; unpubl. reports, Institute for Inland Water Management). An additional problem is that loss of genotoxicity may occur when the sample is filtered for sterilization. Testing of concentrates will limit the scope of a survey to that part of the organic matter that can be recovered by concentration techniques.

New Opportunities with In Vivo Assays in Aquatic Species

Many of the problems encountered with the in vitro assays may be circumvented with direct testing in aquatic organisms. In the past ten years, a number of tests were developed, either with plants or with aquatic animals (annelids, mussels, fish, and amphibians, see the reviews in Ref. 3, 11, and 14, and 4,7-10,12,13,20,26,28,29), which can potentially be used to assess the genotoxic potency of a wastewater. The advantages are clear: there is no need to concentrate a sample or to sterilize it, and the tests can be carried out with intact animals, taking into account uptake and elimination, internal transport, and metabolism.

Until now, only a few attempts have been made to incorporate in vivo genotoxicity assays in the quality assessment of effluents (4,27). Comparisons between different approaches, both from the technical point of view (sensitivity, practical use) and from the economic side, have not yet been made.

Pilot Study for New Approaches

In order to make a comparison between several approaches for the assessment of genotoxins in wastewater, the first results of this pilot study are presented here. A part of this project involved the characterization of an industrial effluent with three different in vivo assays for genotoxicity, including sister chromatid exchange (SCE) induction in the fish Nothobranchius rachowi (25,26), and the formation of micronuclei (MN) in the amphibian Pleurodeles waltl (12) and in the mussel Mytilus edulis (16). The same effluent was tested in the Ames test (17) and in the SOS-chromo-test (21). The genotoxicity assays were also performed in XAD-extracts of this effluent, as well as in the flow-through of the XAD columns. This approach should give a first impression of the possibility to use available XAD-techniques to recover genotoxins from wastewater.

MATERIALS AND METHODS

Sample Treatment

Effluent. The pilot study was carried out with an effluent from a biological wastewater treatment unit receiving effluents from various petrochemical industries.

Sampling. An 80-liter sample of wastewater was taken on October 29, 1988, at the discharge point of the biological wastewater treatment plant. The sample was stored in four stainless steel tanks, cooled to 4°C and transported to the laboratory within 24 hr. Twenty liters of the sample were used for direct in vivo testing with the SCE-assay in Nothobranchius and with the MN-test both in Pleurodeles and in Mytilus. Fif-

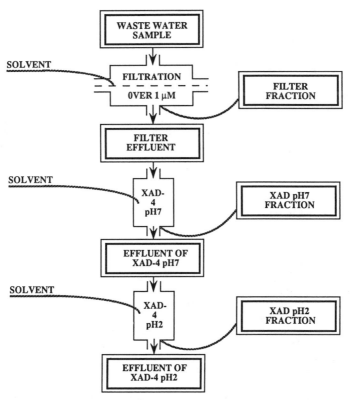

Fig. 1.   The wastewater was first filtered over a 1-μm glass fiber filter. The dissolved organics were then adsorbed on XAD 4, first at ambient pH, then at pH2.   The genotoxicity assays were carried out in all water samples and in all extracts.   The water samples were treated with γ-irradiation before testing in bacterial assays.

teen liters were sterilized with γ-rays (25 kGy) to be tested in the bacterial genotoxicity assays and in the SCE assay with Nothobranchius. Two liters were used for chemical analysis (dissolved organic carbon DOC; petroleum-ether extractable organohalogens, EOX; carbon adsorbable organohalogens, CAOX; and measurement of histidine concentration).

   Isolation and concentration of the organics.   The remaining 45 l were used for the physicochemical isolation and concentration procedure (Fig. 1).   After filtration over a 1 μm glass fiber filter, the sample was passed over XAD 4 first at ambient pH and subsequently at pH2.   The XAD-adsorption was carried out according to Noorsdij et al. (19), with a modified flow (0.5 empty bed volumes/hr) and a modified resin/sample ratio (1 ml of resin/100 ml of sample).   The organics were eluted from the filter and the XAD first with 5 bed volumes of ethanol and then with 5 bed volumes of a mixture of cyclohexane and ethanol.   The samples were concentrated in ethanol by azeotropic destillation (400 ml equivalent of sample/ml ethanol).

Part of the sample was collected after filtration and XAD pH7 adsorption for testing in vivo. After XAD pH2 adsorption, the sample was
neutralized with NaOH for in vivo testing.

Physicochemical Measurements

A number of routine parameters have been determined in the wastewater, and in the effluents of the filter and of both XAD columns. The
dissolved organic carbon content was measured with an auto-analyzer
(Technicon Autoanalyser-II). Organohalogens were measured in the same
samples after EOX, CAOX, and also in the XAD pH7 concentrate (X_7OX)
according to the microcoulometric method (1).

Fingerprints were made of the filter-extract and the XAD pH7 and
pH2 fractions with a reversed phase liquid chromatography (RP-LC, elution gradient from 100% water to 100% acetonitrile).

Genotoxicity Testing

Genotoxic assays. Five different assays for genotoxic effects have
been carried out on the sample: two bacterial assays (the Ames test and
the SOS-chromo-test) and three in vivo tests, i.e., an SCE-assay in gill
cells of Nothobranchius, a MN test in peripheral red blood cells of
Pleurodeles, and a MN test with gill cells from Mytilus.

The SCE assay with Nothobranchius rachowi. The SCE assay was
performed according to van de Kerkhoff and van der Gaag (25). The
fish were obtained from a breeder (de Biezen, Boskoop, The Netherlands), and treated for 5 min in sea water to eliminate possible parasites.
After a quarantine period of 4 wk, groups of three fish were exposed to
concentration 32 resp. 100 ml (or ml/equivalents) of sample (or concentrate) in a total volume of one liter for 96 hr, under gentle aeration.
After 48 hr, the sample was renewed, and 100 mg/l of bromodeoxyundine
(BrdU) was added. Colchicine (50 mg/l) was added to the water 10 hr
before sacrifying the fish. Ten metaphases of gill cells of each fish were
scored for the SCE-frequency.

The newt-micronucleus test. The MN test with Pleurodeles waltl was
carried out according to Jaylet et al. (12). Larvae of the urodele Pleurodeles waltl originated from the breeding stock of the Universite Paul
Sabatier in Toulouse (France), and were transferred to Lelystad (The
Netherlands) for the tests. Groups of 20 larvae (stage 53) were exposed
to 64 resp. 200 ml (or ml/equivalents) of sample (or concentrate) in a
total volume of 2 l of water, under mild aeration. The water with the
sample was renewed every day. All larvae exposed to the unconcentrated
samples were fed with live daphnids from the DBW/Riza breeding stock,
and afterwards with a single batch of chironomid larvae. The assays with
the concentrates were carried out in Toulouse.

Blood smears were prepared from samples taken by heart-punction
after 12 days of exposure to the wastewater. From each animal, 1,000
red blood cells were screened for the presence of micronuclei.

The mussel-micronucleus test. The MN test with Mytilus edulis was
carried out with specimens collected from a natural habitat near Gilleleje,
Denmark. They were allowed to acclimatize in the laboratory 10 days pri-

or to the experiment, in sea water of 28-30% salinity at 16 ± 1°C. All mussels had a shell length between 2 and 3 cm. The mussels were exposed for 10 days at 16°C and were fed with Liquifry marine (Interpet, Dorking, Great Britain) throughout the experiment in order to maintain mitotic activity. The test solution (32 or 100 ml wastewater/liter) was changed every day. At the end of exposure the mussels were sacrificed and the gills were removed and fixed in ethanol-acetic acid (3:1) for a 1/2-hour twice and thereafter kept at 4°C until further processing. Slides were made from a gill cell suspension in 50% acetic acid and stained according to the Feulgen method plus Naphtol Yellow S (24). From each exposure group 4 to 5 mussels were examined for micronuclei by scoring approximately 2,000 cells per mussel. Only cells with an intact nucleus were scored, according to the following criteria (24): (i) diameter smaller than one-third of the nucleus; (ii) nonrefractibility of micronuclei; and (iii) color the same as or lighter than the mucleus.

The Salmonella microsomal mutagenicity assay. This test was carried out by the Netherlands Waterworks Research Institute (KIWA) with strains TA98 and TA100 without and with metabolic activation (17). The ethanol concentrates from filter-extract or XAD were incorporated in the soft agar in the Ames assay, in concentrations ranging from 2-64 ml equivalents per plate. The wastewater as such was tested after sterilization by γ-irradiation, by adding the sample to the bottom-agar. All samples were tested both without and with 0.5-ml rat-liver homogenate per plate (S-9-mix) from male Wistar rats induced with Aroclor 1254.

The SOS-chromo-test. This test was performed by the Compagnie Générale des Eaux (CGE) and used the Escherichia coli strain PQ37 from the Institut Pasteur but without metabolic activation (21). The reagents were obtained from Orgenics .10 µl of sample was added to 100 µl of an overnight culture and incubated for 2 hr at 37°C. The reaction was then stopped by performing the staining reaction.

Procedure blanks. Procedure blanks were tested from all samples concentrated in ethanol with identical batches of chemicals, filters, and XAD-resins. A blank with a higher NaCl content was tested as control for the neutralized XAD pH2 flow-through.

RESULTS

Physicochemical Characteristics

Filtration of the wastewater over a 1 µm glass fiber filter did not substantially affect the concentrations of DOC and of different organohalogen fractions (Tab. 1). Only a minor amount of organohalogen (<2%) was recovered in the filter extract (Tab. 1). A high proportion of nonhalogenated lipophilic material was recovered in the filter extract (Fig. 2). The mean log P_{ow} was estimated at 5 GC and HPLC analysis showed that this material consisted mainly of C_{16} to C_{30} alkanes and C_{14} to C_{18} carbonic acids, and a group of strong UV-adsorbing compounds, probably with an aromatic structure.

Adsorption on XAD-4 at pH7 recovered only 25% of the DOC fraction (Tab. 1). Neutral XAD treatment did retain a major amount of the organohologen fraction. This amount was higher than that found with

Tab. 1.   Dissolved organics and different classes of organohalogens were
          only partly retained by XAD adsorption

	DOC mg/l	EOX μg/l	CAOX μg/l	Frct.OX μg/l
Waste water	140	2000	2600 *	
Filter extract				80
Filter effl.	140	1700	2500 *	
XAD pH7 fract.				4850
XAD pH7 effl.	105	70	1300 *	
XAD pH2 fract.				nd
XAD pH2 effl.	80	25	900 *	
Detection limits	0.2	0.1	1	0.4

nd: not measured because of interference by chloride from the
    hydrochloric acid.

*: underestimation of true concentrations because of the
   presence of high amounts of suspended solids which
   interfere with the adsorption on carbon.

liquid-liquid extraction, but also nearly twice as high as was observed in
the CAOX measurement (Tab. 1). It must be noted, however, that fine
suspended solids interfered with the CAOX measurements, leading to an
underestimation of this figure. The RP-LC fingerprint of the XAD pH7
concentrate demonstrated the presence of a broad spectrum of lipophilic to
moderately hydrophilic compounds (Fig. 2). The mean log P_{ow} of this
fraction ranged from <0 to 4.

Besides 75% of the DOC, a substantial amount of CAOX was still present in the flow-through of the XAD pH7 column. Almost all EOX had
been removed by the XAD.

Adsorption on XAD-4 at pH2 recovered another 18% of the DOC
(Tab. 1). Only a minor part of the remaining CAOX could be adsorbed.
Measurement of the OX content of the XAD pH2 fraction is not possible,
due to the interference of chloride. The RP-LC chromatogram demonstrates a further shift towards moderately hydrophilic compounds in this
fraction (Fig. 2). The compounds isolated at pH2 tend to show a similar
log P_{ow} range (corrected for pH) as in the pH7 fraction, but with a shift
towards lower values.

More than 55% of the DOC was still present in the flow-through of
the XAD pH2 column, together with a substantial amount of the CAOX.
No DOC was retained on the filter, as expected. About a quarter of the
DOC was adsorbed on XAD at pH7.

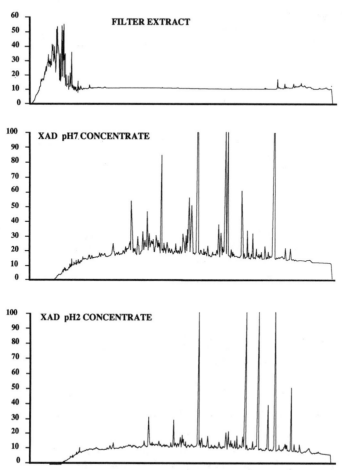

Fig. 2.   The RP-LC chromatogram, as detected with UV shows that the nonsoluble fraction (filter extract) contains mainly lipophilic compounds (top).   The XAD pH7 (middle) and pH2 (below) concentrates contain compounds over a broad range of polarities.

Wastewater Induces Genotoxic Effects In Vivo

   Genotoxic effects were observed in wastewater dilutions with the SCE assay with Nothobranchius and the MN tests with Pleurodeles and Mytilus. The increase in SCE frequency and in incidence of micronucleated cells in Pleurodeles and Mytilus.   The increase in SCE frequency and in incidence of micronucleated cells in Pleurodeles was still highly significant at the lowest concentration tested of 32 ml per liter of dilution water (Tab. 2 and Fig. 3 and 4).   In the mussels, a significant increase only occurred at the highest concentration tested (100 ml/l).

   Filtration did not affect the in vivo genotoxicity.   A significant amount of the genotoxic effect observed in the SCE assay was present in

Tab. 2.   The wastewater sample only induced a significant increase in micronuclei frequency in gill cells of <u>Mytilus edulis</u> at 100 ml/l

Exposure group	Micronucleï (in o/oo)		
Control	0.40	(0	- 1.17)
EMS 100 mg/l	1.24*	(0.34-	1.93)
Waste water			
32 ml/l	0.49	(0	- 0.97)
100 ml/l	2.61*	(0	- 5.34)
XAD pH7 effluent			
100 ml/l	0.47	(0	- 0.95)
XAD pH2 effluent			
100 ml/l	0.80	(0	- 1.66)

* p<0.05

the extract from a 1-μm glass fiber filter that was placed before the XAD columns.   The genotoxicity in the SCE assay was not, however, altered by this filtration, suggesting that a part of the potential genotoxic effect of this effluent was not directly "biologically available" (Fig. 3).

XAD-adsorption removed a major amount of the mutagenicity.   Both in vivo tests, however, still detected a significant genotoxic effect in the flow-through of the XAD pH2 colum (Fig. 3 and 4).   The more hydrophilic part of the genotoxic activity could amount 15 to 40% of the (biologically available) mutagenicity in this sample, according to the estimates from the SCE and MN assays.

γ-Irradiation sterilization significantly reduced the genotoxic activity of the wastewater in the SCE assay and in the Pleurodeles MN test (Tab. 3).   This reduction was not observed consistently at all dosages.   In the SCE assay, only one fish survived the test in the γ-irradiated water at the highest dose.   In the MN test, the difference was not significant at the lowest dose level.

Genotoxic Concentrates in Bacterial Assays

The bacterial mutagenicity assays only detected effects in the different organic concentrates (Fig. 5 and 6).   Direct testing was carried out only in γ-rays sterilized samples, but effects were not observed in the Ames test or in the SOS-chromo-test.

The SOS-chromo-test detected a genotoxic activity in the concentrates without metabolic activation at quite low concentrations (6 ml equivalents per liter in the XAD fractions), while an increase of revertants in the Ames test (TA100 +S-9 mix) was only observed at concentrations over about 100 ml equivalents/l in the XAD fractions.

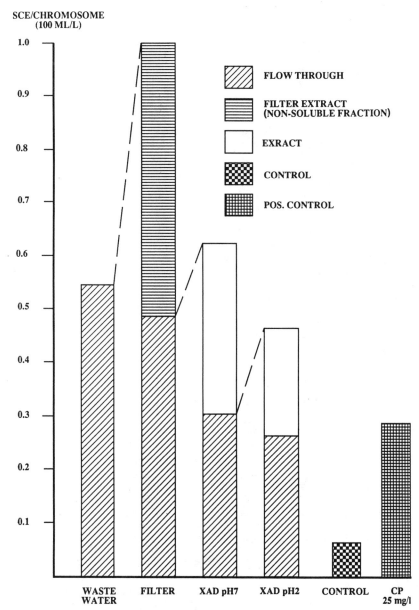

Fig. 3.   The wastewater induced a high number of SCEs in
Nothobranchius. The filter extract contained a large amount of
genotoxic effect that was not directly biologically available.
About 50% of the amount of the original genotoxic effect was
still present in the XAD pH2 flow-though.

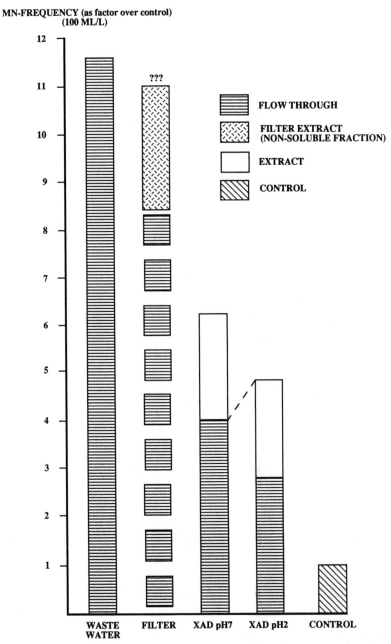

Fig. 4.   The wastewater induced high amounts of micronuclei in Pleuro-
           deles.   A substantial amount of genotoxicity was still observed
           in the flow-through of the XAD pH2 column.   The flow-through
           of the filter was not tested.   The total mutagenicity present in
           the sample therefore remains unknown.

Tab. 3.   Treatment with γ-irradiation reduced the genotoxic activity of
          the wastewater sample (mean frequencies with lowest and high-
          est scores), both in the SCE assay and with the MN-test in red
          blood cells of Pleurodeles

Assay	Frequency of SCE/cell or o/oo of MN/RBC	
	Waste water	γ-irradiated WW
Nothobranchius SCE		
Control	0.06 (0.03 -0.09)	
32 ml/l	0.30 * (0.23 -0.34)	0.16 ** (0.14 -0.18)
100 ml/l	0.54 * (0.51 -0.59)	0.38 *
Pleurodeles MN		
Control	0.004 (.001-.009)	
32 ml/l	0.010 *(.005-.019)	0.008 * (.001-.012)
100 ml/l	0.041 *(.021-.068)	0.013 ** (.003-.025)

*: significantly different from control (p < 0.05)
**: significantly different from non-irradiated waste water

Strain TA100 +S-9 was most sensitive in the Ames test.   No significant
mutagenicity was observed with TA98 and TA100 without S-9 at concentrations up to 64 ml equivalents per liter.   A positive score was also
found with TA98 +S-9 in the filter fraction and in both XAD fractions
(results not shown).

DISCUSSION

XAD: The "Tip of the Iceberg"

This pilot study is a first clear proof that our present analytical
chemical techniques used to monitor wastewaters only cover a part of the
organic components that can be a potential risk to health: 30 to 50% of
the genotoxic effect present in the dissolved fraction of this specific
wastwater was not recovered on XAD.   The yield is even lower if one accounts for the substantial amount of mutagenicity that was found in the
filter extract, a nonsoluble fraction that was not directly biologically
available.

It is not possible to make a final judgment on the suitability of XAD
techniques to concentrate (geno)toxins from wastewater.   The sample that
was selected for this study has its own specific characteristics, which
cannot be translated into more general conclusions about the technique.
The impression is that the dimensions of the columns were sufficient to
prevent a major break-through of the compounds that were meant to be
isolated at ambient pH and at pH2, respectively.   This is supported by

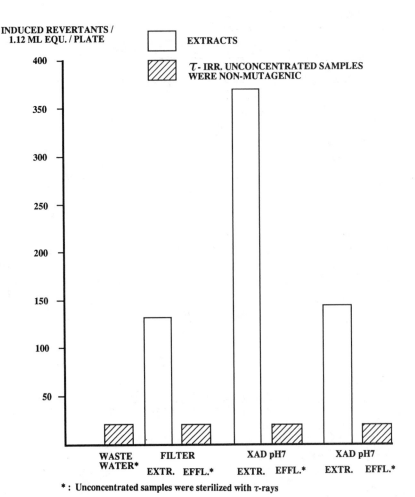

Fig. 5.   γ-Ray sterilized wastewater was not mutagenic in the Ames test.
           The highest mutagenic effect was observed with strain TA100 +
           S-9-mix in the filter extract and in the XAD-concentrates.

an almost total removal of the EOX by XAD pH7 treatment, and a clear
reduction of the DOC, which was also larger than in the following column
at pH2. The reduction of the DOC on the pH7 column was greater than
observed in similar situations with river water or drinking water,
probably because humic acids are only present in a negligible proportion
in this sample (1).

If we assume that the adsorption efficiency of the XAD procedure
was sufficient, it is evident that quite a substantial fraction of relatively
hydrophilic genotoxic compounds were present in this wastewater and
were never detected until now.

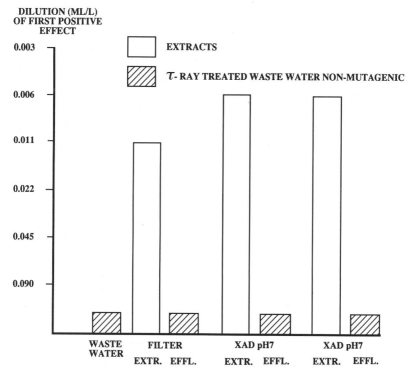

Fig. 6. The γ-ray sterilized wastewater was not genotoxic in the SOS-chromo-test. A high mutagenic effect was observed in the different extracts.

Filtration: The Hidden Fraction

The genotoxic effects observed in the filter extract could not be measured in the direct assay of the sample. The fraction retained by the filter probably contains compounds which are not dissolved in water as well as compounds which are strongly adsorbed to suspended solids. The alkanes and carbonic acids observed in the GC profiles are products originating from the oil industry. The strongly UV-adsorbing fraction that appeared in the RP-LC chromatogram could indicate the presence of poly-aromatic hydrocarbons (PAHs), but this still has to be confirmed. The occurrence of mutagenicity mainly in TA100 and only with S-9 mix and the minor amount of organohalogens found in this fraction can support this assumption.

Although the genotoxins in this fraction are not directly taken up by aquatic animals, it is highly probable that these compounds will finally settle in sediments, and become available for bioconcentration in the long run if they are not degraded.

Overall Impression of the Sampling Method

The overall impression of this method could only be measured with the Nothobranchius SCE assay. With the exception of the nondissolved filter fraction, relatively good recoveries of the mutagenic effect could be noted: the sum of the induced SCE frequencies of the XAD-concentrate and XAD-flow-through were each time only slightly higher than the number of SCEs induced by the XAD-influent. This small increase could be caused by a small increase of the biological availability of the concentrates because of the presence of a solvent. This would indicate that no major loss of genotoxicity occurs by irreversible binding to XAD, nor that any major amount of genotoxicity is generated by the XAD-isolation and concentration procedure, thereby confirming earlier findings on this point (1).

γ-Ray Sterilization: Unexpected Effects

Sterilization with γ-rays not only reduced the genotoxic effect of the wastewater sample in both the SCE and in the MN assay, but also lowered by 30% the toxicity in the Microtox assay, a test for acute toxicity in luminescent bacteria (results not shown). The reason for these effects are unknown. Few data are available on the effect of γ-irradiation on mutagens. Barna (2) mentioned that γ-rays did not induce mutagenicity in treated spices. Whether the reduction of the mutagenicity is due to an effect on molecular structure, or is caused by a reduction of the biological activity in the sample needs further investigation.

It does introduce an uncertain aspect about the results of the bacterial assays: the direct tests were carried out in γ-ray treated wastewater, and did not detect any mutagenicity at all. The testing of the concentrates on the other hand showed that a substantial mutagenic effect could be detected with the bacterial assays at low dilution factors. Additional study is needed to see if the treatment with γ-rays could be responsible for the negative results of the direct assays on the wastewater. Other means of sample sterilization will also be investigated.

No Specific Chemical Characteristics of a Genotoxic Fraction

The best solution would be to find a clear relation between a chemical group of compounds and a specific genotoxic effect. From this study it is clear that genotoxicity is present over a broad range of chemical classes, ranging from a class of highly lipophic compounds (possibly PAHs) retained on the filter, to a wide range of lipophilic and (moderately) hydrophilic substances that is only partly adsorbed on XAD.

The present study did not show any specific relation between the genotoxic effect observed and the presence of organohalogens. It is possible that the effects by small amounts of genotoxic organohalogens could be masked by a bulk of nongenotoxic OX, a fact that has been observed in many other studies (1). This will limit the usefulness of "OX" parameters to monitor genotoxic compounds in wastewater to those situations where a very specific "OX" load is present.

Are In Vivo Tests More Effective?

At first sight the sensitivity of the in vivo assays was higher than that of the bacterial tests. The SCE assay and the MN test could be carried out directly in dilutions of the sample, while a concentrate had to be made of the organics to assess the effect in the Ames test and SOS-chromo-test.

The lower genotoxic response in Mytilus compared to Pleurodeles and Nothobranchius could be interpreted as a general lower sensitivity. Similar differences have been reported earlier in several studies with selected mutagens on the SCE induction in gill cells and larvae of mussels (7,8) compared to corresponding tests in gill cells of fish (28,30). It could be interesting to investigate whether the valve-closure response of adult mussels in the presence of toxicants (15,23) could explain part of the lower response that was observed in this study, as a consequence of a lower exposure.

The difference in sensitivity between bacterial tests and in vivo assays was less clear with the XAD-concentrates. Furthermore, the response of the SOS-chromo-test in filter- or XAD-concentrates was much more sensitive that that of the Ames test, and was observed in a similar range of sensitivity to that of the in vivo tests. Beside possible differences in sensitivity, the difference in mutagenic response between the two bacterial assays may also be caused by a lesser availability of the sample in the Ames test due to the incorporation of the bacteria in agar. Liquid

Tab. 4.   Comparison of costs (in working hours) and speed (in hours) of rapid preliminary screening assays for genotoxins in wastewater in a routine laboratory setting

Test Conditions	Nothobr. SCE three fish		Pleurodeles MN six larvae		Mytilus MN five mussels		Ames-test[1] Two strains ± S9		SOS-chromo test	
	time (hrs)	cost (w.hrs)*	time (hrs)	cost (w.hrs)	time (hrs)	cost (w.hrs)	time (hrs)	cost (w.hrs)	time (hrs)	cost (w.hrs)
Range finding toxicity test	(24)	(2)	(24)	(2)	(24)	(2)	-	-	-	-
Starting mater.	-	3	-	3	-	2-3	-	2	-	2
Sample prepar.	-	-	-	-	-	-	24-72	4-12	24-72	4-12
Test	48-96	2	144-288	3-6	192-288	2-3	48-63	6	24	3
Slide prepar. and staining	24	2	8	2	7	3	-	-	-	-
Counting	1	1	6	6	7-15	7-15	2	2	1	1
Total										
without range find.	73-121	9	158-302	14-20	206-310	14-24	74-137	14-22	49 - 97	10-18
incl. range find.	97-145	10	182-326	16-22	230-334	16-26	74-137	14-22	49 - 97	10-18

*: w.hrs: working hours

phase incubation might therefore increase the sensitivity of the Ames test.

A number of technical modifications of the bacterial assays concerning sterilization and liquid incubation procedures will be needed before final comparisons of sensitivity can be made between the various approaches.

The costs and the duration of the in vivo assays, if carried out on a routine basis, are not very different from those of bacterial assays (Tab. 4). In the present setting the costs for a rapid screening are the lowest with the Nothobranchius SCE test, and are quite similar in the other tests. The SCE test and both bacterial assays are faster than the MN-tests, mainly because of the short exposition periods.

Presently neither costs nor experimental duration should be used, an argument in favor of in vitro bacterial tests. If the problems related to the direct testing in bacterial systems can be solved, the bacterial assays could become faster and cheaper, because there would be no need for concentration procedures.

No Final Conclusions at This Point

The genotoxicity assays in vivo with aquatic species have opened our eyes to the presence of unknown fractions with mutagens. At this time they are the only assays that can be introduced at short notice to get a further impression of the magnitude and consequences of the problems. An additional advantage of the in vivo assays is that they are carried out on whole animals and they score endpoints which are more relevant to eukaryotes than those in bacterial assays.

Whether in vivo or in vitro tests will be used in the long run to monitor effluents for genotoxins will depend on further research with the bacterial systems that have to elucidate the lesser sensitivity of these tests for unconcentrated samples.

REFERENCES

1. AWWA (1988) The search for surrogates. In Cooperative Research Report, KIWA-AWWA-RF, Denver, Colorado.
2. Barna, J. (1983) Ineffectiveness of irradiate spice mixture in DLT. Mutat. Res. 113:231.
3. Choroulinkoff, I., and A. Jaylet (1989) Contamination of aquatic systems and genetic effects (Part 4). In Aquatic Ecotoxicology. Fundamental Concepts and Methodologies, A. Boudo and F.S. Ribeyres, eds. CRC Press Inc. pp. 211-235.
4. Das, R.K., and N.K. Nanda (1986) Induction of micronuclei in peripheral erythrocytes of fish Heteropneustes fossilis by mitomycin C and paper mill effluent. Mutat. Res. 175:67-71.
5. Denkhaus, R., W.O. Grabow, and O.W. Prozesky (1980) Removal of mutagenic compounds in a waste water reclamation system evaluated by means of the Ames Salmonella/microsome assay. Prog. Water Tech. 12:571-589.
6. De Raat, W.K., A.O. Hanstveit, and J.F. de Kreuk (1985) The role of mutagenicity testing in the ecotoxicological evaluation of industrial discharges into the aquatic environment. Fd. Chem. Tox. 23:33-41.

7. Dixon, D.R., and K.R. Clarke (1982) Sister chromatid exchange:
    A sensitive method for detecting damage caused by exposure to envi-
    ronmental mutagens in the chromosomes of adult Mytilus edulis.
    Mar. Biol. Lett. 3:163-172.
8. Harrison, F.L., and I.M. Jones (1982) An in vivo sister-chromatid
    exchange assay in the larvae of the mussel. Mytilus edulis: Re-
    sponse to 3 mutagens. Mutat. Res. 105:235-242.
9. Hooftman, R.N. (1981) The induction of chromosomal aberrations in
    Nothobranchius rachowi after treatment with ethyl methane sul-
    phonate or benzo[α]pyrene. Mutat. Res. 91:347-352.
10. Hooftman, R.N., and W.K. de Raat (1982) Induction of nuclear
    anomalies (micronuclei) in the peripheral blood erythrocytes of the
    eastern mudminnow, Umbra pygmaea, by ethyl methane sulphonate.
    Mutat. Res. 104:147-150.
11. Jaylet, A., and C. Zoll (1989) Tests for the detection of genotoxins
    in fresh water. CRC Critical Reviews in Aquatic Toxicology. CRC
    Press, Inc. (in press)
12. Jaylet, A., P. Deparis, V. Ferrier, and S. Grinfield (1986) A new
    micronucleus test using peripherial blood erythrocytes of the newt
    Pleurodeles waltl to detect mutagens in fresh water pollution. Mutat.
    Res. 164:245-257.
13. Jaylet, A., P. Deparis, and D.Gaschignard (1986) Induction of mi-
    cronuclei in peripherial erythrocytes of axolotl larvae following in
    vivo exposure to mutagenic agents. Mutagenesis 1:211-215.
14. Kligerman, A.D. (1982) Fishes as biological detectors of the effects
    of genotoxic agents. In Mutagenicity, New Horizons in Genetic Toxi-
    cology, J. Heddle, ed., pp. 435-456.
15. Kramer, K.J.M., H.A. Jenner, and D. de Zwart (1989) The valve
    movement response of mussels: A tool in biological monitoring.
    Hydrobiologica
    (in press).
16. Majone, F., R. Brunetti, I. Gola, and A.G. Levis (1987) Persis-
    tence of micronuclei in the marine mussel, Mytilus galloprovincialis,
    after treatment with mitomycin C. Mutat. Res. 191:157-161.
17. Maron, D.M., and B.N. Ames (1983) Revised method for the
    Salmonella mutagenicity test. Mutat. Res. 113:173-215.
18. Meier, J.R., W. Blazak, E.S. Riccio, B.E. Stewart, D.F. Bishop,
    and L.W. Condie (1987) Genotoxic properties of municipal wastewa-
    ters in Ohio. Arch. Env. Contam. Tox. 16:671-680.
19. Noordsij, A., J. van Beveren, and A. Brandt (1983) Isolation of
    organic compounds from water for chemical analysis and toxicological
    testing. Intern. J. Env. and Analyt. Chem. 13:205-217.
20. Pesch, G.G., and C.E. Pesch (1980) Neanthes arenaceodentata
    Polychaeta: Annelida), a proposed cytogenetic model for marine gen-
    etic toxicology. Can. J. Fish. Aquat. Sci. 37:1225-1228.
21. Quillardet, P., and M. Hofnung (1985) The SOS chromotest, a
    colorimetric bacterial assay for genotoxins: Procedures. Mutat.
    Res. 147:65-78.
22. Rappaport, S.M., M.G. Richard, M.C. Hollstein, and R.E. Talcott
    (1979) Mutagenic activity in organic waste water concentrates.
    Env. Sci. Tech. 13:957-961.
23. Slooff, W., and D. de Zwart (1987) Continuous effluent
    biomonitoring with an early warning system. Naturvardsverket
    Rapport 3275:20-21 (Stockholm).
24. Tates, A.D., I. Neuteboom, M. Hofker, and L. den Engelse (1980)
    A micronucleus technique for detecting clastogenic effects of

mutagens/carcinogens (DEN, DMN) in hepatocytes of rat liver in vivo. Mutat. Res. 74:11-20.

25. Van de Kerkhoff, J.F.J., and M.A. van der Gaag (1985) Some factors affecting the differential staining of sister-chromatids in vivo in the fish Nothobranchius rachowi. Mutat. Res. 143:39-43.

26. Van der Gaag, M.A., and J.F.J. van de Kerkhoff (1985) Mutagenicity testing of water with fish: A step forward to a reliable assay. Sci. Total Env. 47:293-298.

27. Van der Gaag, M.A. (1989) Rapid detection of genotoxins in waste water: New perspectives with the sister-chromatid exchange assay in vivo with Nothobranchius rachowi. In Conference Proceedings of the First European Conference on Ecotoxicology, M. Løkke, H. Tyle, and F. Bro-Rasmussen, eds. Copenhagen 259-262.

28. Van der Hoeven, J.C.M., I.M. Bruggeman, G.M. Alink, and J.H. Koeman (1982) The killifish Nothobranchius rachowi, a new animal in genetic toxicology. Mutat. Res. 97:35-42.

29. Van Hummelen, P., C. Zoll, J. Paulussen, M. Kirsch-Volders, and A. Jaylet (1989) The micronucleus test in Xenopus: A new and simple in vivo technique for detection of mutagens in fresh water. Mutagenesis 4:12-16.

30. Vigfusson, N.V., E.R. Vyse, C.A. Pernsteiner, and R.J. Dawson (1983). In vivo induction of sister-chromatid exchange in Umbra limi by the insecticides endrin, chlordane, diazinone and guthion. Mutat. Res., 118:61-68.

31. Weaver, D.L., P.K. Hopke, J.B. Johnston, and M.J. Plewa (1981) Mutagenicity of Chicago municipal sewage sludge in the Salmonella/microsome reverse mutation assay. Env. Muta. 3:350.

DETECTION OF GENOTOXICITY IN CHLORINATED OR OZONATED

DRINKING WATER USING AN AMPHIBIAN MICRONUCLEUS TEST

André Jaylet,[1] Laury Gauthier,[1] and Yves Lévi[2]

[1]Centre de Biologie du Développement
Université Paul Sabatier
31062 Toulouse, Cédex, France

[2]Anjou Recherche
Centre de Recherche de la Compagnie Générale
des Eaux O.T.V.
78600 Maisons-Laffitte, France

INTRODUCTION

Ideally, a detailed chemical analysis of a given sample of water should be able to identify all pollutants which might have a genotoxic action on living organisms. However, a number of considerations effectively rule out such an approach. In the first place, it has to be recognized that current analytic methods are not sufficiently sensitive to enable identification or quantification of all potential micropollutants. Furthermore, in order to evaluate potential risk, detailed understanding of the toxicological properties of all the pollutants known to be present in the sample will be required. Lastly, potential synergistic or antagonistic effects occurring in a complex mixture of pollutants need to be taken into account.

Another approach to the evaluation of water quality is based on tests of mutagenesis. A comprehensive review of this topic and the applications for testing drinking water have been recently published by Meier (26). The main mutagenesis tests applied to various types of water samples have also been discussed by Chouroulinkov and Jaylet (6) and Jaylet and Zoll (21).

The use of short-term tests such as that described by Ames et al. (2) in Salmonella has demonstrated that factory effluents, river water, and drinking water may have genotoxic effects. However, such bacterial tests are not sufficiently sensitive to be used without prior concentration

Genetic Toxicology of Complex Mixtures
Edited by M. D. Waters *et al.*
Plenum Press, New York, 1990

or extraction of micropollutants. Another approach to the study of genotoxic effects is to examine the effects on whole organisms exposed to the
water in question. For example, it has been shown that Rhine water induces chromosome aberrations or sister chromatid exchanges (SCEs) in
fish reared directly in this water (29-37).

Evaluation of mutagenicity in organisms reared directly in unconcentrated water samples has obvious practical application. With this end in
mind, we have developed a micronucleus test using the larvae of three
species   of   amphibian,   Pleurodeles   waltl   (Pleurodele),   Ambystoma
mexicanum (Axolotl), and Xenopus laevis (Xenopus).

Evans (7) first suggested that micronucleus formation in plants could
be used to evaluate genetic damage, although small rodents are commonly
employed for in vivo tests of this type (32). The extensive literature on
this subject has been reviewed by Heddle (12). With respect to aquatic
animals, tests have been described in the fish Umbra pigmea (13) and in
the frog Rana catesbeiana (23). We have developed a micronucleus test
using amphibian larvae. This has been described in detail in a series of
papers devoted to the Pleurodele (8,11,16,18,19,24,28,35,42), the Axolotl
(17), and Xenopus (39), and in a general review comparing the relative
merits of the three species (20).

Amphibian larvae have nucleated red blood cells which actively divide
in circulating blood. The presence of micronuclei in red blood cells is
readily detected on stained blood smears. The principle of the test is to
rear a group of larvae for about ten days in the water to be tested,
while a control group is reared in purified water. The level of micronucleated cells in the treated animals is thus compared to that of the
controls.

Micronuclei are small masses of intracytoplasmic chromatin resembling
small nuclei. They are chromosome fragments or whole chromosomes which
have not migrated to the spindle poles during cell division. They result
from chromosome breaks (chromosome fragments devoid of centromere cannot attach to the spindle), or to a disruption of the mitotic machinery
(whole chromosomes may not migrate normally during anaphase). A substance leading to formation of a micronucleus may either be a clastogen or
an agent which disrupts normal mitosis. The micronculeus test can thus
detect both chromosomal and genomic mutations.

Genotoxic micropollutants in drinking water come from a variety of
origins. They may be the result of pollution of raw water of industrial,
agricultural, or household origin, but also may stem from the treatment
processes used to purify water (see review by Meier, Ref. 26). We were
interested to find out whether the amphibian micronucleus test, which has
been found to be highly sensitive to known genotoxins, could also detect
clastogenic properties of drinking water without the requirement for concentration of the water.

We have included here the results obtained using the Pleurodele
micronucleus test applied to water samples treated with chlorine, monochloramine, or ozone. The results demonstrate that this test can be used
to monitor genotoxicity in water, and may thus be of value in treatment
plants used for the production of drinking water.

TEST PROCEDURES

In this test, the Pleurodele larvae are used when they are approximately two months of age, when they are about 30 mm long. They grow a further 10 mm or so within the next ten days. The rate of cell division in red blood cells is particularly high at this time. The general procedure has been described in detail elsewhere. We will just briefly describe the conditions used for these particular tests.

Statistical analysis is based on the method described by MacGill et al. (25) for median values. In our first publications, we used the relationship:

$$M \pm 1.57 \times IQR/\sqrt{n}$$

to calculate the confidence limit of the median (95%) for a sample ($n \geq 7$), where M is the median and IQR = the interquartile range. After more detailed analysis of the conditions of use of the micronucleus test, we found that this relationship was too restrictive, and we have now modified it to:

$$M \pm 1.51 \times IQR/\sqrt{n}$$

where IQR = interquartile range of the controls and n is the number of animals in the control group. This statistical formula is used in all our subsequent work.

DETECTION OF CLASTOGENIC EFFECTS IN TAP WATER

Procedure

The water to be tested was taken from a tap in our laboratory in Toulouse (France). The water was pumped from the Garonne River to the treatment plant whose daily production ranges from 6,000-9,000 m^3. The raw water is treated with lime and aluminum sulfate or ferric chloride, depending on the time of year. After flotation and filtration over sand, the water is disinfected with chlorine. The mean monthly organic carbon content during treatment ranges from 1.2-2.1 ppm. Seven tests were carried out between October 1985 and May 1986.

In all cases, the level of micronucleus in the red cells of larvae reared in the tap water were compared with levels from larvae reared in the same water that had been filtered over sand and activated carbon.

The chlorine content ranged from 0-0.2 ppm. It should be noted that chlorine levels over 0.4 ppm led to the death of the larvae within a few hours.

For the October and December tests, the larvae were reared in 5-liter vessels containing 2 l of water in aquaria fitted with an overflow. The aquaria were supplied with a trickle of water directly from the tap, while the controls were reared in aquaria supplied with tap water filtered over sand and activated carbon.

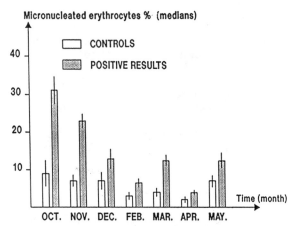

Fig. 1.   Median values obtained for the various tests on laboratory
drinking water samples. White bars: Tests using drinking
water filtered over sand and carbon; Stippled bars: Tests over
the same time period using nonfiltered tap water.

In all cases, the water temperature was maintained at 20 ± 0.25°C.
The Pleurodele larvae were fed on live Chironoma larvae which were
added daily in a single amount.

Results and Discussion

As can be seen in Fig. 1, positive results were obtained in all seven
tests.   This indicated that the tap water contained clastogenic pollutants
which were totally or partially eliminated by filtration over sand and
activated carbon.

The level of micronucleated cells from larvae reared in the nonfiltered tap water varied considerably, indicating that the level of pollution
differed at different times of the year.   However, the level of micronuclei
in controls also varied, in most cases in the same direction as for the
experimental animals.   This tends to indicate differences in sensitivity,
perhaps hatch-related, or that the filter did not retain all the micropollutants.

The micropollutants probably had a variety of origins.   Various possibilities suggest themselves which are not mutually exclusive.   The
micropollutants may be substances present in the raw water, substances
produced by the action of chlorine on organic matter forming halogenated
organic compounds, or even residual chlorine itself.   In order to examine
the latter possibility, further experiments were carried out.

CLASTOGENIC EFFECT OF CHLORINE AND MONOCHLORAMINE

The following experiments were performed to find out whether chlorine or monochloramine had intrinsic clastogenic effects in the absence of
organic matter in the water.

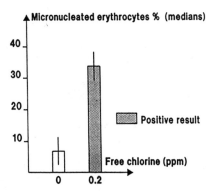

Fig. 2.   Clastogenic effect of chlorine in the presence of food. Median
           values of micronucleated red blood cells (%) of newt larvae
           reared for 12 days in reconstituted water chlorinated (0.2 ppm)
           with sodium hypochlorite.

Procedure

    Water quality.   The water used for rearing the control larvae was
"ultrapure" water prepared by dialysis and deionization in a Milli Q system (Millipore).   This water was enriched in salts normally present in
surface waters.   We called this water "reconstituted ultrapure water," and
it contained per liter the following salts: 294 mg $CaCl_2 \cdot 2H_2$; 123.25 mg
$MgSO_4 \cdot 7H_2O$); 64.75 mg $NaHCO_3$; and 5.75 mg KCl.   Its pH was close to
8.

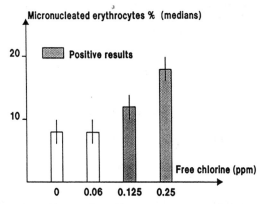

Fig. 3.   Clastogenic effect of different concentrations of chlorine in the
           presence of food: Median values of micronucleated red blood
           cells (%) of newt larvae reared for 12 days in reconstituted
           water chlorinated with sodium hypochlorite.

Organic compounds. Sodium hypochlorite and ammonium hydroxide
were supplied by Aldrich (France). Monochloramine was prepared by reacting solutions of sodium hypochlorite and ammonium hydroxide at 20°C,
pH = 7.7. Under these conditions, monochloramine was formed according
to the following reaction:

$$ClOH + NH_3 \longrightarrow NH_2Cl + H_2O$$

Free and bound chlorine were determined by a colorimetric method
using N,N'-diethyl-p-phenylenediamine.

Test conditions. All tests were carried out in stoppered 5 l vessels
containing 2 l of water and 20 larvae. All tests lasted 12 days. The
animals were fed on live Chironoma larvae.

Effect of Chlorine

Clastogenic effect of chlorine in the presence of food. In two consecutive tests (Fig. 2 and 3), we compared the levels of micronucleated
red cells in a group of larvae reared in reconstituted ultrapure water
chlorinated with sodium hypochlorite, to those of larvae reared in the
same water in the absence of chlorine. The water, the hypochlorite, and
the food were renewed daily at the same time. Under these conditions,
chlorine concentrations of 0.125, 0.2, and 0.25 ppm led to a significant
rise in the number of micronucleated red cells. A concentration of 0.06
ppm had no effect. In these experiments, the clastogenic effects could
not be assumed to be due to the effect of chlorine on organic matter in
the water, although it was possible that the chlorine might have reacted
with substances present in the food producing potential genotoxins. A
further set of experiments was set up to test this possibility.

Effect of chlorine in the absence of food. In order to try to rule
out the possibility of an interaction between chlorine and the food, the
following experiments were carried out.

A group of control larvae was reared in reconstituted ultrapure
water. A second group was reared in the same water containing 0.2 ppm
chlorine to which the Chironoma were added at the time of change of
water as in the previous experiments. In a third group, the larvae were
exposed for the first 3 hr following the change of water to 0.2 ppm chlorine. The water was then replaced with unchlorinated, reconstituted
ultrapure water, and the food was added only at this time.

It can be seen from Fig. 4 that the two groups exposed to chlorine
had similar levels of micronucleated red cells, which were significantly
higher than those of controls. The level in the controls was around 5%,
while in the group exposed to chlorine in the absence of food it was
18.5%, and 17% in the group exposed to food and chlorine at the same
time. This indicated that chlorine had a direct action on the larvae in
the absence of the food.

We have also shown (10) that in the absence of food, the larvae
exposed to 0.2 ppm chlorine absorb all the chlorine within 2 hr. After 2
hr, chlorine can no longer be detected in the water, showing that a relatively short exposure to chlorine can lead to a significant clastogenic
effect.

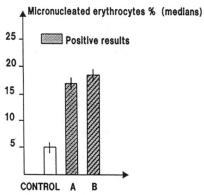

Fig. 4.   Clastogenic effect of chlorine in the presence (A) or absence
(B) of food.   Median values of micronucleated red blood cells
(%) of newt larvae reared for 12 days in reconstituted water
chlorinated with sodium hypochlorite.   Control: Reconstituted
water + food changed daily.   A: Reconstituted water + chlorine
(0.2 ppm) + food, medium changed daily.   B: Reconstituted
water + chlorine (0.2 ppm) without food for first 3 hr, replaced
with nonchlorinated water + food for remaining 21 hr.   Procedure repeated daily for 12 days.

Clastogenic effect of monochloramine.   Further   experiments   were
carried out using chlorine which is sometimes used as a disinfectant.
Various concentrations of chloramine (0.05, 0.1, and 0.125 ppm) were
used (Fig. 5).   A concentration of 0.3 ppm killed the larvae.   The level
of micronuclei increased with increasing concentration of chloramine, although statistically significant results were only obtained at the 0.15 ppm
dose.   This latter result was confirmed in further experiments.

Thus, it appeared that both chlorine and chloramine had a distinct
clastogenic effect under conditions which are likely to be employed in the
treatment of drinking water.

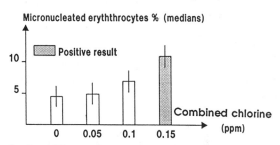

Fig. 5.   Clastogenic effect of monochloramine.   Median values of micronucleated red blood cells (%) of newt larvae reared for 12 days
in reconstituted water chlorinated with different concentrations
of monochloramine.

It has been found that chlorination of water containing organic matter can lead to the formation of halogenated organic compounds (3,30). Subsequent work has shown that this may give rise to numerous micropollutants with potential genotoxic activity. However, it has not generally been recognized that free chlorine (HOCl, OCl⁻) or chloramines themselves might in fact be responsible, and thereby represent a danger to the consumer. Our results tend to confirm studies demonstrating the mutagenic effects of hypochlorite (15,31,41) and chloramines (33,34,36) in bacteria. Moreover, it has been shown that both hypochlorite and chloramine lead to the formation of chromosome aberrations in cell cultures (15). Furthermore, in the mouse, drinking water containing sodium hypochlorite at a dose of 4 and 8 mg per kg body weight per day has been shown to lead to the formation of abnormal spermatozoids within 3 wk (27).

In our initial experiments we found that tap water sampled in our laboratory at all times of year leads to the formation of micronuclei. In this case, it does not seem necessary to invoke the presence of pollutants other than chlorine in order to account for these results.

However, the micronucleus test may also reveal genotoxic effects due to the action of chlorine on water containing organic compounds. We have shown (unpubl. results) that reconstituted ultrapure water artificially enriched in commercially available humic acids has clastogenic properties after chlorination.

Since ozone is also used to disinfect water, we carried out other experiments in which Pleurodele larvae were reared in river water containing different concentrations of ozone.

RESULTS WITH OZONATED SEINE WATER

The use of chlorine for disinfection of water for human consumption can lead to the formation of halogenated organic compounds, some of which have been found to be mutagenic. In an attempt to improve water quality, ozone has been increasingly used in Europe in view of its oxidizing, bleaching, and deodorizing properties.

Various methods have been used to test the mutagenicity of water treated with ozone, although the Ames test has been most widely used after concentration or extraction of micropollutants in the water sample (26).

Since ozone alters the chemical composition of the raw water, differences in observed mutagenicity will thus depend greatly on the quality of the raw water and the actual treatment procedure. Increases or decreases in genotoxicity have been observed depending on the exact conditions (26).

Results of bacterial tests and on cell cultures have indicated that the mutagenicity of the treated water is a function of the ozone concentration (4,5,22,38). We were interested to find out whether the micronucleus test using Pleurodele larvae was sufficiently sensitive to confirm this finding without the requirement for concentration of the water sample.

Procedure

The experiments were carried out on Seine River water sampled daily at Maisons-Lafitte, downstream from Paris. Two tests were carried out, the first in February and the second in June of 1988. Both tests lasted 12 days, and for each test the water was treated immediately after sampling, as follows.

Determination of the ozone demand of the water. Using an ozone demand flask, ozonated air containing a known concentration of ozone was introduced into a sample of the water. A direct reading on the graduations of the flask gave a measure of the ozonated air introduced. Thus, the amount of ozone (mg/l) put in the flask could be calculated. Contact between the water and the gas was facilitated by vigorous shaking for 5 min.

The residual ozone was assayed by a colorimetric method (Cifec-Lovibond, Paris) using N,N'-diethyl-p-phenylenediamine (DPD). The difference between the amount of ozone introduced and the residual ozone enabled calculation of the ozone demand of the water (referred to as dO_3 hereafter).

Ozonation of water under test. For each of the two sets of experiments, the following ozone concentrations were employed:

- ¼ of the ozone demand of the raw water ($dO_3/4$);
- ½ the ozone demand of the raw water ($dO_3/2$);
- the overall ozone demand (dO_3); and
- twice the ozone demand ($2 \cdot dO_3$).

Test conditions. The tests were carried out under the usual conditions in closed 5-l vessels containing 2 l of water and 20 larvae. The first test was carried out from February 17 to 28, 1988 and the second from June 8 to 19, 1988. The larvae were fed on live Chironoma larvae.

Tab. 1. Mean values and standard errors of the ozone concentrations (mg/l) added to the raw water during the tests of February and June 1988. dO_3 = water ozone demand

	1/4 dO_3	1/2 dO_3	dO_3	2 dO_3
February 1988	1.8 ± 0.3	3.5 ± 0.6	7.1 ± 1.1	14.2 ± 2.3
June 1988	1.6 ± 0.1	3.3 ± 0.2	6.6 ± 0.5	13.2 ± 0.9

ozone concentrations (mg/l)

During these tests, the ozone demand varied somewhat. Table 1 shows the mean values and standard errors (for the 12 test days) of the ozone concentrations (mg/l) added to the raw water.

In both experiments, six groups of 20 larvae from the same hatching were reared in separate vessels containing the following media:

- reconstituted ultrapure water;
- Seine River water (raw water);
- raw water + $\frac{1}{4}$ dO_3;
- raw water + $\frac{1}{2}$ dO_3;
- raw water + dO_3; and
- raw water + $2 \cdot dO_3$.

Results and Discussion

Figures 6 and 7 illustrate the results obtained. Histograms for the reconstituted ultrapure water are not shown, as they were essentially the same as those for the nonozonated raw water. Both experiments led to the same results. Larvae reared in the raw water had similar levels of micronuclei in their red cells to the larvae reared in the reconstituted ultrapure water. Seine River water thus does not contain pollutants with genotoxic effects on Pleurodele larvae.

Low levels of ozone ($\frac{1}{4}$ dO_3 or 1.8 mg/l for the first test and 1.6 mg/l for the second) led to a genotoxic effect, whereas higher levels had no significant effect on the level of micronucleated red cells.

These results are in general agreement with those of Bourbigot et al. (5) who showed using the Ames test that Oise River water ozonated with 1.5 mg/l was mutagenic, whereas water containing higher levels of ozone was not.

In addition, using samples of water from a Paris water treatment plant, Bourbigot et al. (4) found similar results in both cell cultures (V79/HGPRT) and a bacterial test (SOS chromo-test). These authors concluded that ozone in an adequate concentration does not increase the mutagenic potential of the original water supply. Used in sufficient concentrations it may even remove mutagenic activity, although low doses appear to increase it.

Kool and Hrubec (22), using the Ames test on water samples from five different sources, showed that treatment with low concentrations of ozone (3 mg/l) increased mutagenicity, whereas higher concentrations (10 mg/l) abolished it.

Van Hoof et al. (38) measured the formation of linear aldehydes and the mutagenic potential of Meuse River water after ozonation. Using the TA98 strain in the Ames test they found no correlation between the two. Increase in ozone concentration and rise in temperature did not increase the proportion of linear aldehydes, although the mutagenic potential fell with increasing ozone concentration (between 1 and 4 mg/l).

Our in vivo results rearing Pleurodele larvae in ozonated water confirm these in vitro results on water extracts. Ozonation of surface water with low doses of ozone may lead to genotoxic effects which are abolished by higher doses of ozone.

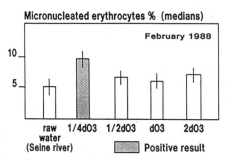

Fig. 6.   Clastogenic effect of raw water treated with ozone. First ex-
           periment, February 1988. Median values of micronucleated red
           blood cells (%) of newt larvae reared for 12 days in Seine River
           water treated with different ozone concentrations. dO_3 = water
           ozone demand.

CONCLUSION

     Using the Pleurodele micronucleus test we found that larvae reared
in drinking water had a higher level of micronucleated red cells than
larvae reared in the same water filtered over sand and activated carbon.
This positive results was found in seven different trials carried out over
a period of one year.   The Pleurodele micronucleus test is thus sufficiently sensitive to detect clastogenic micropollutants in situ in chlorinated
drinking water derived from river water.

     Larvae reared in water devoid of chlorinated organic matter contain-
ing 0.125 and 0.25 ppm sodium hypochlorite were also found to have
higher levels of micronucleated red cells than control larvae.   This was
also true for larvae reared in water containing 0.15 ppm chloramine, but
devoid of organic matter.   It can thus be concluded that both free chlorine or chloramine, which may be present in chlorinated drinking water,
have genotoxic effects.

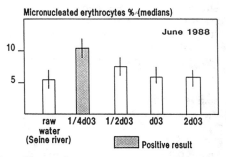

Fig. 7.   Clastogenic effect of raw water treated with ozone. Second Ex-
           periment, June 1988. Median values of micronucleated red blood
           cells (%) of newt larvae reared for 12 days in Seine River water
           treated with different ozone concentrations. dO_3 = water ozone
           demand.

Although it is widely recognized that chlorination of water enriched in organic matter leads to the formation of mutagenic compounds, the potential danger of chlorine itself may tend to be overlooked.

The normal practice of supplying drinking water containing some ppm of chlorine in order to prevent potential bacterial contamination may not be, therefore, as harmless as might be supposed.

Drinking water containing chlorine could lead to the formation of halogenated mutagens by interacting with foods. Chlorine may also act directly on gastric epithelial cells or may even penetrate tissues, especially in the presence of gastric lesions.

Preliminary results using the Pleurodele micronucleus test have also shown that water enriched in humic acids becomes genotoxic after chlorination. These results are in overall agreement with findings (26) demonstrating that chlorination of water containing humic acids leads to the formation of mutagenic compounds.

Treatment of Seine River water with low concentrations of ozone, corresponding to $\frac{1}{4}$ of the ozone demand (1.8 mg/l in the first experiment and 1.6 mg/l in the second) led to a genotoxic effect, whereas ozonation at higher doses (2,4, or 8 times the ozone demand) had no significant effect on the level of micronuclei. These results are in agreement with other work using different systems (4,5,22,38).

Overall, our results demonstrate that the Pleurodele micronucleus test can be used to detect genotoxic potential in water without the need for concentration.

The avoidance of extraction or concentration procedures effectively eliminates the possibility of modification of the chemical composition during these procedures. In addition, the wide variety of methods used for isolation and concentration of micropollutants leads to considerable discrepancies in the actual micropollutants removed and the estimation of the respective risks (21,26,40). Differences in the results of mutagenicity tests may thus be largely due to the different methods used for isolation or concentration, rather than to real differences in the composition of the water samples tested.

Overall, our results are in agreement with the in vitro methods employed on water extracts. However, it should be noted that in most cases the commonly used in vitro tests can detect gene mutations, while the micronucleus test only reveals chromosomal or genomic mutations. Thus, although this in vivo method avoids the requirement for isolation and extraction, it is in fact complementary to the other types of tests.

The experiments described here all used the Pleurodele as the test animal. Results using the larvae of the Axolotl (17) or Xenopus (39) indicate that both these species can be employed for this test system. Since these three amphibian species are widely employed in biological research, and their rearing conditions are well established, they represent valuable systems for in situ monitoring of the genotoxic potential of water (20).

REFERENCES

1. Alink, G.M., E.M.H. Frederix-Wolters, M.A. van der Gaag, J.F.F. van de Kerkhoff, and C.L.M. Poels (1980) Induction of sister chromatid exchanges in fish exposed to Rhine water. Mutat. Res. 78:369-374.
2. Ames, B.N., J. McCann, and E. Yamasaki (1975) Methods for detecting carcinogens and mutagens with the Salmonella/mammalian-microsome mutagenicity test. Mutat. Res. 31:347-363.
3. Bellar, T.A., J.J. Lichtenberg, and R.C. Kroner (1974) The occurrence of organohalides in chlorinated drinking water. J. Am. Water Works Assoc. 66:703-706.
4. Bourbigot, M.M., M.C. Hascoest, Y. Lévi, F. Erb, and N. Pommery (1986) The role of ozone and granular activated carbon in the removal of mutagenic compounds. Environ. Health Perspect. 69:159-163.
5. Bourbigot, M.M., J.L. Paquin, L.H. Pottenger, M.F. Blech, and P.H. Hartemann (1983) Study of mutagenic activity in water in a progressive ozonization unit. Aqua 3:99-102.
6. Chouroulinkov, I., and A. Jaylet (1989) Contamination of aquatic systems and genetic effects (Part 4). In Aquatic Ecotoxicology. Fundamental Concepts and Methodologies, A. Boudou and F. Ribeyre, eds., CRC Press (in press).
7. Evans, H.J., G.J. Neary, and F.S. Williamson (1959) The relative biological efficiency of single doses of fast neutrons and rays on Vicia faba roots and the effects of oxygen. II. Chromosome damage, the production of micronuclei. Int. J. Rad. Biol. 1:216-229.
8. Fernandez, M., and A. Jaylet (1987) An antioxidant protects against the clastogenic effects of benzo(a)pyrene in the newt in vivo. Mutagenesis 2:293-296.
9. Fernandez, M., L. Gauthier, and A. Jaylet (1989) Use of newt larvae for in vivo genotoxicity testing of water: Results on 19 compounds evaluated by the micronucleus test. Mutagenesis 4:17-26.
10. Gauthier, L., Y. Lévi, and A. Jaylet (1989) Evaluation of the clastogenicity of water treated with sodium hypochlorite or monochloramine using a micronucleus test in newt larvae (Pleurodeles waltl). Mutagenesis 4:170-173.
11. Grinfeld, S., A. Jaylet, R. Siboulet, P. Deparis, and I. Chouroulinkov (1986) Micronuclei in red blood cells of the newt Pleurodeles waltl after treatment with benzo(a)pyrene: Dependence on dose, length of exposure, post-treatment time and uptake of the drug. Environ. Mutagen. 8:41-51.
12. Heddle, J.A., M. Hirte, B. Kirkhart, K. Mavournin, J.T. Mac-Gregor, G.W. Newell, and M.F. Salamone (1983) The induction of micronuclei as a measure of genotoxicity. A report of the U.S. Environmental Protection Agency Gene-Tox Program. Mutat. Res. 123:61-118.
13. Hooftman, R.N., and W.K. de Raat (1982) Induction of nuclear anomalies (micronuclei) in the peripheral blood erythrocytes of the eastern mudminnow Umbra pigmaea by ethyl methanesulphonate. Mutat. Res. 104:147-152.
14. Hooftman, R.N., and G.J. Vink (1981) Cytogenetic effects on the eastern mudminnow, Umbra pygmea exposed to ethylmethane sulfonate, benzo(a)pyrene and river water. Ecotox. Environ. Safety 5:261-269.

15. Ishidate, M.J., T. Sofuni, K. Yoshikawa, M. Hayashi, T. Nohmi, M. Sawada, and A. Matsuoka (1984) Primary mutagenicity screening of food additives currently used in Japan. Food Chem. Toxicol. 22:623-636.

16. Jaylet, A., P. Deparis, V. Ferrier, S. Grinfeld, and R. Siboulet (1986) A new micronucleus test using peripheral blood erythrocytes of the newt Pleurodeles waltl to detect mutagens in fresh water pollution. Mutat. Res. 164:245-257.

17. Jaylet, A., P. Deparis, and D. Gaschignard (1986) Induction of micronuclei in the peripheral erythrocytes of the Axolotl larvae following in vivo exposure to mutagenic agents. Mutagenesis 1:211-215.

18. Jaylet, A., L. Gauthier, and M. Fernandez (1987) Détection of mutagenicity in drinking water using a micronucleus test in the newt larvae (Pleurodeles waltl). Mutagenesis 2:211-214.

19. Jaylet, A., L. Gauthier, M. Fernandez, M. Marty, and V. Ferrier (1986) A micronucleus test for the detection of mutagenic activity in drinking water using erythrocytes from newt larvae. In Proceedings of the XVIth Annual Meeting of the European Environmental Mutagen Society, A Léonard and M. Kirsch-Volders, eds., pp. 137-144.

20. Jaylet, A., L. Gauthier, and C. Zoll (1989) Micronucleus test using peripheral red blood cells of Amphibian larvae for detection of genotoxic agents in freshwater pollution. In In Situ Evaluation of Biological Hazards of Environmental Pollutants, S. Sandhu, ed. Plenum Press, New York (in press).

21. Jaylet, A., and C. Zoll (1989) Tests for detection of genotoxins in freshwater. In CRC Critical Review in Aquatic Toxicology, R. Anderson, ed. CRC Press (in press).

22. Kool, H.J., and J. Hrubec (1986) The influence of an ozone, chlorine and chlorine dioxide treatment on mutagenic activity in (drinking) water. Ozone Sci. and Engin. 8:217-234.

23. Krauter, P.W., S.L. Anderson, and F.L. Harrison (1987) Radiation induced micronuclei in peripheral erythrocytes in Rana catesbeiana: An aquatic animal model for in vivo genotoxicity studies. Environ. Molec. Mutagen. 10:285.

24. Marty, J., P. Lesca, A. Jaylet, C. Ardourel, and J.L. Rivière (1989) In vivo and in vitro metabolism of benzo(a)pyrene by the larva of the newt Pleurodeles waltl. Comp. Biochem. Physiol. (in press).

25. Mac Gill, R., J.W. Tuckey, and W.A. Larsen (1978) Variations of box plots. Am. Statist. 32:12-16.

26. Meier, J.R. (1988) Genotoxic activity of organic chemicals in drinking water. Mutat. Res. 196:211-245.

27. Meier, J.R., R.J. Bull, J.A. Stober, and M.C. Cimino (1985) Evaluation of chemicals used for drinking water disinfection for production of chromosomal damage and spermhead abnormalities in mice. Environ. Mutagen. 7:201-211.

28. Normalisation Francaise (1987) Essais des eaux. Détection en milieu aquatique de la génotoxicité d'une substance vis-à-vis de larves de batraciens (Pleurodeles waltl et Ambystoma mexicanum). Essai des micronoyaux. In Fascicule AFNOR T90-325. Association Française de Normalisation, Tour Europe, Paris, France.

29. Prein, A.E., G.M. Thie, G.M. Alink, C.L.M. Poels, and J.H. Koeman (1978) Cytogenetic changes in fish exposed to water of the river Rhine. Sci. Total Environ. 9:287-291.

30. Rook, J.J. (1974) Formation of haloforms during chlorination of natural waters. Water Treat. Exam. 23:234-243.

31. Rosenkranz, H.S. (1973) Sodium hypochlorite and sodium perborate: Preferential inhibitors of DNA polymerase-deficient-bacteria. Mutat. Res. 21:171-174.

32. Schmid, W. (1976) The micronucleus test for cytogenetic analysis. In Chemical Mutagens, Vol. 4, A. Hollaender, ed. Plenum Press, New York, pp. 31-53.

33. Shih, K.L., and J. Lederberg (1976) Chloramine mutagenesis in Bacillus subtilis. Science 192:1141-1143.

34. Shih, K.L., and J. Lederberg (1976) Effects of chloramine on Bacillus subtilis deoxyribonucleic acid. J. Bact. 125:934-945.

35. Siboulet, R., S. Grinfeld, P. Deparis, and A. Jaylet (1984) Micronuclei in red blood cells of the newt Pleurodeles waltl Michah: Induction with X-rays and chemicals. Mutat. Res. 125:275-281.

36. Thomas, E.L., M.M. Jefferson, J.J. Bennett, and D.B. Learn (1987) Mutagenic activity of chloramines. Mutat. Res. 188:35-43.

37. van der Gaag, M.A., G.M. Alink, P. Hack, J.C.M. van der Hoeven, and J.F.J. van de Kerkhof (1983) Genotoxicological study of Rhine water with the killifish Nothobranchius rachowi. Mutat. Res. 113:311 (abstr.).

38. Van Hoof, F. (1982) Formation and removal of mutagenic activity in drinking water by ozonation. Aqua 5:475-478.

39. Van Hummelen, P., C. Zoll, J. Paulussen, M. Kirsch-Volders, and A. Jaylet (1989) The micronucleus test in Xenopus: A new and simple in vivo technique for detection of mutagens in fresh water. Mutagenesis 4:12-16.

40. Wilcox, P., and S. Denny (1985) Effect of dechlorinating agents on the mutagenic activity of chlorinated water samples. In Water Chlorination, Chemistry, Environmental Impact and Health Effects, Vol. 5, R.L. Jolley, R.J. Bull, W.P. Davis, S. Katz, M.H. Roberts, Jr., and V.A. Jacobs, eds. Lewis Publishers, Inc., Chelsea, MI, pp. 1341-1353.

41. Wlodkowski, T.J., and H.S. Rosenkranz (1975) Mutagenicity of sodium hypochlorite for Salmonella typhimurium. Mutat. Res. 31:39-42.

42. Zoll, C., E. Saouter, A. Boudou, F. Ribeyre, and A. Jaylet (1988) Genotoxicity and bioaccumulation of methyl mercury and mercuric chloride in vivo in the newt Pleurodeles waltl. Mutagenesis 3:337-343.

THE SIGNIFICANCE OF MUTAGENICITY AS A CRITERION

IN ECOTOXICOLOGICAL EVALUATIONS

W.K. de Raat, G.J. Vink, and A.O. Hanstveit

Division of Technology for Society TNO
Biological Department
2600 AE Delft, The Netherlands

ABSTRACT

To date environmental mutagenesis has been a predominantly anthropocentric discipline of toxicology. It is ultimately motivated by the fear of human exposure to carcinogens and mutagens. Exposure of and effects in other organisms are regarded as relevant as far as they indicate exposure of and effects in humans. The ecotoxicologically relevant effects, i.e., changes of the structure and function (quality) of ecosystems caused by environmental pollution with mutagens, receive far less attention. Nevertheless, only these effects justify the inclusion of mutagenicity as a criterion in ecotoxicological evaluations. Both these effects and the extent to which they support an ecotoxicological role of mutagenicity tests, are the central themes of this paper.

First, a short treatment of ecotoxicology and the characteristic differences between this branch of toxicology and human toxicology is given; ecotoxicological evaluations for the aquatic environment will be briefly touched upon. Then the ecotoxicologically relevant effects of mutagens in aquatic ecosystems will be discussed. They can be split up into two groups: those that will not and those that will escape detection in an adequate set of conventional ecotoxicity tests. The first includes effects on division rate of unicellular organisms, reproduction of multicellular organisms, and survival; the second, cancer and the long-term (multi-generation) effects of an increased mutational load. It is concluded that, with respect to the first group of effects, mutagenicity tests can only be used as prescreening, alerting, and supporting tests; mutagenicity may play a role in its own right with respect to cancer and increased mutational load.

Additional arguments for the inclusion of mutagenicity in ecotoxicological evaluations are provided for by the actual pollution and effects encountered in the aquatic environment.

Genetic Toxicology of Complex Mixtures   **249**
Edited by M. D. Waters *et al.*
Plenum Press, New York, 1990

ENVIRONMENTAL MUTAGENESIS AS AN ANTHROPOCENTRIC DISCIPLINE

Notwithstanding the adjective used, it is no exaggeration to say that environmental mutagenesis is a predominantly anthropocentric discipline of toxicology. In this respect it does not differ much from most other disciplines of toxicology. This might seem a trivial statement, because all sciences are in one way or another ultimately aimed at the benefit of man. It is, of course, not the intention of this paper to annoy the reader with trivialities. Here we use the word anthropocentric in a stricter sense, i.e., to express that environmental mutagenesis is predominantly aimed at effects on humans. The adjective environmental means that the agents possibly causing these effects find their origin outside man. Pollution of the environment is studied because it may rebound on man, for instance by inhalation of airborne particles (1) or drinking water prepared from polluted surface water (18). So attention is focussed on environmental agents and human health. Other organisms only fulfill the role of test organisms: to detect and identify mutagenic agents that may threaten human health, to elucidate the processes underlying these threats, and to bridge the gap between, on the one hand, the presence of mutagens detected with our sensitive and easy microbial test systems, and, on the other, the deleterious effects in the exposed human. Perhaps the most environmentally directed branch of environmental mutagenesis is the one that is the topic of these proceedings. It is devoted to complex mixtures either emitted to the environment or collected from the environment. It tries to tackle the problems associated with testing complex mixtures and to identify compounds and processes to which the effects of complex mixtures can be attributed. When a clear picture about the effects is obtained the question arises: What does this mean to man?

This chapter is concerned with this important question. The research effort aimed at the answer to this question certainly deserves strong support. However, this paper will not deal with this question. Other questions are its starting point: What about other organisms? What does the presence of mutagens in the environment mean to them? A more anthropocentric one may be added: What does their exposure mean to us? These questions refer to the ecotoxicological aspects of environmental pollution with mutagens. Ecosystems are exposed to the mutagens emitted and may as a consequence undergo changes in their quality. These changes form the central theme of this paper. The paper is anthropocentric in the sense that humans have a negative appreciation of such changes and feel that ultimately they depend on the quality of ecosystems; it is not anthropocentric in the sense of human concern of exposure via the environment.

The paper will start with a short treatment of ecotoxicology and the characteristic differences between this branch of toxicology and human toxicology; ecotoxicological evaluations will be briefly touched upon; more time will be spent on the ecotoxicologically relevant effects of the presence of mutagens in the environment and the possible contribution of mutagenicity tests to ecotoxicological evaluations; the last part of the paper will deal with the actual pollution with mutagens and the effects encountered in situ that may be attributed to this pollution. We will restrict ourselves to the aquatic environment.

ECOTOXICOLOGY VERSUS HUMAN TOXICOLOGY

The clearly observable effects that environmental pollution has on organisms other than man in our environment has led to a separate branch of toxicology or ecotoxicology. In this case, the adjective refers to exposure and effects in organisms other than man, and therefore environmental toxicology and ecotoxicology can, without problem, be regarded as synonyms. The objective of this branch can be defined as follows: To unravel the relations between emissions or discharges of wastes and changes in the quality of the recipient ecosystem.

It differs fundamentally in two aspects from human toxicology:

(i)   Ecotoxicology deals with all the species assembled in the ecosystem while, per definition in the case of human toxicology, only one species is the subject;

(ii) Ecotoxicologists are confronted with the extra dimension of interdependence of the exposed populations.

An ecosystem is not just a collection of species, but numerous interactions determine its composition and functioning. Effects on one species can, and often will, give rise to effects on another. Effects on individual organisms are not important to the ecotoxicologist as long as they do not bear any relevance for the quality, i.e., the composition and functioning of the ecosystem. In human toxicology, any disfunctioning of an individual is, in principle, relevant.

The quality of an ecosystem has to be defined and established. This is one of the major problems of ecotoxicology. Human activities will always interfere with the quality; the definition of the acceptable level of interference is delegated to politics. In contrast, any change in human health is unacceptable. Both branches of toxicology are faced with the problem of extrapolation. Ecotoxicologists try to predict the effects of a complicated system of populations in a certain physicochemical environment from tests with one species or very simple artificial ecosystems. They cannot experiment with the ecosystem itself. Monitoring studies can be added to their laboratory experiments by which the damage caused in situ by actual pollution can be investigated.

Human toxicologists try to predict the effects in one species while only being allowed to carry out experiments in other species. Epidemiology can in this context be regarded as the parallel of environmental monitoring. To summarize, extrapolation from a single species to complex systems, versus extrapolation from one species to another, can be employed.

EVALUATION FOR ECOTOXICOLOGICALLY RELEVANT EFFECTS

Ecotoxicology finds its application of the ever-increasing number of man-made chemicals and waste products and their effects on the recipient ecosystem (8-10). In this it supports authorities who try to regulate and thereby prevent damage to ecosystems (11,29,34), and it enables industries to include ecotoxicity as a quality criterion of products and processes. Because of their importance as recipients, aquatic ecosystems were the first for which tests were developed.

Generally, three main criteria can be discerned in ecotoxicological evaluations for the aquatic environment (4,8,10,82). The first is the fate of the toxicants in the aquatic environment (52). Various physical and chemical processes determine the actual spatial and temporal distribution of the toxicants. Important ones are: evaporation from the water, absorption by the water from the air, adsorption by particulate material, chemical conversion, complexation, and last but not least, biodegradation (50). It is generally impossible to investigate all these processes. In most cases only simple tests for biodegradation and chemical degradation are conducted. Fate models are used to estimate the distribution of compounds in the different compartments of the ecosystem. The processes, however, are important because they determine time span and intensity of exposure as well as the geographic distribution of exposure.

Bioavailability or the extent of actual internal exposure, given certain concentrations in the water, is the second criterion. Bioaccumulation is the important process in this context (77). Lipophylic compounds are less easily excreted and tend to accumulate in the exposed organism. Internal exposure levels are often much higher than would appear from the concentrations in the water. Additional accumulation can take place via the food chain. Often a simple measure is taken for bioaccumulation, i.e., the octanol-water-coefficient. The high correlation between this measure, the bioaccumulation, and effects in the biological tests justify its use.

The third criterion is, of course, the toxicity of the compound(s) given a certain external and internal exposure. Tests with organisms representative of the recipient ecosystem are carried out for effects on survival, reproduction, and growth. Over the last 20 years, a large series of tests have been developed covering these criteria. Tests are selected to compose sets adapted to regulations formulated by authorities,

Tab. 1.   Criteria used in ecotoxicological evaluations

FATE	– chemical degradation – biological degradation – adsorption to particles – evaporation – absorption from air – complexation	determine external dose at certain time and place
BIOAVAILABILITY	– bioaccumulation – either after uptake from water – or via foodchain (biomagnification)	determine internal dose given a certain external dose
EFFECTS	on: – survival – reproduction – division of unicellular organisms	determine structure and function (quality) of eco- system

Tab. 2.   Limitations of ecotoxicological tests

General:
  - lack of knowledge about actual ecotoxicologically relevant effects
  - restricted time and means
  - no experiments possible with real ecosystems

Those of the tests for effects:
  - effects are measured at the organismal level
    (interactions between organisms are largely omitted)
  - limited number of general criteria
    (hardly any sublethal criteria with well-defined endpoints are applied)
  - laboratory conditions differ widely from conditions in ecosystem
    (e.g., food ad libitum versus competition)
  - a limited number of laboratory-resistant species which are easy to
    manipulate are selected
  - most tests have rather short-term nature

Result of these limitations:
  - reduced sensitivity?
  - large safety factors necessary

the type and intensity of the discharge, the amount of chemical produced, the chance of actual exposure, and the properties of the recipient ecosystem (4,10,29,81) (Tab. 1).

The sets of methods currently in use have several shortcomings. They are listed in Tab. 2. These are inevitable consequences of our lack of knowledge about the interactions between toxicants and ecosystems in general, the restricted time and means available for ecotoxicological evaluations, and the fact that no tests can be performed with realistic experimental ecosystems. So at the moment we are seriously hampered if we strive for an accurate prediction of changes in the quality of the ecosystem. Safety factors are used to compensate for our uncertainty (44,74). We will go into more detail as far as the shortcomings of the toxicity tests are concerned. A major one is that effects investigated are largely restricted to the organismal level. Organisms of one species are exposed in the laboratory. Interactions between organisms, (i.e., those processes that make an ecosystem of a collection of populations of different species) are virtually omitted, either because no unambiguous tests are available or because the available ones are too costly and tedious for most routine evaluations.

Furthermore, in most cases only three criteria are used: effects on survival, growth (of a culture of unicellular organisms), and reproduction. Hardly any tests for other, sublethal effects are used, although such tests are available (3). It must be emphasized that ecological interactions may very well be affected by these effects, in particular if they affect the fitness to compete.

In routine ecotoxicological tests, the organisms are kept under stress conditions that differ completely from those in situ. Food is given ad libitum and competition with other species plays no role. It can thus be expected that organisms in routine tests will be less sensitive than those in situ. Other shortcomings are related to the short-term nature of the tests (exposure is short compared to the lifetime of the test organism), the limited number of species that can be handled, and the fact that species are selected for resistance to laboratory conditions and ease of manipulation. We may assume that most of these shortcomings will tend to make an evaluation less sensitive. Ecotoxicity is more likely to be under- than overestimated.

Much ecotoxicological research is aimed at improvement of this situation. General principles underlying the changes in ecosystems due to exposure to toxicants are intensively studied (42,43), and this will improve our insight in the ecological relevance of single species tests, i.e., will allow a more certain extrapolation. Apart from that, attempts are being made to develop practically feasible tests that include more subtle criteria than survival and include interactions (48). Against the background of this research effort, it is surprising that hardly any attention is paid to the ecotoxicological significance of mutagens.

ECOTOXICOLOGICALLY RELEVANT EFFECTS OF THE PRESENCE OF MUTAGENS IN THE AQUATIC ENVIRONMENT

The ecotoxicologically relevant effects of mutagens can be split into two groups (see Tab. 3): effects that can be detected with conventional ecotoxicity tests, and those that cannot. Mutagens can have effects on survival and reproductive capacity of individual organisms and on growth (i.e., division rate) of unicellular organisms. An optimum set of conventional tests will be sensitive to these effects. Other effects are cancer and effects which we have to define in a peculiar way; those effects remaining if the other are insignificant. We will first treat the "conventional" effects.

As we all are very well aware, mutagenicity is the result of the chemical reactivity of a compound towards biological macromolecules. Mutagens, being electrophiles, react with nucleophiles such as proteins and nucleic acids. If there is a preference for certain sites in nucleic acids, the compound is a clear-cut mutagen, i.e., we can easily distinguish mutagenicity from other toxic effects. If other reactions dominate,

Tab. 3.   Ecotoxicologically relevant effects of mutagens

I.   Effects that can be detected with conventional tests:
     - decreased survival
     - decreased reproductive capacity
        - fitness of offspring - mutations
                                - teratogenic effects
        - number of offspring
     - decreased division rate of unicellular organisms
        - algae   - primary production
        - bacteria   - biodegradation

II. Other effects:
     - induction of neoplasms
     - long-term effect of enhanced mutation rate?
        - accumulation of mutations over many generations

---

it is hard to demonstrate mutagenicity. If doses are high enough (and if physical properties allow these high doses to be reached), a mutagen will be picked up with a set of conventional tests just because it induces all sorts of other toxic effects.

We may expect one group of ecotoxicity tests to be particularly sensitive to mutagens, i.e., those in which organisms are exposed during reproduction. Two types of tests can be discerned: those in which more than one generation is exposed, and those in which exposure occurs during ontogenesis. In practice, multigeneration tests are confined to tests with unicellular organisms. Two types of organisms are used: algae, because of their important role as primary producers (78), and bacteria, because of their involvement in biodegradation (5). Reaction of the mutagen with the DNA will result in a hampered DNA replication, will affect the synthesis of products necessary for cell division due to mutations, and will lead to an affected offspring and a reduced survival. Altogether the growth rate of the culture will slow down. Effects on growth are the very effects determined in these tests.

Multigeneration tests with higher organisms are very scarcely applied in evaluations. Some have been developed with crustaceans as test organisms (Daphnids and Artemia), but these are too time-consuming for practical use. Fish would be an obvious choice; however, generation times of these animals are far too long.

Tests in which exposure occurs during ontogenesis are the equivalent of the teratogenicity tests in mammalian toxicology. A fair number of them have been developed. Their criteria range from simply comparing numbers of fertilized eggs and surviving offspring to a thorough observation of ontogenesis. It must be emphasized that these two groups of tests are not solely sensitive to mutagens; other processes besides interactions with the genetic material may affect these criteria as well. No dominant lethal assays are used in human toxicology, i.e., assays in which males are exposed before mating. Various mutagenicity assays with aquatic organisms have been developed; they are as far as we know not frequently used in ecotoxicological evaluations. A number of them are listed in Tab. 4.

Let us now consider the effects not detected in conventional toxicity tests. First, cancer. The relation between carcinogenicity and mutagenicity needs not be elaborated upon (30,59). A major impetus for environmental-mutagenicity studies is the threat of human exposure to carcinogens. The presence of mutagens in the aquatic environment may, or maybe we must say will, inevitably result in tumors in aquatic organisms. Later we will present a short overview of the literature on the occurrence of cancer in fish in situ, and this will show that we are not dealing here with a hypothetical effect. Nevertheless, carcinogenicity tests are not included in ecotoxicological evaluations, the main reason being that they are too laborious and time-consuming.

Are there any other ecotoxicologically relevant effects of mutagens? We have to realize that not all changes in the genetic information of the exposed organisms are drastic enough to be detected by the conventional set of tests in the form of effects on survival or reproductive capacity or by carcinogenicity tests. Many mutations will go unnoticed. If these occur in the somatic cells they will not be relevant because an ideal set of

Tab. 4.   Some mutagenicity tests with aquatic organisms

Organism	Effect	Reference
Umbra limi, central mud minnow	Chromosomal aberrations Sister chromatid exchanges	39 40
Umbra pygmeae, eastern mud minnow	Chromosomal aberrations Sister chromatid exchanges Micronuclei (nuclear anomalies)	25,67 2 26
Notobranchius rachowi, killifish	Chromosomal aberrations	27
Opsanus tau, oyster toad fish	In vitro unscheduled DNA synthesis in hepatocytes Chromosomal aberrations Sister chromatid exchanges	35 51
Caudiverbera caudiverbera, amphibian	Micronuclei	83
Thethya lyncurium, marine sponge	DNA damage and repair	87
Mytilus edulis, mussel	Sister chromatid exchanges in adults in larvae	 13,84 21
Mytilus galloprovincialis, mussel	Sister chromatid exchanges in developing eggs	7
Neanthes arenaceondendata, marine worm	Chromosomal aberrations	66
Pomotoceres triqueter, marine worm	Chromosomal aberrations	14
Two algae and a grass	Micronuclei	17

tests will detect all decreases of fitness of the exposed organism that are regarded as relevant. Furthermore, effects caused by somatic mutations are equivalent to effects with any other mechanistic background from the point of view of the ecotoxicologist. The situation is different if such unnoticed mutations occur in the germ cells. Then the effect of exposure is spread out over the population; it is not restricted to the lifetime of the exposed specimen or by the reduced viability of its offspring. Reproduction transfers effects of mutagens from the organismal level to the population level. If these effects will somehow affect competitiveness within the population or among populations, they represent a temporarily extra stress. Temporarily, because they will disappear due to selection. Such a stress may cause changes of ecosystem quality that are not indicated by conventional ecotoxicological evaluations. Many habitats are characterized by continuous presence of mutagenic pollutants (see the section "An Important Additional Argument: Actual Population and Effects"). We are dealing then with a continuous extra stress. Accumulation of mutations may occur; competition within a population and among populations is continuously affected by an unnaturally high mutational load.

We may also approach this effect of the presence of mutagens in another way. Under natural conditions all organisms are exposed to a background mutagenic stress and this stress, together with sensitivity of the species, determines the background mutation rate. Emission of mutagens means an anthropogenic increase of this mutation rate. We may assume that somehow the quality of ecosystems is determined in the long run by the background mutation rate, and that an increase of the mutation rate will consequently affect the quality in the long run.

Hardly any knowledge is as yet available on the type of changes to be expected and on the increase of the mutagenic stress which is needed for changes to occur that are unacceptable from an ecotoxicological standpoint. If exposure levels are in the same range as those detected in conventional tests, there is no real reason to take this effect into account in our evaluations; if they are lying below these levels, our tests must somehow be supplemented. It is evident that a thorough research effort is needed to fill in this important gap in our knowledge, a new challenge to ecotoxicology as well as to environmental mutagenesis.

ARGUMENTS FOR THE USE OF MUTAGENICITY AS A CRITERION IN ECOTOXICOLOGICAL EVALUATIONS

We may ask whether the considerations of the previous section really urge us to use mutagenicity as a criterion in ecotoxicologic evaluations. Do the possible effects of the presence of mutagens in the environment provide us with arguments for the inclusion of mutagenicity tests in the test sets? The first group of possible effects are those that can be detected in conventional tests. Are mutagenicity tests of any use to detect these effects? The obvious answer to this question seems to be "no." The aim of the evaluations is to gain insight into the effects expected to occur in situ. The effects determined in conventional evaluations must be as similar as possible to those effects in situ. As we have seen in the section entitled "Evaluation for Ecotoxicologically Relevant Effects," the shortcomings of the conventional tests have to do with a lack of similarity. Including mutagenicity tests will not increase the similarity. We only obtain information on the mechanism possibly underlying the effect determined in the conventional tests. In ecotoxicological evaluations the question "Does an Effect Occur?" has a higher priority than "How does it occur?" Clarification of mechanisms is, of course, important in ecotoxicology; research on this topic is aimed at the role of ecological interactions and not so much at molecular-biological backgrounds of effects. As argued earlier, lack of sensitivity might be another shortcoming of the ecotoxicological evaluations. Mutagenicity might lead to a higher sensitivity. However, what is to be done when a compound is mutagenic but not positive in conventional tests? Is it not a better strategy to avoid this problem by improving our conventional tests?

A function mutagenicity tests might have in this context, i.e., in the context of the effects detected with the conventional tests, is alerting the researcher to the fact that he is dealing with a compound or waste product that is reactive to biological macromolecules and may, therefore, have various effects, and may, specifically, due to its preference to the genetic material, affect production. A mutagen not active in a first set of conventional tests might deserve a closer look. The ultimate evaluation has to be based, however, on the end points of the conventional tests. It would be interesting to investigate to what extent clear-cut mutagens escape detection in conventional tests. To this end, we would have to determine the correlation between effects in conventional tests, effects in short-term mutagenicity tests, and mutagenic effects in exposed aquatic organisms. If this were to reveal that mutagenicity is generally accompanied by effects in one of the conventional tests, the alerting function of mutagenicity is certainly justified.

Mutagenicity tests are generally more flexible toxicological tools than the conventional tests. This points to another role they may play. If ecotoxicologically relevant effects can be attributed to the presence of a mutagen, studies on its fate in the environment or its identity can be guided by the effect itself (68).

So far we have restricted ourselves to the effects in conventional tests, i.e., to the contributions of mutagenicity to the detection of these effects. We have seen in the previous section that mutagens can exert two types of effects that will escape conventional tests: cancer and increase of mutational load. It is clear that if we regard these effects as ecotoxicologically relevant, mutagenicity tests may play a role in their own right and not only as supporting or alerting tests.

Cancer can, of course, be investigated with carcinogenicity tests. However, it is hard to see how these laborious and time-consuming tests could find a place in routine evauations. Then the relation between mutagenicity and carcinogenicity may lead to the use of mutagenicity tests instead. Tests with aquatic organisms known to be sensitive, particularly in fish, may possibly be preferred to the usual short-term mutagenicity tests, e.g., the Salmonella/microsome test (see Tab. 4).

However, the question as to whether or not cancer is an important phenomenon in the aquatic environment must be raised. Tumors in feral fish have received much attention, but a link with the presence of mutagens has until now not been unambiguously demonstrated (see the following section, "Tumor-like Lesions in Shellfish and Fish"). Furthermore, it is important to establish whether cancer is a result of very high

pollution levels; whether it is the tip of the iceberg. If so, it will be accompanied by other effects that will <u>not</u> escape the conventional tests and
inclusion of carcinogenicity as a criterion would only complete our picture; it would not lead to detection. In the case of the increase of the
mutational load, only mutagenicity tests will give us the relevant information. However, our knowledge about this effect is so scarce that much
further research is needed before we can answer the question as to
whether the use of mutagenicity tests for this purpose is justified.

AN IMPORTANT ADDITIONAL ARGUMENT:
ACTUAL POLLUTION AND EFFECTS

The arguments discussed in the previous section were all based on
the possible changes of the ecosystem that can be caused by mutagens.
An additional argument is the subject of this section: the actual
anthropogenic pollution of the aquatic environment, and the actual

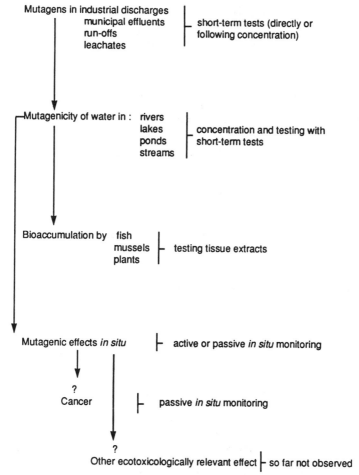

Fig. 1. Schematic summary of evidence for actual pollution and effects.

Tab. 5. Mutagenicity of emissions into the aquatic environment

Sample	Effect	Reference
Wastewater from coal gasification	Mutations in bacteria	80
Wastewater from paper production	Mutations in bacteria and SCE in tissue-culture cells Mutations in bacteria	49,85 47
Wastewater from nitrofurance production	Mutations in bacteria SCE in tissue-culture cells	68
Wastewater from municipal incinerator	Mutations in bacteria	33
Coke-oven effluents	Mutations in bacteria	24
Run-off and leachate from soil polluted with petroleum sludge	Mutations and DNA repair in bacteria	6
Municipal wastewater and effluents of wastewater plants	Mutations in bacteria	28,57,58, 69,70,71
Various effluents and wastewaters	Chromosomal aberrations, micronuclei and SCE in tissue-culture cells	62

mutagenic effects in aquatic environment. Many studies point out that both pollution and effects are not uncommon phenomena (see Ref. 23,68,72,73 for reviews). One could say that the signals from the environment urge us to answer the questions mentioned in the foregoing sections of this paper; detailed review lies beyond the scope of this paper. A short overview will suffice to emphasize the weight of this additional argument. A schematic summary of the available evidence is given in Fig. 1. A more detailed listing of the available evidence is presented in Tab. 5 to 8. Anthropogenic sources of mutagens to rivers, lakes, ponds, and the sea have been detected. Also, here dilution is not the solution to pollution, because our sensitive detection techniques, based on concentration with XAD-resins or otherwise, and the flexible and sensitive short-term tests for mutagenicity (in particular, Salmonella/microsome test) enable us to demonstrate the presence of mutagens in surface water. However, the very sensitivity, the high concentration factors necessary to achieve significant effects, might lead to the question: Do these effects bear any relevance? If we concentrate far enough we might be able to detect mutagenicity everywhere. What is the relevance of the effect induced by the equivalent of a gallon per petri dish?

Tab. 6. Mutagenicity of surface waters detected with short-term tests

Surface Water	Effect	Reference
River Rhine, the Netherlands	Mutations in bacteria	19,45,76
River Rhine, Germany	Mutations in bacteria	58
River Meuse, the Netherlands	Mutations in bacteria	45,76
River Meuse, Belgium	Mutations in bacteria	24
Sediment of Black River, Ohio	Mutations in bacteria UDS in primary rat hepatocarcinoma	86
Nisitakase River, Japan	Mutations in bacteria	54
Creek receiving wastewater from nuclear-fuel processing plant	Mutations in bacteria	64
Katsura River, Japan	Mutations in bacteria	55
Sediment of Tama River, Japan	Mutations in bacteria	79
Cai River, Brazil	Mutations in bacteria	82
Lake Kinnereth, Israel	Mutations in bacteria	20
Two lakes in Illinois	Mutations in bacteria	22
Sediment of Puget Sound, Washington	Chromosomal aberrations	41

The organisms in the field have showed us that these exposure levels are indeed relevant! They themselves may concentrate the mutagens,
as is demonstrated by the fact that the bile fluid of fish and extracts of
mussel tissue are clearly mutagenic in short-term tests. Furthermore, a
number of studies have revealed that the external and/or the internal exposure levels are high enough to induce effects in aquatic organisms.
Suitable indicator organisms can be introduced into the polluted environment for monitoring purposes. This approach can be called "active
biomonitoring." When organisms living in the polluted environment are
compared with their congeners living in clear environments, one speaks of
passive or in situ biomonitoring, because these organisms are not introduced on purpose. We believe that the available evidence for pollution
and effects is quite alarming. Rivers and other surface waters can be so
heavily polluted that the accumulation of genetic damage can be demonstrated in the organisms bound to these habitats. Extrapolation from
high exposure levels in the experiment to low levels in situ is not necessary; it looks as if the toxicological experiment is carried out in situ.

Tab. 7.   Bioconcentration of mutagens by aquatic organisms

Organism	Habitat	Effect	Reference
Mytilus edulis, mussel	Coastal waters, Wales	Mutations in bacteria	32,65
Shrimps, algae, and clams	Coastal waters, Kuwait	Mutations in bacteria	31
Larus argentatus, herring gull, eggs	Great Lakes region	SCE and chromosomal aberrations in tissue culture cells	16
Fish	Sheep River, Alberta in sewage plume	Mutations in bacteria	63
Nibea mitsukunii, spotted sea trout liver	Coastal waters, Japan	Mutations and DNA repair in bacteria	36
Abramis brama, bream bile fluid	River Rhine	Mutations in bacteria	46

TUMOR-LIKE LESIONS IN SHELLFISH AND FISH

The mutagenic effects found in situ do not prove that the presence of mutagensis actually affecting fitness, survival, and reproduction of the exposed organisms and in this way (or in another?) the quality of the ecosystem.   Our experience tells us that we have to expect deleterious effects to occur when a mutagen reaches high concentrations in tissues or when mutations or cytogenic effects are so significantly induced.   Can we indeed find such effects in situ, effects that can be unambiguously at-

Tab. 8.   Mutagenic effects in aquatic organisms exposed in situ

Organism	Habitat	Effect	Reference
Umbra pygmeae, eastern mud minnow	River Rhine, the Netherlands	Chromosomal aberrations SCE	25,67 28
Osmunda regalis, fern	Millers River, Massachusetts	Chromosomal abberrations Mutations	37 38
Mytilus edulis, mussel	Dock of Swansea, Wales	Aneuploidy SCE in developing eggs	15 7

tributed to the presence of mutagens? Here we are confronted with the problem that it will often be difficult to distinguish effects caused by interactions with the genetic material from those based on other mechanisms. How to prove that reduced reproduction rate is not caused by a nonmutagenic compound? To this end we can investigate whether mutagens are present and whether their concentrations or their mutagenic effects correlate with the deleterious effect. Ultimate proof has to come from laboratory experiments. These allow us to vary the compounds that the organism is exposed to in situ, as well as the exposure conditions. Their results will indicate whether or not the effects encountered in situ can indeed be induced by the exposure conditions in situ.

In this last section of the paper, we will touch upon the only deleterious effect encountered in situ that is generally directly associated with the presence of mutagens, i.e., cancer. The phrase "generally directly associated" has to be emphasized because this effect is not necessarily the result of the presence of mutagens. Carcinogenesis is a complex process and the role of mutagens in this process is largely confined to the first step, i.e, the initiation. Nonmutagenic compounds can be involved in other steps. It lies beyond the scope of this paper to go into details on this subject; nevertheless, it may be hypothesized that pollution of the aquatic environment with nonmutagenic compounds may also lead to an increase of tumor incidence. So the tumors and tumor-like lesions found by many investigators in shellfish and fish (see Ref. 12 and 60 for reviews) do not necessarily signify the presence of mutagens. Evidence for such a link can only be obtained with the correlation studies and laboratory experiments mentioned above. In fact, up to now evidence is only scanty.

Mix (60,61) has reviewed the literature on the occurrence of tumors and tumor-like lesions in shellfish (clams, mussels, and oysters). None of the studies demonstrated a clear link between effect and pollution. Often the concentrations of a very limited number of mutagens/carcinogens were determined in the tissues of shellfish; in most cases unsubstituted polcyclic aromatic hydrocarbons (PAHs). Results to date do not point to correlations between these concentrations and tumor incidence.

The studies for fish are reviewed by Mix (60), Malins et al. (53), and McCain (56). In this case, a number of studies do indicate a link between pollution and effect or even between pollution with mutagens/carcinogens and effects. High incidences of hepatic neoplasms have been shown to occur in bottom-dwelling fish living in waters adjacent to various urban areas in the U.S.A. (West coast: Puget Sound near Seattle and Southern California; East coast: Boston Harbor and the Great Lakes Region). The Puget Sound study, which covered a period of five years, showed statistically significant correlations between, on one hand, hepatic neoplasm in the English sole (Parophizus vetulus) and, on the other hand, sediment concentrations of aromatic hydrocarbons (characteristic for creosote) and concentrations of metabolites of aromatic compounds in the bile.

Recently, evidence for DNA adducts (e.g., 6-oxybenzo(a)pyrene) in the liver of the English sole was obtained with ESR spectrometry. Adducts were only found in fish with lesions from Puget Sound. So, convincing and rather complete evidence was obtained in this study for the induction of tumors in fish by matagenic/carcinogenic pollutants. Good evidence is also provided by a number of other studies (see reviews mentioned above). Besides these studies, two other groups can be discerned: studies whose experimental design does not allow unambiguous conclusions, and studies which provide impressive evidence that the presence of pollutants is not associated with tumor growth in fish, even in those habitats known to contain mutagenic substances. Extensive and thorough surveys conducted in heavily polluted aquatic environments in Yugoslavia (Sava River), Germany (Rivers Rhine and Elbe), and Australia (Port Phillip Bay) all gave negative results (60). In The Netherlands, Slooff (75) found tumors in eight fish out of a total of nearly 8,000 collected from the Rivers Rhine and Meuse. Seven (1.7%) came from the Rhine and one (0.7%) from the Meuse which may reflect the fact that the Rhine is much more heavily polluted with mutagens and carcinogens than the Meuse. No tumor-bearing fish were found in a clean control environment (Lake Braassem). Nevertheless, this study shows that even in a river heavily polluted with mutagens and carcinogens, tumor incidence is low. Table 8 shows the concentrations of mutagens to be high enough to induce cytogenic effects in fish. Vethaak (84) collected 15,600 fish from Dutch coastal water. He also found low tumor incidences (0.1 to 1.7% for three different species). Tumor incidence showed no link with sampling sites (extent of pollutions) or with other fish diseases.

Altogether the conclusion seems to be justified that in some cases pollution with mutagens seems to lead to an observable increase of tumor incidence in aquatic animals. However, the levels of pollution necessary to induce a significant increase are so high (see, for instance, the Rhine-Meuse study), that it is questionable whether this deleterious effect provides us with an argument for the inclusion of mutagenicity as a criterion in ecotoxicological evaluations. Such high levels of pollution may also result in effects that can be detected by tests based on criteria other than mutagenicity.

REFERENCES

1. Albert, R.E., J. Lewtas, S. Nesnow, T.W. Thorslund, and E. Anderson (1983) Comparative potency method for cancer risk assessment: Application to diesel particulate emissions. Risk Analysis 3:101-117.
2. Alink, G.M., E.M.H. Frederix-Wolters, M.A. van der Gaag, J.F. van de Kerkhoff and C.L.M. Poels (1980) Induction of sister chromatic exchanges in fish exposed to Rhinewater. Mutat. Res. 78:369-374.
3. Bayne, B.L., D.A. Brown, K. Burns, D.R. Dixon, A. Ivanovici, D.R. Livingstone, D.M. Lower, M.N. Moore, A.R.D. Stebbing, and J. Widdows (1985) The Effects of Stress and Pollution on Marine Animals. Praeger Press, New York.
4. Bergman, H.L., R.A. Kimerle, and A.W. Maki (1986) Environmental Hazard Assessment of Effluents. SETAC Special Publication Series. Pergamon Press, New York.
5. Bitton, G., and B.J. Dutka, eds. (1986) T_____ _____ ___ ___ croorganisms, Vol. 1. CRC Press, Inc., Boca Raton, Florida.
6. Brown, K.W., and K.C. Donnelly (1984) Mutagenic activity of run-off and leachate water from hazardous waste land treatment. Envir. Poll. 35:229-246.

7. Brunetti, R., J. Gola, and F. Majone (1986) Sister-chromatid exchange in developing eggs of Mytilus galloprovincialis Bivalvia. Mutat. Res. 174:207-212.

8. Cairns, J., K.L. Dickson, and A.W. Maki, eds. (1978) Estimating the Hazard of Chemical Substances to Aquatic Life, American Society for Testing and Materials (ASTM), Philadelphia.

9. Connell, D.W. and G.J. Miller (1984) Chemistry and Ecotoxicology of Pollution, Environmental Science and Technology Series, Wiley Interscience, New York.

10. Conway, R.A., ed. (1982) Environmental Risk Analysis for Chemicals. Van Nostr and Reinhold Environmental Engineering Series, Van Nostrand Reinhold, New York.

11. Conway, R.A. (1982) Introduction to environmental risk analysis. In Environmental Risk Analysis for Chemicals, R.A. Conway, ed., Van Nostr and Reinhold, New York.

12. Couch, J.A., W.P. Schoor, L. Courtney, and W. Davis (1984) Effects of Carcinogens, Mutagens and Teratogens on Non-Human Species (Aquatic Animals). Final Report, Project 3, NCI/EPA Collaborative Program, U.S. EPA, Gulf Breeze, Florida.

13. Dixon, D.R., and K.R. Clarke (1982) Sister-chromatid exchange: A sensitive method for detecting damage caused by exposure to environmental mutagens in the chromosomes of adult Mytilus edulis. Mar. Biol. Lett. 3:163-172.

14. Dixon, D.R. (1986) Promotoceres triqueter: A test system for environmental mutagenesis. In The Effects of Stress and Pollution on Marine Animals. Praeger Press, New York.

15. Dixon, D.R. (1982) Aneuploidy in mussel embryos (Mytilusedulis) originating from a polluted dock. Mar. Biol. Lett. 3:155-161.

16. Ellenton, J.A., M.F. McPherson, and K.L. Mans (1983) Mutagenicity studies on herring gulls from different locations on the Great Lakes. 2. Mutagenic evaluation of extracts of herring gull eggs in a battery of in-vitro mammalian and microbial tests. J. Toxic. Env. Health 12:325-336.

17. Fang, T., T.-H. Ma, G. Lin, J. Ho, J. Dai, R. Zhou, D. Chen, Y. Ou, and J. Cui (1983) Preliminary report on the development of a bioassay for detection on mutagens in sea water. Envir. Exp. Bot. 23:303-310.

18. Gaag, M.A. van der, A. Noordsij, and J.P. Oranje (1982) Presence of mutagens in Dutch surface water and effects of water treatment processes for drinking water preparation. In Mutagens in Our Environment, M. Sorsa and H. Vainio, eds. Alan R. Liss, New York.

19. Gaag, M.A. van der, A. Noordsij, and C.L.M. Poels (1982) Mutagenicity of River Rhinewater in The Netherlands and influence of several water treatment processes on the mutagenic effect. Mutat. Res. 97:231-232.

20. Guttmann-Bass, N., M. Bairey-Albuqeurque, S. Ulitzur, A. Chartrand, and C. Rav-Acha (1987) Effects of chlorine and chlorine dioxide on mutagenic activity of Lake Kinnereth water. Env. Sci. Tech. 21:252-260.

21. Harrison, F.L., and J.M. Jones (1982) An in-vivo sister-chromatid exchange assay in the larvae of the mussel Mytilus edulis: Response to three mutagens. Mutat. Res. 105:235-242.

22. Heartlein, M.W., D.M. Demarini, A.J. Katz, J.C. Means, M.J. Plewa, and H.E. Brochman (1981) Mutagenicity of municipal water obtained from an agricultural area. Env. Mut. 3:519-530.

23. Hoffmann, G.R. (1982) Mutagenicity testing in environmental toxicology. Env. Sci. Tech. 16:560A-574A.
24. Hoof, F., van and J. Verheyden (1981) Mutagenic activity in the Meuse River in Belgium. Sci. Total Env. 20:15-22.
25. Hooftman, R.N., and G.J. Vink (1981) Cytogenetic effects on the eastern mudminnow, Umbrapygmeae, exposed to ethylmethanesulphonate, benzo(a)pyrene and river water. Ecotox. Env. Safety 5:261-269.
26. Hooftman, R.N., and W.K. de Raat (1982) Induction of nuclear anomalies (micronuclei) in the peripheral blood erythrocytes of the eastern mudminnow Umbra pygmaeaby ethylmethanesulphonate. Mutat. Res. 104:147-152.
27. Hooftman, R.N. (1981) The induction of chromosome aberrations in Notobranchius rachowi (Pisces: Cyprinodontidae) after treatment with ethylmethanesulphonate or benzo(a)pyrene. Mutat. Res. 191:347.
28. Hopke, P.K., M.J. Plewa, and P. Stapleton (1987) Reduction of mutagenicity of municipal waste waters by land treatment. Sci. Total Env. 66:193-202.
29. Hueck-van derPlas, E.H., ed. (1981) Disposal of chemical waste in the marine environment. Implications of the International Dumping Conventions. Chemosphere 10(6).
30. ICPEMC (1988) Testing for mutagens and carcinogens: The role of short-term genotoxicity assays. ICPEMC Publication No. 16. Mutat. Res. 205:3-12 (and other papers in this issue of Mutation Research).
31. Jabar, M.A., M.A. Salama, and A. Salem (1984) The detection of mutagenic pollutants in marine organisms of Kuwait. Water Air Soil Poll. 22:131-142.
32. Kadhim, M., and J.M. Parry (1984) The detection of mutagenic chemicals in the tissues of shellfish exposed to oil pollution. Mutat. Res. 136:93-105.
33. Kamiya, A., and J. Ose (1987) Study of the behaviour of mutagens in waste water and emission gas from a municipal incinerator evaluated by means of the Ames assay. Sci. Total Env. 65:109-120.
34. Karim Ahmed, A., and G.S. Domiquez (1982) The development of testing requirements under the toxic substances control act. In Environmental Risk Analysis for Chemicals, R.A. Conway, ed., Van Nostr and Reinhold, New York.
35. Kelly, J.J., and M.B. Maddock (1985) In-vitro induction of unscheduled DNA synthesis by genotoxic carcinogens in the hepatocytes of the oyster toadfish Opsanus-tau. Arch. Env. Contam. Tox. 14:555-564.
36. Kinae, N., T. Hashizume, T. Nakita, I. Tomita, I. Kimura, and H. Kanamori (1981) The toxicity of pulp and paper mill effluents. 2. Mutagenicity of the extracts of the liver from spotted sea trout Nibea mitsukurjii. Water Res. 15:25-30.
37. Klekowski, E.J., and D.M. Poppel (1976) Ferns potential in-situ bioassay systems for aquatic-borne mutagens. Am. Fern. J. 66:75-79.
38. Klekowski, E.J. (1976) Mutational load in a fern population growing in a polluted environment. Am. J. Bot. 63:1024-1030.
39. Kligerman, A.D., S.E. Bloom, and W.M. Howell (1975) Umbralimi: A model for the study of chromosome aberrations in fishes. Mutat. Res. 31:225.
40. Kligerman, A.D. (1979) Induction of sister-chromatid exchanges in

the central mudminnow following in vivo exposure to mutagenic agents. Mutat. Res. 64:205.

41. Kocan, R.M., K.M. Sabo, and M.L. Landolt (1985) Cytotoxicity/genotoxicity: The application of cell culture techniques to measurement of marine sediment pollution. AquaticTox. 6:165-177.

42. Kooijman, S.A.L.M. (1985) Toxicity at population level. In Proceedings of the ESA/SETAC Symposium on Multispecies Toxicity Testing, J. Cairns, ed. Pergamon, New York.

43. Koojiman, S.A.L.M. (1984) On the dynamics of chemically stressed populations: The deduction of population consequences from effects on individuals. Ecotox. Env. Safety 8:254.

44. Kooijman, S.A.L.M. (1987) A safety factor for LC50 values allowing for differences in sensitivity among species. Water Res. 3:269-276.

45. Kreijl, C.F. van, H.J. Kool, M. de Vries, H.J. Kranen, and E. de Greef (1980) Mutagenic activity in the Rivers Rhine and Meuse in The Netherlands. Sci. Total Env. 15:137-147.

46. Kreijl, C.F. van, A.C. van den Burg, and W. Sloof (1982) Accumulation of mutagenic activity in bile fluid of River Rhine fish. In Mutagens in Our Environment, M. Sorsa and H. Vainio, eds. Alan R. Liss, New York.

47. Kringstad, K.P., P.O. Ljungquist, F. de Sousa, and L.M. Stromberg (1981) Identification and mutagenic properties of some chlorinated aliphatic compounds in the spent liquor from kraft pulp chlorination. Env. Sci. Tech. 15:562-566.

48. Kuiper, J. (1982) The use of enclosed plankton communities in aquatic ecotoxicology. Thesis, Agricultural University, Wageningen, The Netherlands.

49. Langi, A., and M. Priha (1988) Mutagenicity in pulp and paper mill effluents and in recipient. Water Sci. and Tech. 20:143-152.

50. Leisinger, T., A.M. Cook, R. Hutter, and J. Nuesch, eds. (1981) Microbial Degradation of Xenobiotics and Recalcitrant Compounds. Academic Press, London.

51. Maddock, M.B., H. Northup, and T.J. Ellingham (1986) Induction of sister-chromatid exchanges and chromosomal abberations in hematopoietic tissue of a marine fish following in-vivo exposure to genotoxic carcinogens. Mutat. Res. 172:165-176.

52. Maki, A.W., K.L. Dickson, and J.Cairns, Jr., eds. (1980) Biotransformation and Fate of Chemicals in the Aquatic Environment. American Society for Microbiology, Washington, D.C.

53. Malins, D.C., B.B. McClain, J.T. Landahl, M.S. Myers, M.M. Krahn, D.W. Brown, S.-L. Chan, and W.T. Roubal (1988) Neoplastic and other diseases in fish in relation to toxic chemicals: An Overview. Aquatic Tox. 11:43-67.

54. Maruoka, S., S. Yamanaka, and Y.Yamamoto (1986) Isolation of mutagenic components by high-performance liquid chromatography from XAD extract of water from the Nishitakase River, Kyoto City, Japan. Sci. Total Env. 57:29-38.

55. Maruoka, S., and S. Yamanaka (1982) Mutagenicity in Salmonella typhimurium tester strains of XAD-2 resin ether extracts recovered from Katsura River water in Kyoto City, Japan and its fractions. Mutat. Res. 102:13-26.

56. McCain, B.B., D.W. Brown, M.M. Krahn, M.S. Myers, R.C. Clark, Jr., S.-L. Chan, and D.C. Malins (1988) Marine pollution problems, North American West coast. Aquat. Tox. 11:143-162.

57. Meier, J.R., W.F. Blazak, E.S. Ricio, B.E. Stewart, D.F. Bishop, and L.W. Condie (1987) Genotoxic properties of municipal waste waters in Ohio. Arch. Env. Cont. Tox. 16:671-680.

58. Mersch-Sundermann, V., N. Dickgiesser, K. Koetter, and M. Harre (1988) The mutagenicity of surface water, waste water and drinking water in the Rhine-Neckar region with the Salmonella microsome Ames-test. Zentralbl. Bakteriol. Mikrobiol. Hyg. Ser. B 185:397-410.

59. Miller, E.C., and J.A. Miller, 1972 The mutagenicity of chemical carcinogens: Correlation, Problems and Interpretations. In Chemical Mutagens, Vol. I A. Hollaender, ed., Plenum Publishing Corp., New York.

60. Mix, M.C. (1986) Cancerous diseases in aquatic animals and their association with environmental pollutants: A critical literature review. Mar. Env. Res. (special issue), 20:1-141.

61. Mix, M.C. (1988) Shellfish diseases in relation to toxic chemicals. Aquat. Tox. 11:29-42.

62. Muellerschoen, H., and H.G. Miltenburger (1988) Detection of mutagens in water samples using the cell line V79. Vom Wasser. 71:195-206.

63. Osborne, L.L., R.W. Davies, K.R. Dixon, and R.L. Moore (1982) Mutagenic activity of fish and sediments in the Sheep River Alberta Canada. Water Res. 16:889-902.

64. Pancorbo, O.C., P.J. Lein, and R.D. Blevins (1987) Mutagenic activity of surface waters adjacent to a nuclear fuel processing facility. Arch. Env. Contam. Tox. 16:531-537.

65. Parry, J.M., D.J. Tweats, and MA.J. Al-Mossawi (1976) Monitoring the marine environment for mutagens. Nature 246:538-540.

66. Pesch, G.G., C.E. Pesch, and A.R. Malcolm (1981) Neanthes arenaceondentata, a cytogenetic model for marine genetic toxicology. Aquat. Tox. 1:301-312.

67. Prein, A.E., G.M. Thie, G.M. Alink, J.H. Koeman, and C.L.M. Poels (1978) Cytogenic changes in fish exposed to water of the River Rhine. Sci. Total. Env. 9:287-292.

68. Raat, W.K. de, A.O. Hanstveit, and J.F. de Kreuk (1985) The role of mutagenicity testing in the ecotoxicological evaluation of industrial discharges into the aquatic environment. Food & Chem. Tox. 23:33-41.

69. Rappaport, S.M., M.G. Richard, M.C. Hollstein, and R.E. Talcott (1979) Mutagenic activity in organic waste water concentrates. Env. Sci. Tech. 13:957-961.

70. Reinhard, M., M. Goodman, and K.E. Mortelmans (1982) Occurrence of brominated alkylphenol polyethoxy carboxylates in mutagenic waste water concentrates. Env. Sci. Tech. 16:351-362.

71. Saxena, J., D.J. Schwartz, and M.W. Neal (1979) Occurrence of mutagens/carcinogens in municipal waste waters and their removal during advanced waste water treatment. In Proceedings of Water Reuse Symposium, AWWA Research Foundation, Denver, Colorado, Vol. 3, 2209-2229.

72. Sims, R.C., J.L. Sims, and R.R. Dupont (1988) Human health effect assays. J. Water Poll. Cont. Fed. 16:1093-1106.

73. Sims, R.C., J.L. Sims, and R.R. Dupont (1985) Human health effect assays. J. Water Poll. Cont. Fed. 57:728-742.

74. Slooff, W., J.A.M. van Oers, and D. de Zwart (1986) Margins of uncertainty in ecotoxicological hazard assessment Env. Tox. Chem 5:841-852.

75. Slooff, W. (1983) A study on the usefulness of feral fish as indicators for the presence of chemical carcinogens in Dutch surface waters. Aquat. Tox. 3:127-139.

76. Slooff, W., and C.F. van Kreijl (1982) Monitoring the Rivers Rhine and Meuse in The Netherlands for mutagenic activity using the Ames test in combination with rat or fish liver homogenates. Aquatic Tox. 2:89-98.

77. Spacie, A., and J.L. Hamelink (1985) Bioaccumulation. In Fundamentals of Aquatic Toxicology, G.M. Randand S.R. Petrocelli, eds., Hemisphere, Washington, D.C.

78. Swanson, S., and H. Peterson (1988) Development of Guidelines for Testing Pesticide Toxicity to Non-target Plants. Environment Canada, SRC Publication No. E-901-20-E-88.

79. Suzuki, J., T. Sadamasu, and S. Suzuki (1982) Mutagenic activity of organic matter in urban river sediment. Env. Pollut., Ser. A, Ecol. Biol. 29:91-100.

80. Timourian, H., J.S. Felton, D.H. Stuermer, S. Healy, P. Berry, and M. Tompkins (1982) Mutagenic and toxic activity of environmental effluents from underground coal gasification experiments. J. Tox. Env. Health 9:975-994.

81. TNO (the Dutch Organization for Applied Scientific Research) (1980) Degradability, Ecotoxicity and Bioaccumulation. The Determination of the Possible Effects of Chemicals and Wastes on the Aquatic Environment. Government Publishing Office, The Hague.

82. Vargas, V.M., V.E.P. Motta, and J.A.P. Henriques (1988) Analysis of mutagenicity of waters under the influence of petrochemical industrial complexes by the Salmonella-microsome test. Rev. Bras. Genet. 11:505-518.

83. Venegas, W., I. Hermosilla, J.F. Gavilan, R. Naveas, and P. Carrasco (1987) Larval stages of the anuran amphibian Caudiverbera caudiverbera: A biological model for studies of genotoxic agents. Bol. Soc. Biol. Conception 58:171-180.

84. Vethaak, A.D. (1985) Inventory to the Presence of Dish Diseases in Relation to Pollution in Dutch Coastal Waters (in Dutch). Report of the Netherlands Institute for Fishery Research. IJmuiden, No. CA 85-01.

85. Walden, C.C. (1981) Biological effects on pulp and paper-mill effluents. In Advances in Biotechnology, Vol. 2., M. Moo-Young and C.W. Robinson, eds., Pergamon Press, Canada.

86. West, W.R., P.A. Smith, G.M. Booth, and M.L. Lee (1988) Isolation and detection of genotoxic components in a Black River sediment. Env. Sci. Tech. 22:224-228.

87. Zahn, R.K., G. Zahn-Daimler, W.E.G. Muller, M.L. Michaelis, B. Kurelec, M. Rijavec, R. Batel, and N. Bihari (1983) DNA damage by PAH and repair in a marine sponge. Sci. Total. Env. 26:137-156.

DNA ADDUCTS AND RELATED BIOMARKERS IN

ASSESSING THE RISK OF COMPLEX MIXTURES

F.P. Perera,[1] P. Schulte,[2] R.M. Santella,[1] and
D. Brenner[3]

[1]Columbia University School of Public Health
New York, New York

[2]National Institute of Occupational Safety and Health
Cincinnati, Ohio

[3]Environmental Health Institute
Pittsfield, Massachusetts

INTRODUCTION

Risk assessment for complex mixtures is not merely a theoretical problem but a "real-life" one. Most common human exposures involve multiple chemicals and agents in combination. To illustrate, cigarette smoke contains more than 3,800 different chemicals including at least 40 carcinogens (i.e., known human or animal carcinogens) (41). More than 300 different toxic chemicals, including approximately 50 carcinogens, have been detected both in ambient air (18,56) and in drinking water (29).

Among the 387 pesticides currently registered for use on food, 66 are carcinogenic (17,26). The food supply also contains an unknown number of naturally occurring carcinogenic substances (1). A large proportion of the more than 50 carcinogens identified as having sufficient evidence of carcinogenicity in humans or laboratory animals by the International Agency for Research on Cancer (IARC) have been measured in the workplace, often in combination (40).

There is no question that these real-life mixed exposures contribute significantly to the burden of human cancer--460,000 cancer deaths per year in the United States of which 120,000 are deaths from lung cancer (10). An estimated 80% of these cancers are considered to be preventable in that they are associated with the environmental exposures/factors discussed above. While lifestyle-related environmental exposures are major contributors, involuntary exposures to man-made and industrial chemicals

Genetic Toxicology of Complex Mixtures
Edited by M. D. Waters *et al.*
Plenum Press, New York, 1990

are also significant (64). Therefore, the public health community faces a major challenge: to identify environmental hazards promptly and to more accurately assess their risk to humans. Only in this way can effective preventive strategies be designed to reduce the large burden of cancer.

In this effort, there are several key questions that need answers. First, "do interactions occur between individual components of the mixture?" Despite the importance of this question, in terms of public health, even the most conservative models used in quantitative risk assessment assume an additive relationship between concurrent exposures and do not factor in the possibility of synergism. Second, "what is the extent of individual variation in biological response to complex mixtures--hence in cancer risk?" This, of course, is the crux of cancer prevention--to identify and protect that subset of the population that is most at risk from environmental exposures. Yet again, even so-called conservative models assume that the human population is fairly homogeneous in its biological response to carcinogens. And the related question: "What does the low does-response curve look like for a population? Is it linear as has generally been assumed? Sublinear? Supralinear?"

Conventionally, these questions have been pursued separately by laboratory investigators and epidemiologists. Now, it is becoming possible to use biological markers to bridge the gap between the two disciplines, thus merging sensitive laboratory methods with analytical epidemiology. This approach can provide quantitative measures directly in humans of the biologically effective dose and effect of carcinogens, and even of susceptibility to exposure (through indices of genetic and/or acquired factors which influence an individual's response to carcinogens). Therefore, biological markers and molecular epidemiology have considerable potential as a tool in risk assessment and, ultimately, in cancer prevention.

The purpose of this chapter is to critically evaluate the role of biological markers and molecular epidemiology in the risk assessment of environmental carcinogens, particularly of complex mixtures. Examples of recent studies assessing macromolecular adducts formed by carcinogens will be used to illustrate the potential and the current limitations of this approach. The following two sections present the conceptual framework of biological markers.

CATEGORIES OF BIOLOGICAL MARKERS

This and the following sections are based on a recent review (59).

During the past decade, sophisticated laboratory techniques have been developed that are capable of detecting exposures to very low levels of pollutants and of assessing the behavior, fate, and effect of these xenobiotics at the cellular/molecular level. Their availability has stimulated interest in the incorporation of biological markers into environmental research, particularly the study of somatic effects from exposure to environmental carcinogens and mutagens. (For in-depth discussion of biomarkers, interested readers can turn to Ref. 25, 30, 31, 38, 53, 62, 63, 75, and 76.)

Exposure can be defined as the amount of a pollutant substance with which a subject has contact for a specific time interval. In order for an exposure to result in an adverse health effect, it must generate a chain of subsequent events, each of which can, in theory, be reflected by biological markers. Thus, biological markers are alternations occurring at the biochemical, cellular, or molecular level on the continuum between exposure and disease, which can be measured by assays performed on body fluids, cells, or tissues.

Conventionally, three broad categories of biomarkers have been distinguished: Exposure or dose, effect, and susceptibility. A distinction can be made between a marker of exposure (which gives qualitative information as to whether an individual was exposed) and a marker of dose (which provides a quantitative measure).

Markers of Internal Dose

Markers of internal dose directly measure levels of the parent compound, its metabolites or derivatives in cells, tissues, or body fluids. Sensitive physicochemical and immunological methods are capable of detecting and quantifying very low levels of xenobiotic substances in the body. Measurements are usually made on exhaled air, blood, and urine although other body fluids such as breast milk, semen, and adipose tissue have been utilized as well. Each biological medium has a different and unique relevance both to exposure and to health outcome, which affects interpretation of results.

Biological markers of internal dose can be characterized according to their chemical-specificity/selectivity (48), with selective markers representing measures of unchanged pollutants of their metabolites detected in biological media. Examples of selective internal dosimeters include blood levels of styrene, pesticides, metals, exhaled volatile organic chemicals (VOCs), concentrations of polychlorinated biphenyls (PCBs), DDE (a metabolite of DDT), and TCDD (dioxin) in adipose tissue; and urinary mandelic acid resulting from styrene exposure. Various nonselective markers such as urinary excretion of thioethers and mutagenicity of urine and other body fluids have also been assessed in humans, the latter, fairly widely. Disadvantages of internal dosimeters are that, while the laboratory methods may be highly sensitive, in the absence of bioaccumulation, markers reflect only recent exposure. This can be a limitation in cases where exposure has been interrupted or where information on past exposures is sought. More importantly, internal dosimeters do not reflect actual interactions with critical macromolecules in target cells (50).

Markers of Biologically Effective Dose

Markers in this category measure the amount of pollutant or active metabolites which has interacted with cellular macromolecules at a target site or an established surrogate. Measures of biologically effective dose include DNA and hemoglobin adducts in peripheral blood and other cells and tissues (e.g., lung macrophages, buccal mucosa, bone marrow, placental tissue, lung tissue). Many carcinogens are metabolically activated to electrophilic metabolites that covalently bind to DNA. Adducts on DNA, if unrepaired and occurring at critical sites, can cause gene mutation which has been shown to be a critical initiating step in the multistage carcinogenic process. Several methods to detect DNA-chemical adducts in lymphocytes and target tissues are currently available, including radio- or enzyme-linked immunoassays utilizing polyclonal or monoclonal antibodies, ^{32}P-postlabeling, and synchronous fluorescence spectrophotometry (74).

For example, antibodies were first applied to lung tissue and peripheral white blood cells (WBCs) from lung cancer patients and controls to demonstrate the presence of PAH-DNA adducts (62). Subsequently, they have been used to analyze WBCs and other tissues of individuals exposed to PAHs in cigarette smoke and in occupational settings (e.g., foundries and coke ovens) (30,32,34,60,77). Examples of antibodies available to assess formation of DNA adducts in humans by chemicals include those to aflatoxin B_1, alkylating agents, 4-aminobiphenyl, benzo(a)pyrene (BaP) and PAHs, cisplatinum, and 8-methoxypsoralen. Immunoassays are able to detect as few as one adduct per 10^8 nucleotides. While assays using monoclonal antibodies are highly specific, those with polyclonals, such as the PAHDNA antibody, can show cross-reactivity with multiple structurally related compounds. Immunoassays are dependent on the development of appropriate antibodies, a time-consuming and technically difficult procedure.

The ^{32}P-postlabeling technique is highly sensitive, as it is able to measure one adduct per 10^{10} nucleotides (70). It has been applied to the identification of a range of alkylating and methylating agents (71). The method has also been used to assess BaP/PAH exposure of foundry workers, coke oven workers, and roofers (33,34,36,69). Some investigators have related adducts in placenta and lung measured by ^{32}P-postlabeling to cigarette smoking (20,68); others have not seen smoker/nonsmoker differences in bone marrow, WBC, or buccal mucosa (16,67). While the method produces images that are considered an idiosyncratic "fingerprint" of the exposure, researchers involved with developing this assay have faced difficulties in identifying the adducts formed.

A third approach, synchronous fluoresence spectrophotometry, has recently been applied to coke oven workers with a reported sensitivity of one BaP adduct per 10^7 nucleotides (86). The method has also confirmed the presence of BaP-DNA adducts in placental tissue from smokers (92). High pressure liquid chromatography (HPLC) and fluorescence spectroscopy have been used to detect excised carcinogen-DNA adducts in urine (2).

Assays which measure carcinogen-protein adducts, especially those formed by the binding of metabolites with hemoglobin, can in some cases represent a good surrogate for DNA-adduct measurements. Their use in this capacity is supported by correlations in animal studies between protein and DNA binding by BaP, vinyl chloride, ethylene oxide (EtO), acetylaminofluorene, methyl methanesulfonate, and trans-4-dimethylaminostilbene (39,42,54). Methods available for measuring these adducts include immunoassays, gas chromatography-mass spectrometry (GC/MS), ion-exchange amino acid analysis, and negative chemical ionization mass spectrometry (NCIMS). This last method has been successfully applied to the quantitation of 4-aminobiphenyl-hemoglobin (4-ABP-Hb) adducts in smokers (8,61, 89). Sensitive GC/MS methods exist for measuring protein adducts in individuals exposed to ethylene oxide in cigarette smoke and the workplace (22,57,83). Because of the three month lifespan of hemoglobin, these assays reflect relatively recent exposures, in contrast to DNA adducts which represent exposure integrated over a longer time period.

Markers of Early Biological Effect

These markers indicate an event resulting from a toxic interaction, either at the target or an analogous site, which is known or believed to be a step in the pathogenesis of disease or to be qualitatively or quantitatively correlated with the disease process. Whereas markers of biologically effective dose indicate an interaction with critical macromolecules that might potentially lead to disease but that might also be repaired or otherwise "lost," biomarkers of early biological effect indicate the occurrence of irreversible toxic interactions either at the target or an analogous site. Markers of biologically effective dose and early biological effect provide integrated "black box" measurements indicative of the net result of all the biological processes that occur when the body is exposed to a particular pollutant(s) (37). These include pharmacokinetic events occurring on the cellular or systemic level such as absorption, metabolism, detoxification, and elimination, as well as macromolecular processes such as binding, repair, and immune response.

Cytogenetic techniques provide a direct, though nonspecific method of visualizing changes that occur at the chromosomal level. These changes include alterations in chromosome number, structural chromosomal changes such as breakage and rearrangement, and exchanges between reciprocal portions of a single-chromosome (sister chromatid exchanges or SCEs). Elevated frequencies of chromosomal aberrations have been observed in persons exposed to radiation and in workers exposed to benzene, vinyl chloride, and styrene (87). An increased frequency of SCEs has been found in workers exposed to ethylene oxide, styrene, benzene, arsenic, chloromethylether, chloroprene, organophosphates, and ionizing radiation (19). Although SCEs constitute a biological effect, the significance of this increase in relation to disease outcome is unclear. Micronuclei (MN) (fragments of nuclear material left in the cytoplasm following replication) are generally considered an indication of the prior existence of chromosomal aberrations.

Another approach to assessing genetic effects involves measurement of single-strand breaks in lymphocyte DNA. An elevated frequency of breaks was found in workers exposed to styrene (7,90). In addition, DNA hyperploidy measured in exfoliated bladder and lung cells has been shown to be a biological marker of response to carcinogens. DNA hyperploidy has been found to be highly correlated with degree of exposure in workers exposed to aromatic amines (35).

An important new marker is the activated oncogene and its protein product. During oncogenesis, a normal segment of DNA (termed a protooncogene) is activated to a form which causes cells to become malignant. Activation can occur through several mechanisms, including gene mutation, chromosome breaks, and rearrangements. For example, the ras oncogene was first identified in rat sarcoma, but was subsequently seen in human bladder, colon, and lung cancers (79,80). Activated ras oncogenes containing a point mutation have been produced in vitro, by a number of ambient pollutants including BaP, dimethylbenzanthracene, and N-nitroso compounds (82). Activation of the ras oncogene can be measured by complex immunoblotting techniques. A more convenient and relatively simple method involves measurement of the gene's abnormal protein product, designated P_{21}. P_{21} can be measured in tissue, sputum, blood, or urine using immunoblotting techniques. The assay, however, is still in the validation stage (5). The polymerase chain reaction (PCR) technique is another new technique allowing the investigation of oncogene alterations in DNA from human tissues (44). PCR allows rapid amplification of DNA and has the advantage of requiring small amounts of tissue. It is likely that more than one oncogene needs to be activated to convert a normal cell to a cancerous one (81). More studies will be needed to understand the sequential requirements in terms of oncogene activation.

Markers of Susceptibility

Markers of susceptibility reflect inherent or acquired differences affecting an individual's response to exposure. These differences can serve as effective modifiers and thereby increase or decrease risk at any point on the continuum between exposure and the emergence of symptomatic disease. Markers of susceptibility may indicate the presence of inherited genetic factors that affect the individual or a population of which he/she is a part. They may also reflect certain host factors, such as lifestyle, activity patterns, prior exposures to environmental toxicants, nutritional status, or prior presence of disease.

Individual variability in cytochrome P-450 metabolism, for example, may explain differences in lung cancer risk, although results of various studies have been conflicting (27,45,46). Enzyme activity as measured by metabolism of indicator drugs such as debrisoquine (debrisoquine/4-hydroxy debrisoquine ratio in urine) and antipyrine (urinary clearance) may be useful in identifying genetic polymorphisms which modulate the effects of exposure (14,30). Increased debrisoquine hydroxylation has been found in patients with bronchial cancer compared to controls, irrespective of cigarette consumption (3). N-acetylation rates, which cause differential metabolism of aromatic amines, may also indicate susceptibility. Cartwright et al. (12) have reported a relationship between bladder cancer risk and slow acetylation in workers exposed to aromatic amines.

Markers of susceptibility may also indicate a pre-existing disease condition that could increase an individual's risk. The presence of certain rare hereditary diseases (e.g., ataxia telangiectasia, Fanconi's anemia, and Bloom's syndrome) may reflect heightened susceptibility to potentially genotoxic ambient exposures (11). Persons who are heterozygous for these genetic traits may also be at elevated risk for cancer (61). If so, genetic factors could have an important effect on population-based rates of cancer incidence (38).

Pre-existing conditions may either affect an individual's metabolic status or establish sites within the DNA where initiating events such as point mutations or translocations may be more likely to occur. Restriction enzyme DNA fragment-length analysis of genetic polymorphism (RFLP) is a new approach utilizing recombinant techniques to detect DNA conformations associated with genetic predisposition to cancers, such as retinoblastoma (24). There is also some recent evidence that genetic factors influence BaP-DNA binding. Elevated levels of adducts have been found in individuals with early-age cancers and those with first degree relatives of lung cancer patients (55,73).

ADVANTAGES AND DISADVANTAGES OF BIOMARKERS

Advantages

Biomarkers can permit greater resolution in environmental research. They can provide ordinal and continuous data, rather than the nominal (categorical) and usually dichotomous data that epidemiologists have relied on in the past to characterize exposure, disease, and confounding variables. This fact increases the power of tests for causality and for interactions between variables. Furthermore, by switching the unit of analysis from the whole person to the cell or molecule, the sample size required to see an association may be reduced (31,76). In addition, biomarkers should be able to provide greater understanding of the mechanistic processes involved in the exposure-disease continuum.

Biomarkers can improve exposure assessment. Central to establishing the causal relationship between exposure to a pollutant and disease risk is the characterization of dose-response relationships. The traditional approach in environmental research has been to rely on indirect or imprecise methods (e.g., historical records of ambient or workplace monitoring, mathematical modeling, questionnaires) to classify individuals regarding the environmental exposure of interest. If misclassification (i.e., the incorrect determination of who is exposed) is random, the study will be biased toward the null hypothesis (38,63). Moreover, the underlying assumption that all individuals experiencing comparable exposure will receive an identical biologically effective dose is demonstrably incorrect as has been shown in human studies of DNA and protein adduct formation as discussed below. Uptake, absorption, and distribution of a chemical can vary according to sex, age, health status, diet, hormonal status, and presence of other environmental exposures (53). Biomarkers not only permit individual exposure assessments, but should allow estimation of the actual dose received by the individual. They can thereby reduce misclassification error and increase the power of epidemiological studies to identify causal associations between exposure and disease. In addition, markers could identify individuals who are high responders in terms of either biologically effective dose or early biological effect and who are, therefore, at potentially elevated risk.

Biological markers can establish quantitative relationships between exposure and dose. The best examples come from recent research on carcinogen-DNA and carcinogen-protein adducts. Extensive data on DNA, RNA, and protein binding in experimental systems indicate that these macromolecular effects at the lowest administered doses generally follow first-order kinetics; i.e., that the degree of initial binding in target organs in vivo is usually directly proportional to administered dose, in some cases even at very low levels similar to those which might be encountered by humans as a result of environmental contamination (54,94).

With respect to human data on adducts, a proportional relationship between exposure and adduct levels ("dose-response") has not always been seen. For example, while levels of 4-ABP were significantly elevated in smokers compared to nonsmokers, there was no significant correlation with amount smoked (8,61). This is undoubtedly because of the wide interindividual variation in handling xenobiotics, the inability to precisely determine individual exposure, and the fact that one is dealing with chronic, low, and variable exposures. However, for ethylene oxide and propylene oxide, unlike PAH and 4-ABP which must be metabolically activated and for which greater interindividual variability would be anticipated, the amount of carcinogen-hemoglobin adducts in humans is expected to be reasonably proportional to the estimated dose received. This was indeed the case in the initial study of ethylene oxide-hemoglobin adducts (9), but a subsequent study did not show a correlation between exposure and hemoglobin adduct levels (88) possibly because air levels of EtO were very low (<0.05 ppm). In workers exposed to propylene oxide, good agreement was seen between the degree of hemoglobin alkylation and estimated propylene oxide exposure (57).

Probably the clearest "dose-response" seen to date has been the significant correlation between PAH-DNA adducts measured by immunoassay in peripheral WBCs from Finnish foundry workers and their occupational exposure to PAH (60). Workers were classified as having high (>0.2 microgram per cubic meter or $\mu g/m^3$), medium (0.05-0.2 $\mu g/m^3$) or low (<0.05 $\mu g/m^3$) exposure to BaP as an indicator PAH. The mean adduct levels (femtomole adduct per μg DNA or fmole/μg) were 1.5 (high exposure group), 0.62 (medium), 0.24 (low), and 0.066 (controls). These results were corroborated by the ^{32}P-postlabeling method carried out on WBC DNA from the same worker population (33,69). Despite the correlation between DNA adducts and exposure on the group level, there was significant variation among individuals within the exposed group. Adduct levels measured by immunoassay ranged from nondetectable to 2.8 fmole/μg.

Biomarkers can provide timely identification of individuals or groups at elevated risk of disease. As discussed, markers of dose and biological effect can provide an early warning of hazard or potential risk of disease. In general, they can signal the need for greater surveillance or even action to reduce exposure. Molecular dosimeters can identify the need for surveillance or follow-up studies. Biomarkers of response/effect indicating altered structure and function should trigger immediate remedial or preventive action. By providing this information in a timely fashion, in some cases decades before clinical disease would have appeared, biomarkers can be a valuable tool in disease prevention.

Biomarkers have the potential to improve risk extrapolation between species and between populations. Parallel studies of biological markers in experimental animals and humans can evaluate whether mechanisms or modes of action are similar across species. They can allow calibration of human measurements of biologically effective dose or early response with those measured in laboratory animals being administered the same substance of mixture. Specifically, comparable data on chromosomal effects, DNA adducts, or somatic cell mutations in a human population for whom the relative risks of cancer were unknown and in experimental animals for whom tumor incidences were established could allow better predictions of human risk. Comparative biomonitoring data in several different human populations could allow extrapolation from one whose relative risks of cancer are established historically (e.g., cigarette smokers, coke oven workers) to a population subjected to similar exposure whose relative risks are unknown.

Markers offer improved understanding of the mechanisms of disease causation and progression. Markers, when applied in batteries in parallel animal and human studies or serially to identify a series of events in the continuum, have the potential to elucidate both general mechanisms of disease causation and the way in which specific pollutants may contribute to the manifestation of certain disease endpoints. Markers could help identify and localize critical target sites; for example, specific chromosomal translocations have been associated with certain leukemias as well as with oncogene activation. They might also demonstrate signal events in disease progression, distinguish the effects of long-term chronic vs acute exposures and identify possible interactive effects of pollutants and bioorganisms (e.g., cigarette smoking and papilloma virus in relation to cervical cancer). They should also contribute to an understanding of the effect of marker persistence and should clarify the different roles played by DNA damage and nongenetic factors in disease development.

Biomarkers can improve epidemiological study design and inference. As discussed, accurate exposure assessments can reduce misclassification error and enhance the power of an epidemiological study. Since early biological events are generally more common than the ultimate disease endpoint, markers of effect also expand a study's statistical power (25). Markers also permit more cost-effective studies. Appropriately processed tissue samples can be stored in specimen banks, sometimes for long time periods, and retrieved for assay when needed at a later date. This permits design of a variety of studies to assess exposure, effect, or both (43). Other advantages include shorter follow-up, better characterization of confounders and cofactors, and improved evidence for causal associations.

Disadvantages and Limitations

Most biomarkers lack adequate validation. The principal limitations to the use of biomarkers for exposure/response assessment stem from the fact that most (with the exception of some markers of internal dose and cytogenetic effects) are in a semidevelopmental stage and are not fully validated or field tested. Methodological problems are similar to those faced in the development of any new technology and include limited availability of standardized protocols and reporting criteria, interlaboratory differences (including variability in quantitation methods and sensitivity) and the costliness and labor intensiveness of most procedures.

Markers have not yet been developed to study many significant environmental exposure-response relationships. For example, nearly all existing markers for carcinogens reflect interaction with somatic genetic material and provide information about the initiation or progression stages in the carcinogenic process. Markers relevant to later-stage events important to the promotion and progression of carcinogenesis are not yet available for human studies.

Molecular epidemiological studies are complex and resource-intensive. By virtue of their collaborative and interdisciplinary nature, these investigations are highly demanding in terms of personnel, cost, and effort to develop mutual understanding among researchers in different fields. They also require clinical interaction with subjects. Incorporation of biological markers engenders new and significant methodological problems in study design and in analysis of data. This approach also introduces into health effects research important ethical questions.

OVERVIEW OF RECENT STUDIES INVOLVING COMPLEX MIXTURES

Because of their inability to provide a chemical specific measure of the biologically effective dose of carcinogens, macromolecular adducts have been the subject of considerable research effort. As discussed earlier, carcinogen-DNA adducts are considered to be a necessary although not sufficient event in the initiation of chemical carcinogenesis, while, in certain instances, carcinogen-protein adducts can serve as a feasible surrogate for DNA binding (22,30,51,66,91,94,95). Because they well illustrate the advantages and disadvantages of biological markers, adducts and their relationship to various measures of biological effect will be the focus of this section.

Turning now to the question at hand, i.e., the usefulness of biological markers in assessing the risk of complex mixtures, Tab. 1 summarizes the results of recent studies of macromolecular adducts in humans. Most of the research on macromolecular adducts has been cross-sectional in nature, aimed at elucidating the relationship between carcinogen-DNA or carcinogen-protein adducts on the one hand, and estimated exposure to carcinogens on the other. In many cases, multiple exposures or complex mixtures have been involved, e.g., cigarette smoke, diet, and industrial pollution. The results of various studies of adducts and other markers associated with these complex exposures have been discussed in detail in a recent review (58) and will be only briefly summarized here.

Cigarette smoke is a classical complex mixture which has been the subject of much molecular research. For example, in a study of smoking and nonsmoking volunteers, a battery of markers was evaluated in repeat peripheral blood samples drawn several days apart (61). 4-ABP-Hb adducts were found to be more specific to cigarette smoke than PAH-DNA adducts. Unlike PAH-DNA adducts, for which there was a high and variable "background," 4-ABP-Hb levels were significantly different in smokers and nonsmokers. The levels of ethylene oxide-hemoglobin or EtO-Hb (hydroxyethyl valine) adducts were also significantly higher in a subset of smokers than in nonsmokers (47). SCEs were also significantly elevated in the smokers and were positively correlated with 4-ABP-Hb and EtO-Hb.

In contrast to the smoker/nonsmoker study, a cross-sectional study of PAH-DNA in long-term employees in an iron foundry showed a clear "dose-response" relationship between estimated source-specific exposure to PAH and levels of PAH-DNA adducts after adjusting for cigarette smoking and time since vacation (60). However, here also significant interindividual variation was seen in adduct levels in the exposed workers, ranging from 20-fold in the "medium" exposure group to 29-fold for the low exposure group.

Recently, levels of PAH-DNA adducts (as a marker of biologically effective dose) and activation of oncogenes (as a marker of biological effect) have been examined in exposed workers. Therefore, sera from foundry workers and controls were assayed using a modified Western blotting technique to search for increased levels of oncogene protein products (ras, fes, myc, etc.) as previously described (4). One exposed worker showed elevated levels of the ras oncogene product, while two exposed individuals had significantly elevated levels of the fes oncogene protein product. Samples from the ten unexposed controls were uniformly negative.

Tab. 1. Studies of macromolecular adducts in humans

Compounds Analyzed	Exposure Source	Biological Sample	Population	Reference
Chemical-specific Adducts				
N-3-(2-Hydroxy-histidine; N-(2-Hydroxy-ethyl) valine	Ethylene oxide	RBC	Workers, smokers, unexposed	9,22,47, 83,88
Alkylated Hb	Propylene oxide	RBC	Workers	57
4-Aminobiphenyl-Hb	Cigarette smoke	RBC	Smokers, nonsmokers	8,89
Styrene-Hb	Styrene	RBC	Workers	7
AFB_1-guanine	Diet	Urine	Chinese and Kenyans residing in a high-exposure and high- and low-risk area, respectively	2,28
3-Methyladenine	Methylating agents	Urine	Unexposed	78
PAH-DNA	PAH in cigarette smoke, workplace	WBC, lung tissue, placenta	Lung cancer patients, smokers, workers	7,19,20,30, 34,60-62,68, 77,86,92
Antibodies to PAH-DNA	PAH in workplace	Serum	Workers	32
O^6-Methyldeoxy-guanosine	Nitrosamines in diet	Esophageal and stomach mucosa	Chinese and European cancer patients	84,93
Cisplatinum-DNA, Cisplatinum-protein	Cisplatin based chemotherapy	WBC, Hb, plasma proteins	Chemotherapy patients	15,23,51,58
8-Methoxypsoralen-DNA	8-Methoxypsoralen	Skin	Psoriasis patients	74
Styrene oxide-DNA	Styrene	WBC	Workers	49
Nonchemical-specific Adducts				
Multiple carcinogens by ^{32}P	Betel and tobacco chewing, smoking, industrial exposures, wood smoke	Placenta, lung tissue, oral mucosa, WBC, bone marrow, colonic mucosa	Smokers, workers, volunteers	13,16,20,34, 67-70

Abbreviations: 4-ABP, 4-aminobiphenyl; PAH, polycyclic aromatic hydrocarbons; RBC, red blood cells; WBC, white blood cells; ^{32}P, ^{32}P-postlabelling.

As with all cross-sectional studies, which provide a "snapshot" at a particular point in time, this research cannot establish temporal or causal relationships between exposure, adduct formation, and oncogene activation. For this reason, longitudinal studies in individuals whose exposure changes significantly over time are preferable in terms of establishing the relationship between exposure and biological markers such as macromolecular adducts. There have been few such studies to date but several natural experiments have presented themselves. These include studies of individuals in smoking cessation programs (7), workers who are sampled before beginning employment in a coke oven plant (85), workers who are sampled after an interruption in exposure (32), and chemotherapy patients from whom pretreatment and serial post-treatment samples are obtained (23,52, 58,72).

A "natural experiment" was provided by the plant-wide annual vacation taken by the Finnish foundry workers during the month of July. White blood cell DNA from nine individuals was assayed for PAH-DNA adducts after the workers returned from their four-week vacation and six weeks following their return to work. The levels of adducts were significantly higher in the post-work samples compared to the post-vacation samples, showing a clear biological effect of exposure (60).

The first step in exploring the role of various biomarkers as risk factors for cancer has been taken in several modified case-control studies. Recently a series of cases with primary carcinoma of the lung and controls have been evaluated for levels of PAH-DNA adducts and SCEs in order to determine whether, when exposure to PAHs and other mutagens/carcinogens was comparable, cases had higher levels of biomarkers than controls (61). Such a finding would suggest that the ability to efficiently activate and bind carcinogens may have been a risk factor in their disease. SCEs in lymphocytes did not differ between cases and controls. As in the smokers/nonsmokers study, PAH-DNA adducts were not specific to cigarette smoking but apparently reflected the numerous background sources of PAH, such as diet and ambient air. However, among current smokers, WBC DNA from lung cancer cases had significantly higher levels of PAH-DNA adducts, consistent with a possible genetic/metabolic predisposition to lung cancer.

As with the foundry workers' study, Western blotting techniques were used to screen sera of lung cancer cases and controls for increased concentrations of peptide sequences of nine different oncogene-encoded proteins (4). All 18 cases were positive for at least one oncogene protein (mainly ras and fes), in contrast to two of the 20 healthy controls.

Case-control studies have the serious limitation that if exposure or pharmakokinetic processes have altered between the time of cancer induction and diagnosis, current levels of macromolecular adduct levels may bear little relationship to cancer risk. Thus, the nested case-control is the optimal design for evaluating the relationship between biological markers and cancer risk, provided, of course, that the marker is stable in samples collected at the outset of a prospective study.

CONCLUSION

Returning to the assessment of complex mixtures and multiple exposures, in light of the above discussion, recommendations are as follows:

(i) Chemical-specific or "selective" disometers (DNA and protein adducts) should be used in conjunction with nonspecific markers such as SCEs, micronuclei, gene mutation, and oncogene activation which give an integrated estimate of the genotoxic/procarcinogenic effect of exposure.

(ii) In order to tease apart the effect of specific constituents of the mixture, a step-wise evaluation can be carried out, as illustrated by the reverse pyramid in Fig. 1. First, external exposure would be charac-

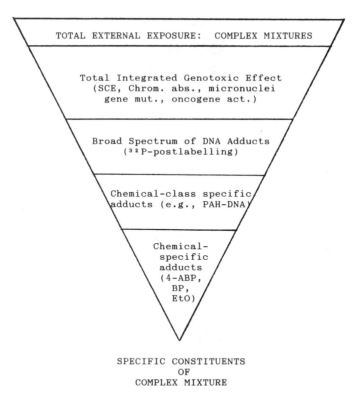

TOTAL EXTERNAL EXPOSURE: COMPLEX MIXTURES

Total Integrated Genotoxic Effect
(SCE, Chrom. abs., micronuclei
gene mut., oncogene act.)

Broad Spectrum of DNA Adducts
(^{32}P-postlabelling)

Chemical-class specific
adducts (e.g., PAH-DNA)

Chemical-
specific
adducts
(4-ABP,
BP,
EtO)

SPECIFIC CONSTITUENTS
OF
COMPLEX MIXTURE

Fig. 1.   Assessment of the potential carcinogenic risk of complex mixtures.

terized as completely as possible through ambient or personal monitoring
and questionnaires. This would provide an estimate of the level and pattern of exposure, both to the mixture and to its individual components.
The next step would be to analyze the relationship between integrated
and chemical-specific exposure variables on the one hand, and the various
biomarkers on the other. These would progress from the general to the
specific. For example, one might sequentially evaluate markers of total
genotoxic/procarcinogenic effect, then a spectrum of DNA adducts by the
postlabelling assay, then chemical class-specific adducts (e.g., PAH-DNA
by immunoassay) and finally, individual chemical chemical-specific adducts
(e.g., 4-ABP-Hb or BaP-DNA). Correlations between biological markers
would also be examined. The question of interest: What proportion of
the total genotoxic/procarcinogenic effect of exposure to complex mixtures
is attributable to specific constituents in the mixture. Also, is there apparent interaction between individual constituents, or do the effects appear to be additive? In this way, it may be possible to identify the major
pathogenic agents present in a chemical mixture and to determine the
shape of the low dose-response curve.

(iii) For purposes of evaluating the role of complex mixtures in cancer
risk, the nested case-control study is the optimal design. Model exposures include cigarette smoking, occupational exposures to fossil fuel combustion products, and combination chemotherapy.

In conclusion, these early studies are encouraging in that macromolecular adducts and related biomarkers can ultimately be useful in identifying the operational carcinogenic constituents of complex mixtures and in estimating whether their effect is additive or interractive. They can also provide valuable insights into the nature and magnitude of interindividual variability and the need to factor this into quantitative risk assessment. To make this happen, well-designed molecular epidemiological studies in human populations with model exposures to single agents and to complex mixtures are essential.

ACKNOWLEDGEMENTS

We gratefully acknowledge the significant contribution of Dr. I.B. Weinstein to these projects. We also thank colleagues at Columbia University, Dr. P. Brandt-Rauf, Dr. S. Smith, Dr. A. Jeffrey, Dr. L. Latriano, T.L. Young, and Dr. B.M. Lee for their contribution to the laboratory assays of oncogene activation and adducts. We also thank Dr. J. Mayer for his contribution to the molecular epidemiologic studies and Dr. D. Warburton, Dr. M. Toor, and Dr. H.K. Fischman for cytogenetic analyses. We are grateful to Dr. S. Tannenbaum and M. Bryant for analysis of 4-aminobiphenyl adducts, and Dr. K. Hemminki for collaboration on the foundry workers' study. This work was supported by Public Health Service Grants 5 R01CA-35809, 5 R01CA-39174, and R01CA-47351-01, and grant P01CA-21111 from the Division of Extramural Activities, National Cancer Institute; by an NIEHS grant ES03881, and by grant 41387 from the American Cancer Society.

REFERENCES

1. Ames, B.N., R. Magaw, and L.S. Gold (1987) Ranking possible carcinogenic hazards. Science 236:271-280.
2. Autrup, H., K.A. Bradley, A.K.M. Shamsuddin, J. Wakhisi, and A. Wasunna (1983) Detection of putative adduct with fluorescence characteristics identical to 2,3-dihydro-2-(7-guanyl)-3-hydroxyaflatoxin B1 in human urine collected in Murang'a District, Kenya. Carcinogenesis 4:1193-1195.
3. Ayesh, R., J. Idle, J. Richie, M. Crothers, and M. Hetzel (1984) Metabolic oxidation phenotypes as markers for susceptibility to lung cancer. Nature 312:169-170.
4. Brandt-Rauf, P.W., S. Smith, and F.P. Perera (1989) Molecular epidemiology and environmental carcinogenesis of the lung. In Current Issues in Respiratory Public Health, T.J. Witek and E.N. Schacter , eds. J.B. Lippincott, NY (in press).
5. Brandt-Rauf, P.W. (1988) New markers for monitoring occupational cancer: The example of oncogene proteins. J. Occup. Med. 30:399-404.
6. Brenner, D., A. Jeffrey, L. Latriano, D. Warburton, M. Toor, and F.P. Perera (1989) Biologic markers in styrene exposed boatworkers (in preparation).
7. Brenner, D., R. Santella, B.M. Lee, D.W. Warburton, M. Toor, and F. Perera (1989) Biologic markers in smokers following cessation of smoking (in prep.).
8. Bryant, M.S., P.L. Skipper, S.R. Tannenbaum, and M. Maclure

(1987) Hemoglobin adducts of 4-aminobiphenyl in smokers and nonsmokers. Cancer Res. 47:612-618.

9. Calleman, C.J., L. Ehrenberg, B. Jansson, S. Osterman-Golkar, D. Segerback, K. Svensson, and C.A. Wachtmeister (1978) Monitoring and risk assessment by means of alkyl groups in hemoglobin in persons occupationally exposed to ethylene oxide. J. Env. Path. Tox. 2:427-442.

10. Cancer statistics (1988) Ca: A Cancer Journal for Clinicians 38:1-23.

11. Carrano, A.V., and A.T. Natarajan (1988) Considerations for population monitoring using cytogenetic techniques. Mutat. Res. 204:379-406.

12. Cartwright, R.A., R.W. Glashhan, H.J. Rogers, et al., (1982) Role of N-acetyltransferase phenotypes in bladder carcionogenesis: A pharmacokinetic epidemiological approach to bladder cancer. Lancet 2:842-846.

13. Chacko, M., and R.C. Gupta (1987) Evaluation of DNA damage in the oral mucosa of cigarette smokers and nonsmokers by ^{32}P-adduct assay. Proc. Am. Assoc. Cancer Res. 28:101.

14. Conney, A.H. (1982) Induction of microsomal enzymes by foreign chemicals and carcinogenesis by polycyclic aromatic hydrocarbons. Cancer Res. 42:4875-4917.

15. Den Engelse, L., P.M.A.B. Terhaggen, and A.C. Begg (1988) Cisplatin-DNA interaction products in sensitive and resistant cell lines and in buccal cells from cisplatin-treated cancer patients. In Proceedings of the American Association for Cancer Research 29:339 (abstract No. 1348).

16. Dunn, B.P., and H.F. Stich (1986) ^{32}P-postlabelling analysis of aromatic DNA adducts in human oral mucosal cells. Carcinogenesis 7:1115-1120.

17. EPA (1988) Regulation of Pesticides in Food: Addressing the Delaney Paradox. Notice, 53 Fed. Reg. 41101; Oct. 19, U.S. Environmental Protection Agency, Washington, DC.

18. EPA (1989) The Toxics Release Inventory: A National Perspective. U.S. Environmental Protection Agency, Office of Toxic Substances, Economics and Technology Division, Washington, DC, June.

19. Evans, H.J. (1982) Cytogenetic studies on industrial populations exposed to mutagens. In Indicators of Genotoxic Exposure, Banbury Report #19, B.A. Bridges, B.E. Butterworth, and I.B. Weinstein, eds. Cold Spring Harbor Laboratory, Cold Spring Harbor, NY.

20. Everson, R.E., E. Randerath, R.M. Santella, R.C. Cefalo, T.A. Avitts, and K. Randerath (1986) Detection of smoking-related covalent DNA adducts in human placenta. Science 231:54-57.

21. Everson, R.E., E. Randerath, R.M. Santella, T.A. Avitts, I.B. Weinstein, and K. Randerath (1988) Quantitative association between DNA damage in human placenta and maternal smoking and birth weight. JNCI 80:567-576.

22. Farmer, P.B., E. Bailey, S.M. Gorf, M. Tornqvist, S. Osterman-Golkar, A. Kautiiainen, and D.P. Lewis-Enright (1986) Monitoring human exposure to ethylene oxide by the determination of hemoglobin adducts using gas chromatography-mass spectrometry. Carcinogenesis 7:637-640.

23. Fichtinger-Schepman, A.M.J., A.T. Van Oosterom, P.H.M. Lohman, and F. Berends (1980) cis-Diamminedichloroplatinum (II)-induced DNA adducts in peripheral leukocytes from seven cancer patients:

Quantitative immunochemical detection of the adduct induction and removal after a single dose of cis-Diamminadichloroplatinum(11). Cancer Res. 47:3000-3004.

24. Francomano, C., and H.H. Kazazian, Jr. (1986) DNA analysis in genetic disorders. Ann. Rev. Med. 37:377-395.

25. Gann, P.H., D.L. Davis, and F. Perera (1985). Biologic markers in environmental epidemiology: Constraints and opportunities. In Fifth Workshop of the Scientific Group on Methodologies for the Safety Evaluation of Chemicals (SGOMSEC). Mexico City, August 12-16.

26. GAO (1989) Statement of P.F. Guerrero, General Accounting Office, before the Subcommittee on Toxic Substances, Environmental Oversight, Research and Development Committee on Environmental and Public Works, U.S. Senate, May 15.

27. Gonzalez, F.J., A.K. Jaiswal, and D.W. Nebert (1986) P-450 genes: Evolution, regulation and relationship to human cancer and pharmacogenetics. In Cold Spring Harbor Symposium Quantitative Biology 51(Pt.2), pp. 879-890.

28. Groopman, J.D., P.R. Donahue, J. Zhu, J. Chen, and G.N. Wogan (1985) Aflatoxin metabolism and nucleic acid adducts in urine by affinity chromatography. Proc. Natl. Acad. Sci., U.S.A. 82:6492-6496.

29. Harris, C.C., A. Weston, J. Willey, G. Trivers, and D. Mann. (1987) Biochemical and molecular epidemiology of human cancer: Indicators of carcinogen exposure, DNA damage and genetic predisposition. Env. Health Persp. 75:109-119.

30. Harris, C.C., K. Vahakangas, N.J. Newman, G.E. Trivers, A. Shamsuddin, N. Sinopoli, D.L. Mann, and W. Wright (1985) Detection of benzo(a)pyrene diol epoxide-DNA adducts in peripheral blood lymphocytes and antibodies to the adducts in serum from coke oven workers. Proc. Natl. Acad. Sci., USA 82:6672-6676.

31. Hattis, D. (1987) The value of molecular epidemiology in quantitative health risk assessment. In Environmental Impacts on Human Health: The Agenda for Long-Term Research and Development, S. Draggen, J.J. Cohrssen, and R.E. Morrison, eds. New York: Praeger.

32. Haugen, A., G. Becher, C. Benestad, K. Vahakangas, G.E. Trivers, M.J. Newman, and C.C. Harris (1986) Determination of polycyclic aromatic hydrocarbons in the urine, benzo(a)pyrene diol epoxide-DNA adducts in lymphocyte DNA, and antibodies to the adducts in sera from coke oven workers exposed to measured amounts of polycyclic aromatic hydrocarbons in the work atmosphere. Cancer Res. 46:4178-4183.

33. Hemminki, K., F.P. Perera, D.H. Phillips, K. Randerath, M.V. Reddy, and R.M. Santella (1988) Aromatic DNA adducts in white blood cells of foundry and coke oven workers. Scand. J. Work Env. Health 14 (Suppl. 1):55-56.

34. Hemminki, K., K. Twardowska-Saucha, J.W. Scroczynski, E. Grzybowka, M. Chorazy, K.L. Putman, K. Randerath, D.H. Phillips, A. Hewer, R.M. Santella, T.L. Young, and F.P. Perera (1989) DNA adducts in humans environmentally exposed to aromatic compounds in an industrial area of Poland (sub. for pub.).

35. Hemstreet, G.P., P.A. Schulte, K. Ringen, W. Stringer, and E.B. Altekruse (1989) DNA hyperploidy as a marker for biological response to bladder carcinogen exposure. Int. J. Cancer (in press).

36. Herbert, R., M. Marcus, M.S. Wolff, F.P. Perera, L. Andrews, J.H. Godbold, M. Rivera, M. Stefanides, X.Q. Lu, P.J. Landrigan, R.M. Santella (1989) A pilot study of detection of DNA adducts in

white blood cells of roofers by ^{32}P-postlabelling. In Experimental and Epidemiologic Approaches to Cancer Risk Assessment of Complex Mixtures. IARC, Lyon, France (in press).

37. Hoel, D.G., N.L. Kaplan, and M.W. Anderson (1983) Implication of non-linear kinetics on risk estimation on carcinogenesis. Science 219:1032-1037.

38. Hulka, B.S., and T. Wilcosky (1988) Biological markers in epidemiologic research. Arch. Env. Health 43:83-89.

39. IARC (1989) Monitoring of humans with exposures to carcinogens, mutagens and epidemiological applications. In Cancer Occurrence, Causes and Control, L. Tomatis, ed. IARC Sci. Publ. Ser., Sec. 6.3, Lyon, France (in press).

40. IARC (1987) Overall Evaluations of Carcinogenicity: An Updating of IARC Monographs, Vol. 1-42: Suppl. 7. International Agency for Research on Cancer, Lyon, France.

41. IARC (1986) Tobacco Smoking. IARC Monographs on the Evaluation of the Carcinogenic Risk of Chemicals to Humans, Vol. 38. International Agency for Research on Cancer, Lyon, France.

42. IARC (1988) Methods for Detecting DNA Damaging Agents in Humans: Applications in Cancer Epidemiology and Prevention, H. Bartsch, K. Hemminki, and I.K. O-Neill, eds. IARC Sci. Publ. No. 89, Lyon, France.

43. Jellum, E., A. Andersen, H. Orjasaeten, O.P. Foss, L. Theodorsen, and P. Lund-Larsen (1987) The Janus serum bank and early detection of cancer. Biochemica Clinica 11:191-195.

44. Jiang, W., S. Kahn, J. Guillem, S. Lu, and I.B. Weinstein (1989) Rapid detection of rat oncogenes in human tumors: Application to colon, esophageal and gastric cancer. Oncogene 4:923-928.

45. Karki, N.T., R. Pokela, L. Nuutinen, and O. Pelkonen (1987) Aryl hydrocarbon hydroxylase in lymphocytes and lung tissue from lung cancer patients and controls. Int. J. Cancer 39:565-570.

46. Khoury, M.J., C.A. Newill, and G.A. Chase (1985) Epidemiologic evaluation of screening. Am. J. Pub. Health 75:1204-1208.

47. Latriano, L., F.P. Perera, D. Brenner, and A.M. Jeffrey (1988) Comparison of ethylene oxide and 4-aminobiphenyl hemoglobin adducts in cigarette smokers. Extended abstract presented at the American Chemical Society Meeting, Los Angeles, California.

48. Lauwerys, R., and A. Bernard (1985) The place of biological monitoring for assessing exposure to industrial chemicals: Presenting status and future trends. Annals Amer. Conf. Indust. Hyg. 12:327-330.

49. Liu, S.F., S.M. Rappaport, K. Pongracz, and W.J. Bodell (1988) Detection of Styrene Oxide-DNA Adducts in Lymphocytes of a Worker Exposed to Styrene. IARC, Lyon, France, pp. 217-222.

50. Lucier, G.W., and C.L. Thompson (1987) Issues in biochemical applications to risk assessment: When can lymphocytes be used as surrogate markers? Env. Health Persp. 76:187-191.

51. Miller, E.C., and J.A. Miller (1981) Mechanisms of chemical carcinogenesis. Cancer 47:1055-1064.

52. Mustonen, R., K. Hemminki, A. Alhonen, P. Hietanen, and M. Kiilunen (1988) Determination of Cisplatin in Blood Compartments of Cancer Patients. IARC, Lyon, France, pp. 329-332.

53. National Academy of Sciences, National Research Council, Committee on Biologic Markers (1987) Biological markers in environmental health. Env. Health Pers. 64:3-9.

54. Neumann, H.G. (1984) Analysis of hemoglobin as a dose monitor for alkylating and arylating agents. Arch. Tox. 56:1-6.

55. Nowak, D., U. Schmidt-Preuss, R. Jorres, et al. (1988) Formation of DNA adducts and water-soluble metabolites of benzo(a)pyrene in human monocytes is genetically controlled. Int. J. Cancer 41:169-173.

56. NWF (1989) Danger Downwind: A Report on the Release of Billions of Pounds of Toxic Air Pollutants, J. Poje, N.L. Dean, R.J. Burke, eds. National Wildlife Federation, Washington, DC, March 22.

57. Osterman-Golkar, S., E. Bailey, P.B. Farmer, S.M. Gorf, and J.H. Lamb (1984) Monitoring exposure to propylene oxide through the determination of hemoglobin alkylation. Scand. J. Work Env. Health 10:99-102.

58. Perera, F., J. Mayer, R.M. Santella, D. Brenner, A. Jeffrey, L. Latriano, S. Smith, D. Warburton, T.L. Young, W.Y. Tsai, K. Hemminki, and P. Brandt-Rauf (1989) Biologic markers in risk assessment for environmental carcinogens. Env. Health Persp. (in press).

59. Perera, F.P., P.A. Schulte, and D. Brenner (1988) Biologic Markers in Assessing Human Exposures to Airborne Pollutants. Working Paper for the Committee on Advances in Assessing Human Exposure to Airborne Pollutants. National Academy of Sciences, Washington, DC.

60. Perera, F.S., K. Hemminki, T.L. Young, D. Brenner, G. Kelly, and R.M. Santella (1988) Detection of polycyclic aromatic hydrocarbon-DNA adducts in white blood cells of foundry workers. Cancer Res. 48:2288-2291.

61. Perera, F.P., R.M. Santella, D. Brenner, M.C. Poirier, A.A. Munshi, H.K. Fischman, and J. Van Ryzin (1987) DNA adducts, protein adducts and sister chromatid exchange in cigarette smokers and nonsmokers. JNCI 79:449-456.

62. Perera, F.P., and I.B. Weinstein (1982) Molecular epidemiology and carcinogen-DNA adduct detection: New approaches to studies of human cancer causation. J. Chron. Dis. 35:581-600.

63. Perera, F.P. (1987) Molecular cancer epidemiology: A new tool in cancer prevention. JNCI 78:887-898.

64. Perera, F., and P. Boffetta (1988) Perspectives on comparing risks of environmental carcinogens. JNCI 80:1282-1283.

65. Perera, F.P., M.C. Poirier, S.H. Yuspa, J. Nakayama, A. Jaretzki, M.C. Curnen, D.M. Knowles, and I.B. Weinstein (1982) A pilot project in molecular cancer epidemiology: Determination of benzo(a)pyrene-DNA adducts in animal and human tissues by immunoassay. Carcinogenesis 3:1405-1410.

66. Perera, F.P. (1988) The significance of DNA and protein adducts in human biomonitoring studies. Mutat. Res. 205:255-269.

67. Phillips, D.H., A. Hewer, and P.L. Grover (1986) Aromatic DNA adducts in human bone marrow and peripheral blood leukocytes. Carcinogenesis 7:2071-2075.

68. Phillips, D.H., A. Hewer, C.N. Martin, R.C. Garner, and M.M. King (1988) Correlation of DNA adducts levels in human lung with cigarette smoking. Nature 336:790.

69. Phillips, D.H., K. Hemminki, A. Alhonen, A. Hewer, and P.L. Grover (1988) Monitoring occupational exposure to carcinogens: Detection by ^{32}P-postlabelling of aromatic DNA adducts in white blood cells from iron foundry workers. Mutat. Res. 204:531-541.

70. Randerath, K., R.H. Miller, D. Mittal, and E. Randerath (1988) Monitoring Human Exposure to Carcinogens by Ultrasensitive Postlabelling Assays: Application to Unidentified Genotoxicants. IARC, Lyon, France, pp. 361-367.

71. Reddy, M.V., and K. Randerath (1987) ^{32}P-postlabelling assay for carcinogen-DNA adducts: Nuclease Pl-mediated enhancement of its sensitivity and applications. Env. Health Persp. 76:41-47.

72. Reed, E., S. Yuspa, L.A. Zwelling, R.F. Ozols, and M.C. Poirier (1986) Quantitation of cis-diamminedichloroplatinum (II) (cisplatin)-DNA-intrastrand adducts in testicular and ovarian cancer patients receiving cisplatin chemotherapy. J. Clin. Invest. 77:545-550.

73. Rudiger, H.W., D. Nowak, K. Hartmann, and P. Cerrutti (1985) Enhanced formation of benzo(a)pyrene: DNA adducts in monocytes of patients with a presumed predisposition to lung cancer. Cancer Res. 45:5890-5894.

74. Santella, R.M. (1988) Application of new techniques for detection of carcinogen adducts to human population monitoring. Mutat. Res. 205:271-282.

75. Schulte, P.A. (1989) A conceptual framework for the validation and use of biologic markers. Env. Res. 48:129-144.

76. Schulte, P.A. (1987) Methodologic issues in the use of biologic markers in epidemiologic research. Am. J. Epidem. 126:1006-1016.

77. Shamsuddin, A.K., N.T. Sinopoli, K. Hemminki, R.R. Boesch, and C.C. Harris (1985) Detection of Benzo(a)pyrene DNA adducts in human white blood cells. Cancer Res. 45:66-68.

78. Shuker, D.E.G., and P.B. Farmer (1988) Urinary Excretion of 3-Methyladenine in Humans as a Marker of Nucleic Acid Methylation. IARC, Lyon, France, pp. 92-96.

79. Slamon, D.J., J.B. de Kernian, I.M. Verma, and M.J. Cline (1984) Expression of cellular oncogenes in human malignancies. Science 224:256-262.

80. Spandidos, D.A., and I.B. Kerr (1984) Elevated expression of the human ras oncogene family in premalignant and malignant tumors of the colorectum. Br. J. Cancer 49:681-688.

81. Stowers, S.J., R.R. Maronpot, S.H. Reynolds, and M.W. Anderson (1987) The role of oncogenes in chemical carcinogenesis. Env. Health Persp. 75:81-86.

82. Sukumar, S., S. Pulciani, J. Doniger, J.A. DiPaolo, C.H. Evans, B. Zbar, and M. Barbacid (1984) A transforming ras gene in tumorigenic guinea pig cell lines initiated by diverse chemical carcinogens. Science 223:1197-1199.

83. Tornqvist, M., S. Osterman-Golkar, A. Kautiainen, P.B. Farmer, S. Jensen, and L. Ehrenberg (1986) Tissue doses of ethylene oxide in cigarette smokers determined from adduct levels in hemoglobin. Carcinogenesis 7:1519-1521.

84. Umbenhauer, D., C.P. Wild, R. Montesano, R. Saffhill, J.M. Boyle, N. Huh, U. Kirstein, J. Thomale, M.F. Rajewsky, and S.H. Lu (1985) O^6 Methyldeoxyguanosine in oesophageal DNA among individual at high risk of oesophageal cancer. Int. J. Cancer 36:661-665.

85. Vahakangas, K., et al. (1989) Symchronous fluorescence spectrophotometry of benzo(a)pyrene diolepoxide-DNA adducts in workers exposed to polycyclic aromatic hydrocarbons. In Experimental and Epidemiologic Approaches to Cancer Risk Assessment of Complex Mixtures. IARC, Lyon, France (in press).

86. Vahakangas, K., G. Trivers, A. Haugen, et al. (1985) Detection of benzo(a)pyrene diol-epoxide-DNA adducts by synchronous fluorescence cence spectrophotometry and ultrasensitive enzymatic radioimmunoassay in coke oven workers. J. Cell Biochem. Suppl. 9C:1271.

87. Vainio, H. M. Sorsa, and K. Falck (1984) Bacterial Urinary Assay in Monitoring Exposure to Mutagens and Carcinogens. IARC, Lyon, France, pp. 247-258.

88. van Sittert, N.J., G. de Jong, M.G. Clare, R. Davies, B.J. Dean, L.J. Wren, and A.S. Wright (1985) Cytogenetic, immunological, and hemotological effects in workers in an ethylene oxide manufacturing plant. Br. J. Ind. Med. 42:19-26.

89. Vineis, P., J. Esteve, P. Hartge, R. Hoover, D.T. Silbergman, and B. Terracini (1988) Effects of timing and type of tobacco in cigarette-induced bladder cancer. Cancer Res. 48:3849-3852.

90. Walles, S.A.S., H. Norppa, S. Osterman-Golkar, and J. Maki-Paakkanen (1988) Single-strand Breaks in DNA of Peripheral Lymphocytes of Styrene-exposed Workers. IARC, Lyon, France, pp. 226.

91. Weinstein, I.B., S. Gattoni-Celli, P. Kirschmeier, M. Lambert, W. Hsiao, J. Backer, and A. Jeffrey (1984) Multistage carcinogenesis involves multiple genes and multiple mechanisms. Cancer cells 1. In The Transformed Phenotype, Cold Spring Harbor Laboratory, New York, pp. 229-237.

92. Weston, A., J.C. Willey, D.K. Manchester, V.L. Wilson, B.R. Brooks, J.-S. Choi, M.C. Poirier, G.E. Trivers, M.J. Newman, D.L. Mann, and C.C. Harris (1988) Dosimeters of Human Carcinogen Exposure: Polycyclic Aromatic Hydrocarbon Macromolecular Adducts. IARC, Lyon, France, pp. 181-189.

93. Wild, C.P., D. Umbenhauer, B. Chapot, and R. Montesano (1986) Monitoring of individual human exposure to aflatoxins (AF) and N-nitrosamines (NNO) by immunoassays. J. Cell Biol. 30:171-179.

94. Wogan, G.N., and N.J. Gorelick (1985) Chemical and biochemical dosimetry of exposure to genotoxic chemicals. Env. Health Persp. 62:5-18.

95. Yuspa, S.H., and M.C. Poirier (1989) Chemical carcinogenesis: From animal models to molecular models in one decade. Adv. Cancer Res. (in press).

IMMUNOLOGICAL METHODS FOR THE DETECTION OF POLYCYCLIC

AROMATIC HYDROCARBON-DNA AND PROTEIN ADDUCTS

Regina M. Santella, You Li, Yu Jing Zhang,
Tie Lan Young, Marina Stefanidis, Xiao Qing Lu,
Byung Mu Lee, Maria Gomes, and Frederica P. Perera

Cancer Center
Division of Environmental Science
Columbia University School of Public Health
New York, New York 10032

ABSTRACT

Sensitive immunological methods are now available for the detection and quantitation of carcinogen-DNA and protein adducts. Both monoclonal and polyclonal antibodies have been developed against DNA modified by benzo(a)pyrene diol epoxide-I (BPDE-I). These antisera recognize a number of structurally related diol epoxide-DNA adducts, and have been utilized in enzyme-linked immunosorbent assays (ELISAs) to determine adduct levels in DNA isolated from humans with several different environmental and occupational exposures to polycyclic aromatic hydrocarbons (PAHs). Populations studied include cigarette smokers, foundry workers, coke oven workers, and coal tar treated psoriasis patients. Since these populations are exposed to complex mixtures of PAHs, multiple adducts will be formed and determined by the ELISA. In addition, the antibodies can be used in immunohistochemical studies to investigate localization of adducts to specific cell and tissue types.

Also methods recently have been developed for the immunological measurement of BPDE-I-protein adducts. For increased sensitivity, proteins are first enzymatically digested to peptides and amino acids before adduct quantitation by ELISA. This assay has been utilized to measure albumin adducts in roofers and controls. In a small number of subjects, no significant differences were seen between the two groups. Further studies are ongoing.

INTRODUCTION

Immunological methods are now available for the sensitive detection and quantitation of carcinogen-DNA adducts. A number of antibodies have been developed against alkylated adducts, oxidative- or UV-damaged

DNA, as well as bulky aromatic adducts (12,17). These antisera can be used in radioimmunoassays (RIA), or enzyme-linked immunosorbent assays (ELISA) to detect adducts with sensitivities in the range of one adduct per 10^8 nucleotides. This sensitivity has allowed the detection of adducts in humans following environmental or occupational exposure to carcinogens.

While DNA is believed to be a critical target for chemical carcinogens, they also bind to RNA and protein. Measurement of protein adducts in humans has been utilized as an alternate method for biological monitoring because of the large amounts of protein, either albumin or hemoglobin, available from blood. While not as well established as immunoassays for DNA adduct detection, methods are available for quantitation of aflatoxin-albumin and ethylene oxide-globin adducts (2,23). We recently have developed an immunoassay for BPDE-I-protein adducts (5).

MEASUREMENT OF PAH-DNA ADDUCTS

To measure DNA adducts formed by benzo(a)pyrene (BaP), both monoclonal and polyclonal antibodies have been developed against BPDE-I-DNA (11,13). When initially characterized, they were found to be highly specific for the modified DNA not recognizing BaP itself or nonmodified DNA. In addition, there was no crossreactivity with several other carcinogen modified DNAs, including acetylaminofluorene and 8-methoxypsoralen-DNA. More recently, the reactive diol epoxide metabolites of several other PAHs became available and were used to modify DNA. Both the monoclonal (15) and polyclonal antisera showed significant crossreactivity with the diol epoxide-DNA adducts of several other PAHs having similar stereochemistry as the BPDE-I adduct. Antibody #29, the polyclonal antisera routinely used for adduct detection in human samples, has significant crossreactivity with DNAs modified by two different benz(a)anthracene diol epoxides (50% inhibition at 42 and 114 fmol, Fig. 1). In addition, this antisera recognizes chrysene diolepoxide modified DNA with higher sensitivity (50% inhibition at 18 fmol) than it recognizes the original antigen BPDE-I-DNA (50% inhibition at 30 fmol). Similar crossreactivities have been found by others with another antibody against BPDE-I-DNA (21). Since humans are exposed to BP in complex mixtures containing a number of other PAHs, multiple adducts may be detected by the ELISA. The identity of the adducts present cannot be determined, and thus absolute quantitation is not possible. However, since a number of PAHs in addition to BP are carcinogenic, the ELISA provides a biologically relevant general index of DNA binding by this class of compounds. Measured values are expressed as fmol equivalents of BPDE-I-DNA adducts which would cause a similar inhibition in the assay.

More recently, we have determined that antisera #29, used for the detection of DNA adducts in biological samples, recognizes the adducts with different sensitivities dependent on the modification level of the DNA (16). Adducts in highly modified DNA (approximately 1.2 adduct/100 nucleotides) are recognized more efficiently than in low modified DNA ($1.5/10^6$ nucleotides). This is not true for all antibodies since a monoclonal antibody recognizing 8-methoxypsoralen modified DNA had similar reactivity with the adduct in high- and low-modified DNA samples (24). These results demonstrate the importance of thorough characterization of the antisera before application to unknown samples and,

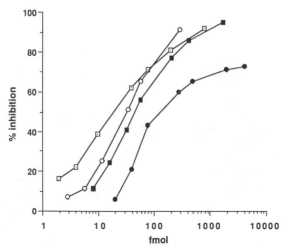

Fig. 1.   Competitive inhibition of polyclonal anti BPDE-I-DNA antibody #29 binding to BPDE-I-DNA. The competitors are BPDE-I-DNA (0, 50% inhibition at 30 fmol), chrysene-1,2-diol-3,4-oxide-DNA (□, 50% inhibition at 18 fmol), benz(a)anthracene-8,9-diol-10,11-oxide-DNA (■, 50% inhibition at 42 fmol), and benz(a)anthracene-3,4-diol-1,2-oxide-DNA (●, 50% inhibition at 114 fmol). Detailed procedures for the ELISA are given in Ref. 13.

where necessary, the utilization of a standard modified in the same range as the unknown samples.

The BPDE-I-DNA antisera has been used to monitor adducts in placental and white blood cell DNA of smokers and nonsmokers (1,7). Adducts levels in placental DNA of smokers were not significantly higher than in nonsmokers (1.9 adducts/10^7 nucleotide in smokers and 1.2/10^7 in nonsmokers). White blood cell DNA adducts were lower but also not significantly different between smokers and nonsmokers (0.15/10^7 in smokers and 0.12/10^7 in nonsmokers). More recently, we have found a seasonal variation in adduct formation which may be related to previously observed variations in aryl hydrocarbon hydroxylase activity (9).

Adduct levels measured in white blood cell DNA from foundry workers and controls showed a clear dose-response relationship when workers were classified into high (5 adducts/10^7 nucleotides), medium (2.1/10^7), or low (0.80/10^7) exposure to BaP (8). Controls, not occupationally exposed to PAHs, had a mean adduct level of 0.22/10^7. Adducts in these same samples were also analyzed by two other laboratories by ^{32}P-postlabeling (4,10). While adduct levels were lower in the postlabeling assay there was a good correlation between the immunoassay and postlabeling (correlation coefficient 0.70).

Coal tar treated psoriasis patients have also been useful as a model population for occupational exposure through skin absorption. These patients are treated by skin application of crude coal tar preparations and have been shown to have elevated risk of skin cancer (19). We have carried out a small pilot study to determine adduct levels in white blood

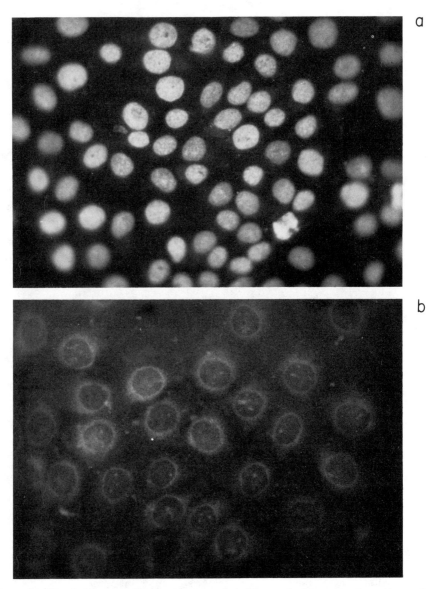

Fig. 2.   Indirect immunofluorescene staining of human keratinocytes
treated with 10 μg/ml BaP and control untreated cells.   Slides
were treated with RNase, proteinase K, and HCl before staining
with antibody 5D11 (13) (1:10 dilution) and goat antimouse IgG
conjugated with fluorescein (1:30).   (a) Cells treated with BaP;
(b) control cells.

cell samples from patients and controls. Preliminary data on 22 patients gave a mean adduct level of $1.7/10^7$. Only one of five controls was positive.

Adduct specific antibodies can also be utilized in immunohistochemical studies to investigate adduct localization in particular cell and tissue types. Initial work with the antibodies against BPDE-I-DNA has been in human keratinocytes treated in culture with 10 µg/ml BaP. Ethanol-fixed cells were treated with RNase to eliminate potential crossreactivity with RNA adducts and with proteinase K to release proteins from the DNA to enhance antibody binding. This was followed by 1N HCl treatment to denature the DNA and further increase sensitivity. Specific nuclear staining could be detected in treated samples but not in untreated control cells (Fig. 2). Other controls, including DNase treatment, staining with nonspecific antiserum, and preabsorption of the antisera with BPDE-I-DNA, were all negative. While adduct levels were not quantitated in these cells, previous studies by ourselves and others with adduct-specific antisera suggest that adduct levels must be in the range of $1/10^6$ nucleotides for detection with immunofluorescence methods (6,24). Initial studies on skin biopsies from coal tar treated psoriasis patients suggest that this sensitivity will be sufficient for detection of adducts in these samples.

DETECTION OF BPDE-I-PROTEIN ADDUCTS

Quantitation of protein adducts on either hemoglobin or albumin has been used as an alternate marker of exposure to environmental carcinogens. In contrast to DNA adduct measurements, large amounts of protein can be obtained from blood samples and no repair occurs, suggesting that chronic low levels of exposure may be measurable. Red blood cells have an average lifespan of four months, while albumin has a half-life of 21 days. Thus, only relatively recent exposure will be measurable. To determine BaP-protein adduct levels, we have utilized a monoclonal antibody, 8E11, developed from animals immunized with BPDE-I-G coupled to bovine serum albumin (13). This antibody recognizes BPDE-I-modified dG, DNA, and protein as well as BPDE-I-tetraols. More recently, we have further characterized this antibody in terms of crossreactivity with a number of other BaP metabolites, other PAHs, and other PAH diol epoxide modified DNAs. The 8E11 crossreacts with a number of BaP metabolites (Tab. 1). Higher sensitivity is for the diols with weak recognition of phenols. There is also some weak crossreactivity with other PAHs, including pyrene, aminopyrene, and nitropyrene. It also recognizes the diolepoxide adducts of several other PAHs (Tab. 1). This antibody will thus recognize a number of PAHs and their metabolites with variable sensitivities.

Direct quantitation of adducts on protein cannot be carried out sensitively due to the low affinity of the antibody for the adduct in intact protein, probably due to burying of the adduct in hydrophobic regions of the protein (Tab. 1). Others have suggested that release of tetraols with acid treatment is a sensitive method for determination of protein adducts (18,22). However, our initial studies in mice treated with radiolabeled BaP indicated that only low levels of radioactivity could be released from globin by acid treatment (20). For this reason, we used an alternate approach for measurement of protein adducts. Globin was enzy-

Tab. 1.   Competitive inhibition of antibody 8E11 binding to BPDE-I-BSA

Competitiors	fmol Causing 50% Inhibition
BPDE-I-tetrol	350
BPDE-I-BSA digested	400
BPDE-I-BSA nondigested	1,450
BaP-7,8-diol	250
BaP-9,10 diol	150
4-OH-BaP	42,700
5-OH-BaP	$>1 \times 10^5$
BaP	6,000
1-Aminopyrene	70,000
1-Nitropyrene	16,000
Dimethylbenz(a)anthracene	$>1 \times 10^6$
1-OH-pyrene	3,400
Pyrene	16,200
BPDE-I-DNA	350
Chrysene diol epoxide DNA	160
Benz(a)anthracene diol epoxide DNA	1,350

Plates were coated with 25 ng BPDE-1-bovine serum albumin. Antibody 8E11 was used at a 1:2,500 dilution and goat antimouse IgG-alkaline phosphatase at a 1:500 dilution.

matically digested to peptides and amino acids before ELISA. When tested on protein modified in vitro with BPDE-I, a three- to four-fold increase in sensitivity resulted (Tab. 1). This assay was validated using globin isolated from animals treated with radiolabeled BaP. The ELISA was able to detect 90-100% of the adducts measured by radioactivity (5). These animal studies also indicated that adduct levels were about ten-fold higher in albumin than in globin. For this reason, our initial work on human samples has been with albumin isolated from workers occupationally exposed to PAHs. Albumin was isolated by Reactive blue 2-Sepharose CL-4B affinity chromatography and enzymatically digested with insoluble protease coupled to carboxymethyl cellulose which could be easily removed by centrifugation. Samples were then analyzed by competitive ELISA with

Tab. 2. Highest serum dilution showing antibodies against BPDE-DNA*

| | No Samples with Positive Titer | | |
	Negative	1:125	1:625
Foundry workers			
High	1	1	1
Medium	2	2	3
Low		1	3
Controls	4	3	3
Roofers	8	1	3
Controls	6	1	5

*Serum was serially diluted and tested by noncompetitive ELISA on plates coated with calf thymus DNA and BPDE-I-DNA. Criteria for positive titer are taken from C.C. Harris et al. (3). Titers are given as the highest dilution of serum which gave absorbance values for BPDE-I-DNA coated wells greater than two standard deviations above that for nonmodified DNA.

antibody 8E11. Initial studies have been carried out on a small number of roofers occupationally exposed during the removal of an old pitch roof and application of new hot asphalt. These studies were carried out in collaboration with Dr. Robin Herbert, Mt. Sinai Medical Center (New York). Seventy percent of the roofers samples (n = 12) were positive with a mean level of 5.4 fmol/μg, while 62% of the controls (n = 12) had detectible adduct levels (mean of 4.0 fmol/μg) (unpubl. data). In this small number of subjects there was a trend but no significant difference between roofers and controls. However, we are continuing studies of PAH-albumin adducts in a larger sample.

DETERMINATION OF SERUM ANTIBODIES AGAINST DNA ADDUCTS

Harris and co-workers have demonstrated the presence of antibodies against PAH diol epoxide-DNA adducts in the sera of individuals with various exposures to PAHs (3,21). They suggested that the presence of such antisera indicated adduct formation in vivo and might be useful as an alternate marker of exposure to PAHs. To test this hypothesis, we screened the sera of Finnish foundry workers, roofers, and controls. Plates were coated with 50 ng of BPDE-I-DNA or nonmodified calf thymus DNA. Sera was diluted and added to three wells: those coated with the modified and nonmodified DNAs and a noncoated well. After incubation, the presence of bound antisera was determined with goat antihuman IgG-alkaline phosphatase using p-nitrophenylphosphate as the substrate. Criteria for positive titer were those defined by Harris and co-workers (3) as an absorbance value for the modified DNA greater than two standard deviations above that for the control, nonmodified DNA. The results are given in Tab. 2. No significant difference was seen between the number of samples with positive titer in the sera from control individuals versus those with high occupational exposure. In contrast, white blood cell DNA adduct measurements in the exposed populations were significantly elevated compared to controls (8,10). Therefore, serum antibody measurement does not appear to be a useful marker of exposure to PAHs.

DETERMINATION OF MULTIPLE ADDUCTS

Human exposure to environmental carcinogens usually occurs in the form of complex mixtures. To monitor exposure to these mixtures, we would ideally like to perform multiple ELISAs on DNA samples from the same individual to determine all adducts present on DNA or protein. In order to get around the problem of limited availability of peripheral blood cell DNA, we have initially tested whether two different DNA adducts can

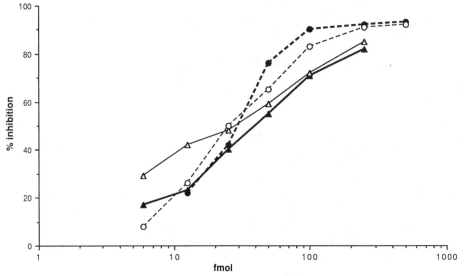

Fig. 3.   Multiple adduct analysis by competitive ELISAs. BPDE-I-DNA was mixed with 8-MOP-DNA, serially diluted and 50 µl mixed with 50 µl of a mixture of BPDE-I-DNA antibody #29 (final dilution 1:30,000) and 8-MOP-DNA antibody 8G1 (1:5,000). This mixture was sequentially incubated on plates coated with BPDE-I-DNA and 8-MOP-DNA. Both antibodies were also used in standard assays. The curves are BPDE-I-DNA in the conventional assay (△), BPDE-I-DNA in the mixture assay (▲), 8-MOP-DNA in the conventional assay (O), and 8-MOP-DNA in the mixture assay (●). Details for the standard ELISA for BPDE-I-DNA are given in Ref. 13 and for 8-MOP-DNA in Ref. 14.

be measured in a DNA sample with specific antibodies recognizing the individual adducts. DNA modified in vitro with BPDE-I was mixed with that modified by 8-methoxypsoralen and UVA light. A mixture of the antibodies recognizing both adducts, each at the appropriate final dilution for the ELISA, was then made. Serial dilutions of the mixed DNAs (50 μl) were then added to 50 μl of mixed antibodies. This competitive mixture was first added to plates coated with BPDE-I-DNA and after a 90 min incubation transferred to plates coated with 8-methoxypsoralen (8-MOP-DNA). Each plate was then incubated with the appropriate alkaline phosphatase conjugate as in the standard assay. These assays were compared to the standard assay for both antibodies (Fig. 3) which indicated that when BPDE-I-DNA is analyzed alone or in the presence of 8-MOP-DNA similar 50% inhibition values result. Similarly for 8-MOP-DNA, no difference is seen when analyzed alone or in the presence of BPDE-IDNA. These results suggest that it may be possible to make a cocktail of antisera to specific DNA adducts and, by sequential transfer to plates coated with the appropriate antigen, quantitate a number of different DNA adducts. Since the DNA adducts are not destroyed by incubation with antibodies in the ELISA, it can be recovered from the competitive mixture on the microwell plate and repurified with phenol/chloroform extraction. The DNA could then be utilized for additional analysis by alternate methods such as postlabeling or fluorescence.

DISCUSSION

Monitoring exposure to complex mixtures with immunological methods has a number of advantages as well as several disadvantages. The ease and rapidity of immunoassays make them ideal for epidemiological studies where large numbers of samples must be analyzed. In addition, no radioactivity is utilized. In general antibodies, even monoclonals, will recognize a number of structurally related adducts. Such chemical class specific antisera cannot give absolute quantitation of adducts. However, even with this limitation, we and others have demonstrated that immunoassays can be useful for the sensitive detection of carcinogen adducts in a number of human populations.

ACKNOWLEDGEMENTS

This work was supported by grants from NIH OHO2622, ESO3881, CA21111, CA35809, CA39174, and ACS41387.

REFERENCES

1. Everson, R.B., E. Randerath, R.M. Santella, R.C. Cefalo, T.A. Avitts, and K. Randerath (1986) Detection of smoking-related covalent DNA adducts in human placenta. Science 231:54-57.
2. Gan, L.-S., P.L. Skipper, X. Peng, J.D. Groopman, J.-S. Chen, G.N. Wogan, and S.R. Tannebaum (1988) Serum albumin adducts in the molecular epidemiology of aflatoxin carcinogenesis: Correlation with aflatoxin B_1 intake and urinary excretion of aflatoxin M_1. Carcinogenesis 9:1323-1325.
3. Harris, C.C., K. Vahakangas, J.M. Newman, G.E. Trivers, A. Shamsuddin, N. Sinopoli, D.L. Mann, and W.E. Wright (1985) De-

tection of benzo(a)pyrene diol epoxide-DNA adducts in peripheral blood lymphocytes and antibodies to the adducts in serum from coke oven workers. Proc. Natl. Acad. Sci., USA 82:6672-6676.

4. Hemminki, K., F.P. Perera, D.H. Phillips, K. Randerath, M.V. Reddy, and R.M. Santella (1988) Aromatic DNA adducts in white blood cells of foundry workers. In Methods for Detecting DNA Damaging Agents in Humans: Applications in Cancer Epidemiology and Prevention, H. Bartsch, K. Hemminki, and I.K. O'Neill, eds. IARC, Lyon, France, Vol. 89, pp. 190-195.

5. Lee, B.M., and R.M. Santella (1988) Quantitation of protein adducts as a marker of genotoxic exposure: Immunologic detection of benzo(a)pyrene-globin adducts in mice. Carcinogenesis 9:1773-1777.

6. Menkveld, G.J., C.J. VanDerLaken, T. Hermsen, E. Kriek, E. Scherer, and L.D. Engelse (1985) Immunohistochemical localization of O^6-ethyldeoxyguanosine and deoxyguanosin-8-yl(acetyl) aminofluorene in liver sections of rats treated with diethylnitrosamine, ethylnitrosourea, or N-acetylaminofluorene. Carcinogenesis 6:263-270.

7. Perera, F.P., R.M. Santella, D. Brenner, M.C. Poirier, A.A. Munshi, H.K. Fischman, and J. van Ryzin (1987) DNA adducts, protein adducts and SCE in cigarette smokers and nonsmokers. JNCI 79:449-456.

8. Perera, F.P., K. Hemminki, T.L. Young, R.M. Santella, D. Brenner, and G. Kelly (1988) Detection of polycyclic aromatic hydrocarbon-DNA adducts in white blood cells of foundry workers. Cancer Res. 48:2288-2291.

9. Perera, F., J. Mayer, A. Jaretzki, S. Hearne, D. Brenner, T.L. Young, H.K. Fischman, M. Grimes, S. Grantham, M.X. Tang, W.-Y. Tsai, and R.M. Santella (1989) Comparison of DNA adducts and sister chromatid exchange in lung cancer cases and controls. Cancer Res. 49:4446-4451.

10. Phillips, D.H., K. Hemminki, A. Alhonen, A. Hewer, and A. Grover (1988) Monitoring occupational exposure to carcinogens: Detection by ^{32}P-postlabelling of aromatic DNA adducts in white blood cells from iron foundry workers. Mutat. Res. 204:531-541.

11. Poirier, M.C., R. Santella, I.B. Weinstein, D. Grunberger, and S.H. Yuspa (1980) Quantitation of benzo(a)pyrene-deoxyguanosine adducts by radioimmunoassay. Cancer Res. 40:412-416.

12. Poirier, M.C. (1984) The use of carcinogen-DNA adduct antisera for quantitation and localization of genomic damage in animal models and the human population. Env. Muta. 6:879-887.

13. Santella, R.M., C.D. Lin, W.L. Cleveland, and I.B. Weinstein (1984) Monoclonal antibodies to DNA modified by a benzo(a)pyrene diol epoxide. Carcinogenesis 5:373-377.

14. Santella, R.M., N. Dharamaraja, F.P. Gasparro, and R.L. Edelson (1985) Monoclonal antibodies to DNA modified by 8-methoxypsoralen and untraviolet A light. Nucl. Acids Res. 13:2533-2544.

15. Santella, R.M., F.P. Gasparo, and L.L. Hsieh (1987) Quantitation of carcinogen-DNA adducts with monoclonal antibodies. Prog. in Exp. Tumor Res. 31:63-75.

16. Santella, R.M., A. Weston, F.P. Perera, G.T. Trivers, C.C. Harris, T.L. Young, D. Nguyen, B.M. Lee, and M.C. Poirier (1988) Interlaboratory comparison of antisera and immunoassays for benzo(a)pyrene-diol-epoxide-I-modified DNA. Carcinogenesis 9:1265-1269.

17. Santella, R.M. (1988) Application of new techniques for the detection of carcinogen adducts to human population monitoring. Mutat. Res. 205:271-282.
18. Shugart, L. (1986) Quantifying adductive modification of hemoglobin from mice exposed to benzo(a)pyrene. Analyt. Biochem. 152:365-369.
19. Stern, R.S., S. Zierler, and J.A. Parrish (1980) Skin carcinoma in patients with psoriasis treated with topical tar and artificial ultraviolet radiation. Lancet 2:732-733.
20. Wallin, H., A.M. Jeffrey, and R.M. Santella (1987) Investigation of benzo(a)pyrene-globin adducts. Cancer Lett. 35:139-146.
21. Weston, A., G. Trivers, K. Vahakangas, M. Newman, and M. Rowe (1987) Detection of carcinogen-DNA adducts in human cells and antibodies to these adducts in human sera. Prog. in Exp. Tumor Res. 31:76-85.
22. Weston, A., M.I. Rowe, D.K. Manchester, P.B. Farmer, D.L. Mann, and C.C. Harris (1989) Fluorescence and mass spectral evidence for the formation of benzo(a)pyrene anti-diol-epoxide-DNA and -hemoglobin adducts in humans. Carcinogenosis 10:251-257.
23. Wraith, M.J., W.P. Watson, C.V. Eadsforth, N.J. van Sittert, and A.S. Wright (1988) An immunoassay for monitoring human exposure to ethylene oxide. In Methods for Detecting DNA Damaging Agents in Humans: Applications in Cancer Epidemiology and Prevention, H. Bartsch, K. Hemminki, and I.K. O'Neill, eds. IARC Publications, Lyon, France, Vol. 89, pp. 271-274.
24. Yang, X.Y., V. DeLeo, and R.M. Santella (1987) Immunological detection and visualization of 8-methoxypsoralen-DNA photoadducts. Cancer Res. 47:2451-2455.

^{32}P-POSTLABELING DNA ADDUCT ASSAY: CIGARETTE SMOKE-INDUCED

DNA ADDUCTS IN THE RESPIRATORY AND NONRESPIRATORY

RAT TISSUES

R.C. Gupta[1,3] and C.G. Gairola[2,3]

[1]Department of Preventive Medicine and
Environmental Health

[2]Tobacco and Health Research Institute

[3]Graduate Center for Toxicology
University of Kentucky
Lexington, Kentucky 40536

INTRODUCTION

Several epidemiological studies have strongly implicated cigarette smoking with higher incidence of respiratory tract cancer including those of larynx, the oral cavity, and the lung (4). In addition, smokers face an increased risk for cancer of the pancreas and bladder, among other organs (14,20).

The smoking of a cigarette leads to the formation of two types of smoke: (i) mainstream cigarette smoke, which is generated during the puff drawing and is voluntarily inhaled by smokers; and (ii) sidestream cigarette smoke, which is generated between the puffs and is the main constituent of environmental tobacco smoke to which both smokers and nonsmokers may involuntarily be exposed.

Bioassays have shown that cigarette smoke and its condensates are carcinogenic in experimental animals (3,22). In the rat, nose-only chronic exposure to cigarette smoke has been found to induce a variety of hyperplastic and metaplastic changes in the upper and lower respiratory tracts of animals (21). Daily long-term exposure of rats to mainstream cigarette smoke has been reported to increase the incidence of tumors in the nasal and pulmonary regions of the respiratory tract (2). By what mechanism(s) the cigarette smoke induces such precancerous alterations in the respiratory tract is not clearly understood.

An interaction between carcinogens and macromolecules, particularly DNA to form addition products (adducts), is believed to be a key event in the initiation of chemical carcinogenesis. Recent development of a

Genetic Toxicology of Complex Mixtures
Edited by M. D. Waters *et al.*
Plenum Press, New York, 1990

highly sensitive and versatile technique, the ^{32}P-postlabeling assay
(10,11,17), now makes it possible to detect and quantitate such DNA
lesions induced by the reactive constituents of cigarette smoke. Here we
have extended our past studies (12) and determined how chronic exposure
to cigarette smoke affects the distribution of DNA adducts in selected
respiratory and nonrespiratory tissues.

MATERIALS AND METHODS

Animals and Treatment Groups

Male Sprague-Dawley rats (9 to 10 weeks of age; 200 g) were pur-
chased from Harlan/Sprague-Dawley and housed in quarantine rooms for
two weeks. Animals showing no abnormal signs were randomly selected
for the study and divided into the following three groups: (a) room
controls (RC), handled once a week during weekly cage cleaning; (b)
sham controls (SH), given daily treatment identical to those of the smoke-
exposed group, but in the absence of smoke; and (c) smoke-exposed
groups (SM), exposed to freshly generated mainstream smoke from a high
tar/high nicotine University of Kentucky reference cigarette (2R1) for 4
to 40 weeks.

Exposure System and Indicators of Smoke Exposure

The smoke generation and exposure system has been described in
detail (7,8). Briefly, the mainstream smoke is generated once each
minute by a two-second 35-ml puff from the cigarette. An average of ten
puffs are generated from each 2R1 cigarette and the smoke is transferred
to a mainstream smoke exposure chamber, into which the animals' noses
protrude. This exposure system allows exposure of animals to fresh main-

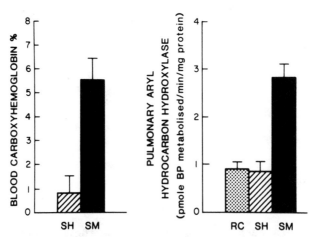

Fig. 1.   Blood carboxyhemoglobin values (mean ± SE) of rats measured
          biweekly over the entire duration of exposure, and aryl hydro-
          carbon hydroxylase activity measured in lung tissue 9,000 x g
          supernatants.   ▦, room control (RC); ▨, sham control (SH);
          and ■, smoke-exposed (SM); blood carboxyhemoglobin levels in
          RC group were not measured.

stream smoke intermittently in a fashion analogous to human smoking. Exposure of animals to cigarette smoke was ascertained by measuring inhaled total particulate matter (TPM), carboxyhemoglobin (COHb) levels and pulmonary aryl hydrocarbon hydroxylase (AHH) activity (5,8).

Isolation of DNA and Measurement of DNA Adducts

DNA was isolated from 0.1 to 0.4 g of individual tissues (lung, larynx, trachea, heart, and bladder), and from 50 to 90 mg of pooled (from four to five rats) nasal mucosa, using a rapid solvent extraction procedure (9) in which protein and RNA are removed by digestion with proteinase K and RNases A + T1, respectively, followed by solvent extractions and ethanol precipitation. DNAs were obtained essentially free of RNA and protein but in varying yields (1.8 to 2.3 mg/g lung, nasal mucosa, and heart; 0.3 to 0.6 mg/g bladder, larynx, and trachea).

DNA adducts were analyzed by the ³²P-postlabeling assay (11) after enhancement of the sensitivity of detection (10,17) as outlined: After

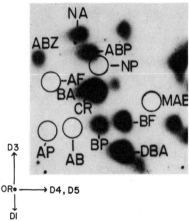

Fig. 2.   ³²P-Postlabeling analysis of calf thymus DNA reacted in vitro with active metabolites of various aromatic amines and polycyclic hydrocarbons as indicated below. After mixing the various DNAs (about 1 to 4 adducts/10⁷ nucleotides, each), the mixed DNA was enzymatically digested and adducts were extracted in 1-butanol, 5'-³²P-labeled, separated by multidirectional PEI-cellulose TLC and detected by autoradiography. Conditions were essentially as described (10,11), except that the eluting solvents in directions 4 and 5 were isopropanol:4N ammonia, 1.2:1 and 1.7 M Na phosphate, pH 6.0, respectively. Circles represent location of adducts that were analyzed in a separate experiment. OR = origin. ABZ = N-OH-N'-acetylbenzidine; AF = N-OH-2-aminofluorene; ABP = N-OH-4-aminobiphenyl; NA = N-OH-β-naphthylamine; AP = N-OH-2-aminophenanthrene; NP = N-OH-1-aminopyrene; MAB = N-benzoyloxy-methyl-4-azoaminobenzene; AB = N-benzoyloxy-4-azoaminobenzene; BP = benzo(a)pyrene diolepoxide; BA = benz(a)anthracene diolepoxide; CR = chrysene diolepoxide; BF = benzo(k)fluoranthene diolepoxide; DBA = dibenz(a,c)anthracene diolepoxide.

enzymatic digestion of DNA (5 to 20 μg), adducts were enriched by extraction in 1-butanol (11) or treatment with nuclease P1 (17), $5'-^{32}P$-
labeled, separated by multidirectional PEI-cellulose TLC, detected by
autoradiography and quantitated by measurement of the adduct radioactivity. To calculate adduct levels, total nucleotides were also $5'-^{32}P$-labeled
in parallel and analyzed by 1-directional TLC (10). Adduct levels were
then evaluated by calculating relative adduct labeling as (cpm in adduct/-
cpm in total nucleotides) x (1/dilution factor) which were then translated
into attomole of adducts per μg DNA (10).

RESULTS

Smoke Exposure

An estimation of the TPM intake (1.46 ± 0.56 mg/session/rat) during
the exposure of rats to cigarette smoke suggested a dose of 5 to 5.5
mg/kg body weight. The smoke-exposed group had significantly elevated
levels of both COHb (Fig. 1a) and pulmonary AHH activity (Fig. 1b) as
compared to the control, suggesting the animals were exposed to cigarette
smoke effectively.

Separation of a Complex Mixture of DNA Adducts

Before the ^{32}P-postlabeling assay was applied to detect multiples of
adducts that can be expected to be induced by a complex mixture such as
cigarette smoke, chromatography conditions were standardized. Numerous
aqueous salt solutions and organic solvent mixtures were screened. Figure 2 shows an optimal resolution of about 18 major adducts produced by
reaction of calf thymus DNA in vitro with active metabolites of various
aromatic amines and polycyclic aromatic hydrocarbons. All but two adducts [benz(a)anthracene- and chrysene-adducts] were clearly resolved,
suggesting that this solvent system was suitable to resolve a complex mixture of adducts.

Adduct Analysis of Respiratory and Nonrespiratory Tissue DNAs

Butanol extraction- and nuclease P1-versions of the ^{32}P-postlabeling
procedures have frequently been used to enrich DNA adducts prior to
^{32}P-labeling in order to achieve a sensitivity of detection of 1 adduct/10^{10}
nucleotides. Preliminary analysis of nasal mucosal and lung DNAs from
SM animals showed three to five times better recovery of lung DNA adducts by nuclease P1-mediated enrichment than butanol procedure, although nasal DNA adducts were recovered similarly in the two procedures
(data not shown). Therefore, all tissue DNAs were analyzed using the
enzymatic enrichment assay, except nasal DNA that was analyzed using
the extraction procedure.

^{32}P-Postlabeling adduct analysis of the nasal mucosal DNA from SM
rats revealed the presence of five adducts (Fig. 3a), while only one
adduct was detectable in the SH (Fig. 3b) and RC (not shown) control
DNAs. The control DNA adduct was chomatographically identical to spot
N1 of SM DNA. Quantitation of the adducts showed that the major spots
N2 and N3 corresponded to 46% and 26% of the total adduct radioactivity
and their levels increased with the duration of the exposure from four
weeks to 40 weeks (data not shown). The levels of spots N2 (14%) and

N5 (13%) also varied during the exposure. Spots N2 and N5 were not detectable after four weeks of the exposure. These results demonstrate that exposure to cigarette smoke induces covalent DNA modifications and that the alterations are dependent on the duration of the exposure.

      ³²P-Postlabeling analysis of lung DNAs from the SH and SM groups both showed at least eight distinct spots (Fig. 3c and 3d). Co-chromatography experiments established that the SH and SM DNA adducts were qualitatively identical. Analysis of several control samples in which DNA and the hydrolyzing or the phosphorylating enzyme(s) were individually omitted did not reveal any detectable amounts of these spots, indicating that the adducts indeed were DNA derived. Rechromatography of the major adducts #3 and #8 in three different solvents (0.6 M lithium chloride/0.5 M Tris·HCl/7 M urea, pH 8; 0.2 M sodium bicarbonate/6 M urea; and 0.7 M sodium phosphate/7 M urea, pH 6.4) failed to demonstrate any unresolved components, suggesting that these spots are single entities. As is also evident from the spot intensities in Fig. 3c and 3d, the levels of SM DNA adducts were an average of 15 times higher than in SH DNA adducts (Fig. 4). The qualitative profiles of the adducts were identical in every member of the SH and SM groups, but quantitatively a signifi-

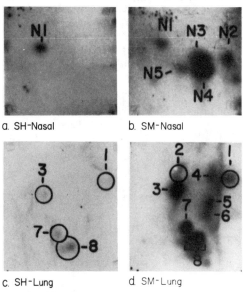

a. SH-Nasal   b. SM-Nasal

c. SH-Lung   d. SM-Lung

Fig. 3.   ³²P-Adduct maps of nasal mucosal and lung DNAs of male Sprague-Dawley rats exposed once a day, seven days a week, for 23 weeks to freshly generated mainstream cigarette smoke. Animals receiving treatment identical to that of the smoke-exposed group, except in the absence of cigarette smoke, served as sham controls. DNAs were analyzed as outlined in the Fig. 2 legend, except that the development in direction 4 for maps a and b was with isopropanol:4N ammonia, 1.3:1. Adducts N3 and N4 were sufficiently resolved and could be located easily after a shorter autoradiographic exposure than used for the present maps.

cant interanimal variation was observed (Fig. 4). Since low levels of adducts were also present in the SH groups, the results indicate that inhalation of cigarette smoke simply accelerates the formation of the pre-existing DNA adducts.

The adduct profiles seen in the SH and SM groups of other respiratory (larynx and trachea) and nonrespiratory (heart and bladder) tissues was essentially identical to the lung DNA adduct pattern, except that in trachea, larynx, and bladder DNAs of the SH group, only spots 1 and 8 were detectable (not shown). Quantitation of the adduct radioactivity revealed that adduct levels in the SM heart DNA were an average of 25-fold higher than in the corresponding SH DNA. A significant (5- to 10-fold) enhancement of DNA adducts in the SM as compared to SH group was also found in the larynx, trachea, and bladder (Fig. 4).

To determine the chemical nature of the smoke-related adducts, the lung DNA adducts were co-chromatographed under conditions outlined in Fig. 2 and compared with reference carcinogen adducts derived from N-OH derivatives of 4-aminobiphenyl, β-naphthylamine, 2-aminofluorene, 1-nitropyrene, and N'-acetylbenzidine, and diolepoxides of benz(a)anthracene, chrysene, dibenz(a)anthracene, and benzo(a)pyrene. None of the lung DNA adducts co-migrated with any of these compounds, except the predominant adduct #8 which co-migrated with benzo(a)pyrene diolepoxide adduct (not shown). However, as shown in Fig. 5, adduct #8 separated from this reference adduct in two of the other three solvents employed, indicating that the lung DNA adduct was unrelated to the benzo(a)pyrene diolepoxide adduct. Likewise, none of the nasal DNA adducts were found to be related to the reference adducts.

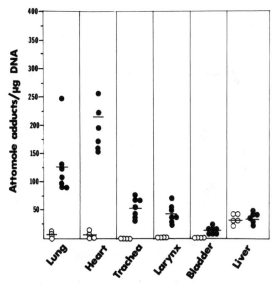

Fig. 4.   Total DNA binding levels in the various tissues of the sham and smoke-exposed rats. Experimental conditions of the exposure and adduct measurements were as described in text.

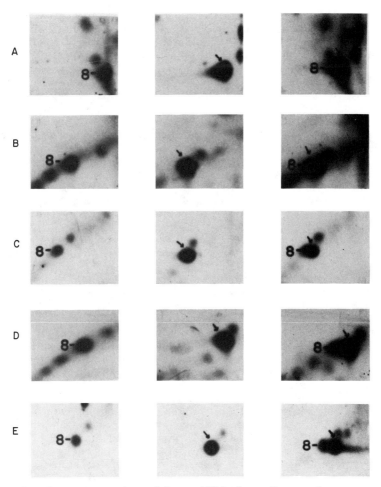

Fig. 5.   Co-chromatography of lung DNA from the smoke-exposed group
and reference calf thymus DNA modified with benzo(a)pyrene
diolepoxide.   The DNAs were mapped essentially as described in
the Fig. 2 legend, except that indicated solvents were used in
direction 4:   A, isopropanol:4N ammonia, 1:1; B, 0.7 M sodium
phosphate, 7 M urea, pH 8.0; C and D, 0.8 M lithium chloride,
0.5 M Tris·HCl, 7 M urea, pH 8.0; and E, 0.2 M sodium bicar-
bonate, 6 M urea.   The development was 1 cm (panel E), 4 cm
(panels A and C), or 8 cm (panels B and D) onto a Whatman 1
wick attached to the top of the sheet.   Spots 1 to 6 in the lung
DNA maps (see Fig. 2d) were lost due to over running the
chromatograms in directions 3 and/or 4.   Spot marked with an
arrow = dG-N²-7β,8α,9α-trihydroxy-7,8,9,10-tetrahydro benzo-
(a)pyrene; other spots, unknown.

DISCUSSION

Exposure to mainstream cigarette smoke significantly elevated the levels of blood COHb and pulmonary AHH. The TPM data showed an average dose of 5.2 mg smoke particulates/kg body weight of animal/day. This is similar to an estimated dose of 6.8 mg/kg/day for human exposure of an adult smoking 20 cigarettes/day (1). These observations demonstrated that animals effectively inhaled both gaseous as well as particulate constitutents of cigarette smoke, but received a relatively milder dose of smoke TPM.

An analysis of the tissue DNA adducts in these animals by the sensitive ^{32}P-postlabeling assay, which is capable of detecting structurally diverse aromatic and/or hydrophobic DNA adducts at a level of 1 adduct/-10^{10} nucleotides, showed one to eight detectable DNA adducts in lung, trachea, larynx, heart, and bladder of the sham controls. Chronic exposure of animals to mainstream cigarette smoke showed a remarkable enhancement of most adducts in the lung and heart DNA. The enhancement was less pronounced in the other respiratory and nonrespiratory tissues. In contrast, four of the five nasal DNA adducts formed in the smoke-exposed rats were undetectable in control animals. Since cigarette smoke contains several thousand chemicals and a few dozen of them are known or suspected carcinogens, the difference between the DNA adducts of nasal and the other tissues may reflect the diversity of reactive constituents and their differential absorption in different tissues.

In previous studies (12), we showed the presence of mainly three adducts (corresponding to spot #1, #2, and #3 in Fig. 3c and 3d) in lung DNA of control and smoke-exposed rats. However, in the present investigation, several additional spots were also detected in both the control and SM animals (Fig. 3c and 3d). To determine whether the additional adducts were induced due to some biological factors or some experimental variables during analysis, a few lung DNAs from SM animals stored at -80°C for over two years were reanalyzed. Two major and several minor adducts were detected in these DNAs, essentially as seen in the present investigation. This result and the fact that no change in the smoke exposure protocol was made suggest instability of the adducts. It may be pointed out, however, that all tissue DNAs analyzed in this investigation repeatedly gave identical spot patterns.

The smoke-related tissue DNA adducts were aromatic and/or hydrophobic in nature, as reflected by their extractability in 1-butanol (10) and selective chromatography (10,11). In comparison to the lung DNA adducts, the adducts in nasal DNA were less hydrophobic as suggested by their faster chromatographic mobility and requirement for less polar developing solvent in direction 4 (Fig. 3). Identity of the predominant adduct #3 and #8 was further investigated by comparison with several reference DNA adducts prepared by reacting calf thymus DNA with active metabolites of 4-aminobiphenyl, 2-naphthylamine, N'-acetylbenzidine, 2-aminofluorene, 2-aminophenanthrene, benzo(a)pyrene, benz(a)anthracene, dibenz(a,c)anthracene, benz(k)fluoranthene, and chrysene. Since some of these chemicals are present in cigarette smoke, our results suggest that these constituents of cigarette smoke may not be directly responsible for formation of DNA adducts in the lungs and heart of the smoke-exposed animals. Similar chromatography experiments indicated that the smoke-induced adducts in nasal DNA might also be unrelated to the above carcinogens in the cigarette smoke. Since cigarette smoke contains many other DNA-reactive constituents, the formation of new nasal DNA adducts and enhancement of the other tissue DNA adducts may therefore occur as a result of the interaction of some other smoke constituents with DNA, thus suggesting that the adducts detected in this study possibly represent only a fraction of the total DNA-damaging potential of the cigarette smoke. However, it is also possible that some of the smoke carcinogens, when present with other fresh smoke constituents, e.g., CO may inhibit their interaction with cellular DNA because of either the inability of cytochrome P-450 enzymes to activate them to reactive species or inactivation of the reactive metabolites before their interaction with DNA.

Aging has been shown to induce structural changes in DNA (reviewed in Ref. 18 and 19), accumulation of chromosomal mutations (15) and generation of DNA adducts (6,13,16). Lung cancer and heart disease are both primarily diseases of aging, and it is conceivable that cigarette smoke simply accelerates the age-related DNA damage, thereby predisposing smokers to lung and heart diseases.

ACKNOWLEDGEMENTS

Expert technical assistance of Karen Earley and Judy Glass is gratefully acknowledged. This work was supported by the EPA Co-operative Agreements CR 813840 and CR 816185 (R.C.G.) and by grants from USPHS CA 30606 (R.C.G.) and KTRB 41031 (C.G.G.).

The research described has been reviewed by the Health Effects Research Laboratory, U.S. Environmental Protection Agency, and approved for publication. Approval does not signify that the contents necessarily reflect the view and policies of the Agency, nor does mention of trade names or commercial products constitute endorsement or recommendation for use.

REFERENCES

1. Binns, R. (1977) Inhalation toxicity studies on cigarette smoke IV. Expression of the dose of smoke particulate material to the lungs of experimental animals. Toxicology 7:189-195.
2. Dalbey, W.E., P. Nettesheim, R. Griesemer, J.E. Caton, and M.R. Guerin (1980) Chronic inhalation of cigarette smoke by F344 rats. JNCI 64:383-390.
3. Dontenwill, W.P. (1974) Tumorigenic effect of chronic cigarette smoke inhalation on Syrian golden hamster. In Experimental Lung Cancer. Carcinogenesis and Bioassays, E. Karbe and E. Park, eds. Springer-Verlag, New York, pp. 331-368.
4. Fielding, J.E. (1985) Smoking--Health effects and control, Part II. New Eng. J. Med. 313:555-561.
5. Gairola, C. (1987) Pulmonary aryl hydrocarbon hydroxylase activity of mice, rats, and guinea pigs following long-term exposure to mainstream and sidestream cigarette smoke. Toxicology 45:177-184.
6. Garg, A., and R.C. Gupta (1988) Tissue-specific DNA modifications in untreated, aged rats as detected by ³²P-adduct assay. Proc. Am. Assoc. Cancer Res. 29:104.
7. Griffith, R.B., and R. Handcock (1985) Simultaneous mainstream-

sidestream smoke exposure systems I. Equipment and procedures. Toxicology 34:123-138.

8. Griffith, R.B., and S. Standafer (1985) Simultaneous mainstream-sidestream smoke exposure systems II. The rat exposure system. Toxicology 35:13-24.

9. Gupta, R.C. (1984) Nonrandom binding of the carcinogen N-hydroxy-2-acetylaminofluorene to repetitive sequences of rat liver DNA in vivo. Proc. Natl. Acad. Sci., USA 81:6943-6947.

10. Gupta, R.C. (1985) Enhanced sensitivity of ^{32}P-postlabeling analysis of aromatic carcinogen-DNA adducts. Cancer Res. 45:5656-5662.

11. Gupta, R.C., M.V. Reddy, and K. Randerath (1982) ^{32}P-Postlabeling analysis of non-radioactive aromatic carcinogen-DNA adducts. Carcinogenesis (Lond.) 3:1081-1092.

12. Gupta, R.C., M.L. Sopori, and C.G. Gairola (1989) Formation of cigarette smoke-induced DNA adducts in the rat lung and nasal mucosa. Cancer Res. 49:1916-1920.

13. Gupta, R.C., K. Earley, J. Locker, and B. Lombardi (1987) ^{32}P-Postlabeling analysis of liver DNA adducts in rats chronically fed a choline-devoid diet. Carcinogenesis (Lond.) 8:187-189.

14. International Agency for Research on Cancer Tobacco Smoking (1986) IARC Monograph 38:199-293.

15. Martin, G.M., A.C. Smith, D.J. Ketterer, C.E. Ogburn, and C.M. Disteche (1985) Increased chromosomal aberrations in first metaphases of cells isolated from the kidneys of aged mice. Israel J. Med. Sci. 21:296-301.

16. Randerath, K., M.V. Reddy, and R.M. Disher (1986) Age- and tissue-related DNA modifications in untreated rats: Detection by ^{32}P-postlabeling assay and possible significance for spontaneous tumor induction and aging. Carcinogenesis (Lond.) 7:1615-1617.

17. Reddy, M.V., and K. Randerath (1986) Nuclease P1-mediated enhancement of sensitivity of ^{32}P-postlabeling test for structurally diverse DNA adducts. Carcinogenesis (Lond.) 7:1543-1551.

18. Schneider, E.L. (1985) Cytogenetics of aging. In Handbook of the Biology of Aging, C.E. Finch and E.L. Schnieder, eds. Van Nostrand Reinhold, New York, pp. 357-373.

19. Scmookler-Reiss, R.J., and S. Goldstein (1985) Genetic diversification and the clonality of senscence. In Review of Biological Research in Aging, Vol. 2, M. Rothstein, ed. Alan R. Liss, New York, pp. 115-139.

20. U.S. Surgeon General (1982) The Health Consequences of Smoking--Cancer (PHS) 82-50179:5-8.

21. Wehner, A.P., G.E. Dagle, E.M. Milliman, P.W. Phelps, D.B. Carr, J.R. Decker, and R.E. Filpy (1981) Inhalation bioassays of cigarette smoke in rats. Tox. Appl. Pharm. 61:1-17.

22. Wynder, E.L., and D. Hoffman (1967) Tobacco and Tobacco Smoke-studies in Experimental Carcinogenesis. Academic Press, New York, pp. 730-752.

POSTLABELING ANALYSIS OF POLYCYCLIC AROMATIC HYDROCARBON:

DNA ADDUCTS IN WHITE BLOOD CELLS OF FOUNDRY WORKERS

K. Hemminki,[1] K. Randerath,[2] and M.V. Reddy[2]

[1]Institute of Occupational Health
Topeliuksenkatu 41 a A, 00250 Helsinki, Finland

[2]Department of Pharmacology
Baylor College of Medicine
Houston, Texas 77030

ABSTRACT

Blood samples were obtained from 61 volunteers working in a Finnish iron foundry who were occupationally exposed to polycyclic aromatic hydrocarbons (PAHs) from 19 control subjects not known to be occupationally exposed to this class of chemical carcinogens. Foundry workers were classified as belonging to high, medium, or low exposure groups according to their exposure to airborne benzo(a)pyrene (BaP) (high > 0.2, medium 0.05-0.2, low < 0.05 µg BaP/m^3 air). Aromatic adducts were determined in white blood cell DNA from the subjects using the ^{32}P-postlabeling technique. There was a high correlation between the estimated exposure and adduct levels.

Differences in the level of adducts were also noted between job categories in the foundry. This study suggests that the ^{32}P-postlabeling assay may be useful in monitoring human exposure to known and previously unidentified environmental genotoxic agents.

INTRODUCTION

Determination of DNA adducts of carcinogenic agents has emerged as a possible dose monitor in exposed humans (2,4,6,14,15). The methods available include the ^{32}P-postlabeling method (7,21,23) immunoassay (13,17,18) scanning fluorescence spectroscopy (29) and mass spectroscopy (12,30). Qualitatively, the methods have been validated by comparing exposed populations to nonexposed controls. The exposures studied in humans include tobacco smoke (3,13,17,22), PAHs in the occupational environment (8,9,12,28,30), and cisplatin in cancer chemotherapy (5,19). Validity of the ^{32}P-postlabeling study has been demonstrated in an inter-

Genetic Toxicology of Complex Mixtures
Edited by M. D. Waters *et al.*
Plenum Press, New York, 1990

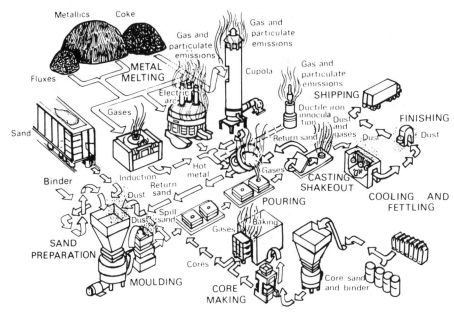

Fig. 1.   Iron foundry processes indicating emission sources from Ref. 1
           and 11.

Tab. 1.   Concentrations of BaP in air samples of iron foundries by work-
           place (coal-tar pitch method).   From two foundries given in
           Refs. (1,11)

Operation	Mean BaP (μgm^3)
Melting	0.2
Moulding	2.2
Casting	2.1
Shake-out	12.6
Fettling	0.5
Transport and others	1.4

laboratory comparison (25). In order to demonstrate the applicability of
the methods in dose monitoring, dose-response studies should be carried
out. We have previously participated in such a study on foundry workers using BaP-DNA-antibodies (17) and, to a limited extent, using
[32]P-postlabeling (20). Here we report results of a large coded study
among foundry workers to establish dose-response relationships using the
[32]P-postlabeling technique.

EXPOSURES IN IRON FOUNDING

Iron founding processes have been reviewed by the International
Agency for Research on Cancer (11). Founding comprises the basic operations that are illustrated in Fig. 1: patternmaking, moulding, core-
making, melting, casting, shake-out, and fettling (1). However, variations and modern technologies have affected processes, particularly moulding and the binding materials involved. PAHs are formed during thermal
decomposition of the organic ingredients of the mould material, and the
levels of PAHs formed are dependent on the types of moulds used
(1,11,26). Schimberg et al. (27) measured the concentrations of BaP in
air samples from two foundries using coal-tar pitch, and from four foundries using coal powder sand additives. BaP levels were more than one
order of magnitude higher in the foundries using coal-tar pitch binders.

PAHs being predominantly formed in the casting process and released
into the foundry air at the time of casting and shaking-out of the moulds,
the levels of PAHs measured differ extensively by workplace (Tab. 1).
Pattern making, melting, and fettling processes usually involve low levels
of exposure to PAHs. By contrast, processes around molding, casting,
and shake-out involve high levels of exposure. Transports in that area
are also associated with at least intermediate levels of exposure. Because
PAHs are absorbed to the mould sand, reuse of sand may also entail exposure to PAHs (11).

MATERIALS AND METHODS

Blood samples were obtained from healthy volunteers from a Finnish
iron foundry. Industrial hygiene measurements for PAHs were carried
out in the foundry in the years 1978-1980; as the work processes have
remained practically unchanged since then, these measurements were used
by two industrial hygienists familiar with the foundry to grade the volunteers for daily exposure by job description (see section on "Exposures in
Iron Founding"). BaP levels in the workplace atmosphere were used as
guidelines to assign the exposure to PAHs as high (> 0.2 μg BaP/m^3),
medium (0.05-0.2 μg/BaP/m^3) and low (< 0.05 μg BaP/m^3). Control
samples were obtained from individuals coming from different parts of
Finland to the Institute of Occupational Health for examination. Their job
titles did not indicate occupational exposure to PAHs. Information on
current smoking (cig./day) was obtained for all subjects.

Blood (20-50 ml) was withdrawn into heparinized tubes and transported on ice to Helsinki. Cells were collected by centrifugation and red
blood cells were lysed by washing twice with 0.15 M NH_4Cl, each time followed by centrifugation at 1,000 g for 5 min. Nuclei were isolated by
washing the cells in 0.1 M NaCl, 0.5 % Tritox X, and 10 mM Tris pH 7.4,

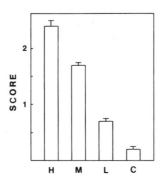

Fig. 2.   DNA adduct score in white blood cell of foundry workers in
           high (H), medium (M), and low (L) exposure category, and in
           the controls (C). The bars show weighed means ± SEM from 5
           (H), 32 (M), 24 (L), and 19 (C) individuals. It is estimated
           that the adduct score of 0 is < 5.0; 1, 5.0-10; 2, 10-20, and 3>
           20 adducts/10^8 nucleotides.

followed by centrifugation. DNA was isolated from nuclei by treatment
with pancreatic RNase and proteinase K followed by extraction with phenol
and chloroform : isoamyl alcohol (24:1), as described (20). Dried DNA
samples were shipped to Houston for adduct analysis.

DNA samples (5 µg) were assayed by the nuclease P_1 enhanced ver-
sion of ^{32}P-postlabeling as described by Reddy and Randerath (24) except that the amounts of nuclease P_1 and (γ-^{32}P) ATP were 7.5 µg and
100 µCi, respectively. ^{32}P-labelled adducts were resolved by the standard PEI-cellulose thin layer chromatography (TLC) techniques as detailed
previously (7,23,24) with contact transfer of adducts after D1 development. Solvents were: 1.0 M sodium phosphate, pH 6.8 (D1); 4.2 M lithium formate, 7.5 M urea, pH 3.35 (D3); and 0.7 M sodium phosphate, 7.0
M urea, pH 6.4, or 0.8 M lithium chloride, 0.5 M Tris-HCl, pH 8.0 (D4).
In some experiments, radioactive background material was removed by development in 1.7 M sodium phosphate, pH 6.0 (24). Adduct spots were
detected by autoradiography with intensifying screens and quantified by
Cerenkov counting of excised areas of the chromatograms. The determinations were carried out in duplicate. For statistical evaluation, DNA
adduct scores were employed. One sample was drawn from most individuals. However, for five individuals, two samples were drawn, and for
one individual, three samples were drawn at different times. The mean
adduct scores were weighed by the number of specimens drawn from the
individuals. The adjusted results were assayed for by variance analyses
using a general linear model.

RESULTS

Figure 2 shows the weighted adduct scores from the foundry workers
and controls. The scores were 2.4, 1.7, 0.7, and 0.2 adducts/10^8
nucleotides in high, medium, and low exposure, and control populations,
respectively. The differences between each group are statistically significant.

Tab. 2   Adduct scores by main job descriptions.   For definition of the adduct score, see Fig. 2

Operation	Mean Score	(N^1)
Pattern making	0.0	(2)
Sand preparation	1.7	(3)
Melting	0.2	(16)
Moulding	2.0	(11)
Casting	1.8	(13)
Shake-out	2.0	(2)
Fettling	0.5	(6)
Transport	2.2	(6)

[1]N refers to different samples (repeated samples were obtained from six individuals, cf. Tab. 3).

Table 2 gives the adduct scores by main job description.   The scores in sand preparation, moulding, casting, shake-out, and transport are about 2.0.   These are also the operations where highest air concentrations of BaP have been measured according to Tab. 1.   By contrast, pattern making, melting, and fettling are associated with low levels of adducts, also in agreement with industrial hygienic measurements.

Tab. 3.   Postlabeling scores from individuals sampled in different time periods

Individual	Exposure	Scores measured
1	High	3;2
2	Medium	3;2
3	Medium	2;2
4	Medium	1;2
5	Medium	2;2
6	Medium	2;0;2

Tab. 4.   Adduct scores at work and after a four-week vacation

Group	Mean adduct score	(N)
At work[1]	2.1	(8)
After vacation	1.0	(7)

[1]The workers were employed three months after vacation.

Repeated samples, drawn in different time periods, were obtained from six individuals. Their adduct scores are shown in Tab. 3. The scores were reasonably uniform from the same individuals. However, for individual No. 6, one low result intervened between two higher scores.

The foundry workers had an annual summer vacation of four weeks. Samples were drawn immediately after they returned from vacation and three months later. The adduct score was 1.0 after vacation and 2.1 after three months of resumed work. The sampling was from the same individuals belonging to the high- and medium-exposure categories.

DISCUSSION

This is the largest coded study published so far to investigate dose-response relationships in humans with the postlabeling assay. The data show a high correlation between the estimated exposure to PAH and presence of aromatic adducts in white blood cells. This reinforces the finding of a smaller study, based on workers from the same foundry (20). The postlabeling assay appears to differentiate the exposure levels equally well as the immunoassay carried out with an antibody raised against BaP-modified DNA (17).

The mean adduct score was 2.4 for the highly exposed foundry workers and 0.2 for the controls. It is estimated that this relates to approximately 20 adducts/10^8 nucleotides in the foundry workers and some 5 adducts/10^8 nucleotides in the controls. These are higher than previous postlabeling data on foundry workers (10,20) because recent improvements in the postlabeling technique have increased the measured adduct levels by about one order of magnitude (25). Even though the present postlabeling data on foundry workers give adduct levels quite similar to those detected by immunoassay (17), and the results from the two techniques correlate (10), it is not clear whether the same types of adducts are being measured.

Most of the adducts detected in foundry worker DNA appear to be related to occupational exposure because of the extensive difference between the highly exposed and the controls. Furthermore, the doubling of the adduct score three months after a four-week vacation reinforces work-related effects.

The coded study reported here suggests that the postlabeling assay can be used as a dosimeter of human exposure to carcinogenic compounds in the occupational setting.

ACKNOWLEDGEMENTS

We thank Kim L. Putman for able assistance in some of the analyses and in manuscript preparation. Pentti Kyyronen's contribution to statistical analysis is gratefully acknowledged. This study was supported by grants from the Finnish Work Environment Fund and from the United States National Cancer Institute (Grant CA 43263).

REFERENCES

1. Baldwin, V.H., and C.W. Westbrook (1983) Environmental Assessment of Melting, Pouring, and Inoculation in Iron Foundries, Contract No. 68-02-3152 and 68-02-3170, U.S. Environmental Protection Agency, Research Triangle Park, NC.

2. Bartsch, H., K. Hemminki, and I.K. O'Neill, eds. (1988) Methods for Detecting DNA Damaging Agents in Humans: Applications in Cancer Epidemiology and Prevention. IARC Scientific Publications, Lyon, France, No. 89, pp. 1-518.

3. Everson, R.B., E. Randerath, R.M. Santella, R.C. Cefalo, T.A. Avitts, and K. Randerath (1986) Detection of smoking related covalent DNA adducts in human placenta. Science 231:54-57.

4. Farmer, P.H., H.-G. Neumann, and D. Henschler, (1987) Estimation of exposure of man to substances reacting covalently with macromolecules. Arch. Tox. 60:251-260.

5. Fichtinger-Schepman, A.-M.J., A.T. van Oosterom, P.H.M. Lohman, and F. Berends (1987) cis-Diamminedichloroplatinum (II)-induced DNA adducts in peripheral leukocytes from seven cancer patients: Quantitative immunochemical detection of the adduct induction and removal after a single dose of cis-diamminedichloroplatinum (II). Cancer Res. 47:3000-3004.

6. Garner, R.C. (1985) Assessment of carcinogen exposure in man. Carcinogenesis 6:1071-1078.

7. Gupta, R.C., M.V. Reddy, and K. Randerath (1982) ^{32}P-post-labeling analysis of non-radioactive aromatic carcinogen-DNA-adducts. Carcinogenesis 3:1081-1092.

8. Harris, C.C., K. Vahakangas, M.J. Newman, G.E. Trivers, A. Shamsuddin, N. Sinopoli, D.L. Mann, and W.E. Wright (1985) Detection of benzo(a)pyrene diol epoxide-DNA adducts in peripheral blood lymphocytes and antibodies to the adducts in serum from coke oven workers. Proc. Natl. Acad. Sci., USA 82:6672-6676.

9. Haugen, A., G. Becher, C. Bemestad, K. Vahakangas, G.E. Trivers, M.J. Newman, and C.C. Harris, (1986) Determination of polycyclic aromatic hydrocarbons in the urine, benzo(a)pyrene diol-epoxide-DNA adducts in lymphocyte DNA and antibodies to the adducts in sera from coke oven workers exposed to measured

amounts of polycyclic aromatic hydrocarbons in the work atmosphere. Cancer. Res. 46:4178-4183.

10. Hemminki, K., F.P. Perera, D.H. Phillips, K. Randerath, M.V. Reddy, and R.M. Santella (1988) Aromatic DNA adducts in white blood cells of foundry workers. In Methods for Detecting DNA Damaging Agents in Humans: Applications in Cancer Epidemiology and Prevention, H. Bartsch, K. Hemminki, and I.K. O'Neill, eds. IARC Scientific Publications No. 89, pp. 190-195, Lyon, France.

11. IARC (1984) Polynuclear Aromatic Compounds. Part 3: Industrial Exposure in Aluminum Production, Coal Gasification, Coke Production and Iron and Steel Foundring. Monographs on the Evaluation of the Carcinogenic Risk of Chemicals to Humans, IARC, Vol. 34, Lyon, France.

12. Manchester, D.K., A. Weston, J.-S. Choi, G.E. Trivers, P.V. Fennessey, E. Quintana, P.B. Farmer, D.L. Mann, and C.C. Harris (1988) Detection of benzo(a)pyrene diol epoxide-DNA adducts in human placenta. Proc. Natl. Acad. Sci., USA 85:9243-9247.

13. Perera, F.P., M.C. Poirier, S.H. Yuspa, J. Nakayama, A. Jaretzki, M.M. Curren, D.M. Knowles, and I.B. Weinstein (1982) A pilot project in molecular cancer epidemiology: Determination of benzo(a)pyrene-DNA adducts in animal and human tissues by innonassays. Carcinogenesis 3:1405-1410.

14. Perera, F., and I.B. Weinstein (1982) Molecular epidemiology and carcinogen-DNA adduct detection: New approaches to studies of human cancer causation. J. Chronic Dis. 3:581-600.

15. Perera, F.P. (1987) Molecular cancer epidemiology: A new tool in cancer prevention. JNCI 78:887-898.

16. Perera, F.P., R.M. Santella, D. Brenner, M.C. Poirier, A.A. Munshi, H.K. Fischman, and J. Van Ryzin (1987) DNA adducts, protein adducts and sister chromatid exchange in cigarette smokers and nonsmokers. JNCI 79:449-456.

17. Perera, F.P., K. Hemminki, T.L. Young, D. Brenner, G. Kelly, and R.M. Santella (1988) Detection of polycyclic aromatic hydrocarbon-DNA adducts in white blood cells of foundry workers. Cancer Res. 48:2288-2291.

18. Poirier, M.C., R. Santella, I.B. Weinstein, D. Grunberger, and S.H. Yuspa (1980) Quantitation of benzo(a) pyrene-deoxyguanosine adducts by radioimmunoassay. Cancer Res. 40:412-416.

19. Poirier, M.C., E. Reed, L.A. Zwelling, R.F. Ozols, C.L. Litterst, and S.H. Yuspa (1985) Polyclonal Antibodies to quantitate cis-diamminedichloroplatinum (II)--DNA adducts in cancer patients and animal models. Env. Health Persp. 62:89-94.

20. Phillips, D.H., K. Hemminki, A. Alhonen, A. Hewer, and P.L. Grover (1988) Monitoring occupational exposure to carcinogens: Detection by ^{32}P-postlabeling of aromatic DNA adducts in white blood cells from iron foundry workers. Mutat. Res. 204:531-541.

21. Randerath, K., M.V. Reddy, and R.C. Gupta (1981) ^{32}P-postlabeling test for DNA damage. Proc. Natl. Acad. Sci, USA 78:6126-6129.

22. Randerath, E., R.H. Miller, D. Mittal, T.A. Avitts, H.A. Dunsford, and K. Randerath (1989) Covalent DNA damage in tissues of cigarette smokers as determined by ^{32}P-postlabeling assay. JNCI 81:341-347.

23. Reddy, M.V., R.C. Gupta, E. Randerath, and K. Randerath (1984) ^{32}P-postlabeling test for covalent DNA binding of chemicals in vivo:

Application to a variety of aromatic carcinogens and methylating agents. Carcinogenesis 5:231-243.

24. Reddy, M.V., and K. Randerath (1986) Nuclease Pl-mediated enhancement of sensitivity of ^{32}P-postlabeling test for structurally diverse DNA adducts. Carcinogenesis 7:1543-1551.

25. Savela, K., K. Hemminki, A. Hewer, D.H. Phillips, K.L. Putman, and K. Randerath (1989) Interlaboratory comparison of the ^{32}P-postlabeling assay for aromatic DNA adducts in white blood cells of iron foundry workers. Mutat. Res. (in press).

26. Schimberg, R.W., P. Pfaffli, and A. Tossavainen (1978) Profile analysis of polycyclic aromatic hydrocarbons in iron foundries (German). Staub-Reinhalt. Luft, 38:273-276.

27. Schimberg, R.W., P. Pfaffli, and A. Tossavainen (1980) Polycyclic aromatic hydrocarbons in foundries. J. Tox. Env. Health 6:1187-1194.

28. Shamsuddin, A.K.M., N.T. Sinopoli, K. Hemminki, R.R. Boesch, and C.C. Harris (1985) Detection of benzo(a)pyrene: DNA adducts in human white blood cells. Cancer Res. 45:66-68.

29. Vahakangas, K., A. Haugen, and C.C. Harris (1985) An applied synchronous spectrophotometric assay to study benzo(a) pyrene-diolepoxide-DNA adducts. Carcinogenesis 6:1109-1116.

30. Weston, A., M.L. Rowe, D.K. Manchester, P.B. Farmer, D.L. Mann, and C.C. Harris (1989) Fluorescence and mass spectral evidence for the formation of benzo(a)pyrene anti-diol epoxide -DNA and -hemoglobin adducts in humans. Carcinogenesis 10:251-257.

PROTEIN ADDUCTS AS BIOMARKERS FOR CHEMICAL CARCINOGENS

Paul L. Skipper and Steven R. Tannenbaum

Division of Toxicology
Massachusetts Institute of Technology
Cambridge, Massachusetts 02139

INTRODUCTION

Protein adducts are one of several types of biomarkers which are valuable to the study of chemical carcinogenesis (8). Over the last 15 years, a great deal of developmental work has gone into making this particular biomarker a useful one. Much of the effort has been focused on hemoglobin as the target protein, but serum albumin has also proven interesting. Enough laboratory experiments and studies of human subjects have now been conducted that it is worthwhile to reflect on the results to attempt an assessment of the role protein adducts can play in the study of chemical carcinogenesis. This chapter specifically addresses issues related to epidemiological studies.

The National Research Council (NRC) has defined three classes of biomarkers, according to whether they are indicators of exposure, effect, or susceptibility (13). Although it is conceivable that protein adducts could be used to detect susceptibility, they are principally of value as biomarkers of exposure and effect. Systemic dose and biologically effective dose are the parameters most closely associated with protein adducts in the NRC breakdown of classes of biomarkers.

Protein adducts are best characterized as biomarkers of exposure. Indeed, the studies of human populations conducted thus far have mostly been analyzed in this context (1,2,7). It should be noted, though, that this may be largely for practical reasons. It is still much easier to design a study in which exposure is assessed in an independent fashion to develop a relationship to adduct levels than one in which adducts are related to endpoints such as mutation or cancer.

Even though protein adducts do fall more to the exposure end in the sequence of biomarkers, there are likely to be some combinations of carcinogen/target organ for which they are good markers of both effect and exposure. Such cases would be characterized by the absence of pharmacological compartmentalization of the carcinogen and by relatively simple or nonexistent metabolic conversions. Ethylene oxide is a good ex-

Genetic Toxicology of Complex Mixtures
Edited by M. D. Waters *et al.*
Plenum Press, New York, 1990

ample (5). In most cases, however, the preferred application is not readily apparent. When metabolism is complex and/or takes place in one or more compartments which are distant from the target, or when other physiological variables have a significant impact on adduct formation, it becomes difficult to predict the relationships between exposure and adducts or adducts and effects.

Study group size is also an important determinant of whether protein adducts should be treated as biomarkers of exposure or of effect. The protein adduct level in an individual, when there exists large interindividual variability in metabolism or other factors affecting the relationship between the intake of a carcinogen and the formation of protein adducts, is unlikely to be a good determinant of environmental contamination or systemic dose. However, it may be closely related to effect, in which case the determination of the level in an individual takes on more significance. Adduct levels in an individual can also very sensitively reveal changes in exposure experienced by that individual (2). Mean levels of an adduct measured in a cohort will accurately reflect environmental contamination, provided the cohort size is adequate. Likewise, better correlations between adduct levels and effects may be expected in group studies in which aggregate observations are made.

4-Aminobiphenyl (4-ABP) is a human carcinogen which forms hemoglobin adducts in greater yield than any other carcinogen which has been investigated thus far. Its metabolism is complex. Furthermore, its target organ, the urinary bladder, is remote from the site where most metabolism occurs. Its reactivity toward DNA within the target organ is influenced by factors which are different from those that govern hemoglobin adduct formation. All these properties combine to make this a nearly ideal compound to use to investigate the different characteristics of the biomarkers described above. The remainder of this chapter will describe various studies which have been performed using hemoglobin adducts of 4-ABP, and will survey additional problems we have approached using protein adducts of other carcinogens.

BIOMARKER OF EXPOSURE

Environmental Monitoring

One of the fundamental advantages of using protein adducts as biomarkers for assessing the environment is that they are integrating markers which sum the exposure from all sources over a relatively long time frame, typically several weeks. This property is extremely useful in that the contribution of any selected source or potential source to the total can be readily judged if that source can be controlled or manipulated. 4-ABP was originally chosen for study because it was suspected that there existed a predominant source of this carcinogen which could ethically be manipulated.

It has now been demonstrated that smokers are predominantly exposed to 4-ABP as the result of cigarette smoking. This conclusion may be reached from the results of studies of markedly different populations composed of smokers and nonsmokers. These include an all male cohort from Turin, Italy (3); students and employees of Columbia University, New York (2); smoking cessation classes conducted by the Stop Smoking Clinic (Danvers, MA) at seven surburban hospitals in eastern

Massachusetts (12); and pregnant women at term (unpubl. results). Mean adduct levels of 4-ABP in each of the groups of smokers are remarkably similar. Likewise, the nonsmokers in each group display quite constant adduct levels, which are typically 20-30% of the level in smokers.

Nonsmokers also appear to be exposed to 4-ABP. One obvious possibility for the source is environmental tobacco smoke (ETS). A recently concluded study was conducted to test this hypothesis by measuring 4--ABP adducts in subjects who could be categorized according to their degree of exposure to ETS (11). 3-ABP adducts were also measured since the previously mentioned studies of smokers had indicated that adducts of this isomer are more specific for exposure to tobacco smoke.

The subjects were divided into two groups (group 1 and group 2) according to whether there was detectable cotinine in their plasma. The two groups were then subdivided, according to the level of exposure reported in their interviews, to create a total of five groups (1a, 1b, 1c, and 2a, 2b). Levels of 4-ABP were marginally higher in group 2 and the difference was of borderline statistical significance: the median two-sample test statistic was 1.64 ($p=0.05$) and the Wilcoxon two-sample test statistic was 1.08 ($p=0.14$). A more significant difference in the hypothesized direction was found for 3-ABP levels. Group 2 had significantly higher levels than group 1. The median and Wilcoxon tests resulted in $p=0.027$ and $p=0.11$, respectively, using the lowest detectable level for subjects with undetectable adduct levels. When undetectable levels were omitted, p values were reduced to 0.017 and 0.05, respectively.

From this study it may be concluded that ETS is an identifiable source of 3- and 4-ABP and that persons with significant exposure to it constitute a distinguishable population. It should also be noted that the increases in adduct level in exposed as compared to nonexposed subjects is actually quite small and could be detected only because of the large number of subjects involved. Thus, although ETS is an identifiable source of aminobiphenyls in the environment, it does not account for much of the total. The origin of the remainder continues to be a perplexing issue.

Estimation of Systemic Dose

In a study of 40 Italian smokers of blond tobacco (3) it was observed that 4-ABP adduct levels were highly correlated with self-reported cigarette consumption. Linear regression analysis yielded a value of 5 pg adduct/g Hb for each cigarette smoked per day. The relationship between daily dose and adduct level resulting from chronic exposure has previously been modeled (17). Applying this model with the results of the Italian study yields a value for the systemic dose of 4-ABP per cigarette in the range of 0.5-1.2 ng. Chemical analysis of cigarette smoke has indicated that 4-ABP is present in the mainstream smoke at the level of 1-5 ng per cigarette (14). Thus, it appears that a considerable fraction, but probably not all, of the 4-ABP inhaled by a smoker is absorbed.

Only seven of the 40 subjects who participated in this study reported smoking more than 20 cigarettes per day (CPD). More recently, a study of more than 70 smokers, none of whom smoked less than 20 CPD, was conducted (unpubl. results). No significant correlation between adduct level and cigarette consumption was observed. An upward trend is apparent in the data, but it depends heavily on a small number of points.

Two explanations may be offered for the difference between the two studies. One is that heavy smokers modify their smoking patterns relative to lighter smokers so that they actually inhale similar amounts of smoke. This explanation is not supported by measurements of carboxyhemoglobin, a reliable index of inhalation, which increases in proportion to the number of cigarettes smoked (9). Neither is it supported by a recent report of ethylene oxide adducts in smokers in which the scatter plot of adducts vs cigarette consumption shows no evidence of plateau of high CPD (1).

An alternative explanation is that while heavy smokers may experience a greater systemic dose of 4-ABP than light smokers, they metabolize relatively less of the amine to the form which binds to hemoglobin as the result of induction of competing metabolic pathways. Changes in metabolic profile as the result of cigarette smoking have certainly been documented (4). Whether these would result in reduced formation of 4-hydroxylaminobiphenyl relative to other metabolites is not known.

Biologically Effective Dose

Since 4-ABP is a urinary bladder carcinogen, this dose may be defined as the amount of electrophilic DNA adduct-forming metabolites which actually reach the genomic material in bladder urothelium. Bladder DNA adducts of 4-ABP are repaired very slowly. It is thus possible in an experimental situation to measure the extent of DNA adduct formation as a means of approximating the effective dose resulting from a known systemic dose. Hemoglobin adducts can be determined simultaneously to assess to what extent they reflect the DNA adducts.

Such an experiment was conducted with six female beagle dogs which were treated with 4-ABP over a period of two weeks (10). Hemoglobin adducts were determined on eight different days of the study, beginning on the day before dosing. DNA adducts were analyzed at the end of the period after the animals were sacrificed. The results are presented in Tab. 1. Hemoglobin adducts are listed as the average of the four measurements made during the second week of the study, at which time the levels had become relatively constant.

No clear pattern emerges from these data. Both hemoglobin and DNA adducts varied over a range of about three-fold. Both Spearman rank analysis and linear regression analysis revealed no significant association of the two different types of adducts. Mechanistic studies are under way to investigate the basis for this apparent lack of association. For the present, it may be said that hemoglobin adduct levels of 4-ABP appear not to reveal the relative extent of bladder DNA adduct formation in individuals experiencing the same exposure to this amine.

BIOMARKER OF EFFECT

Whether carcinogen-protein adducts can be predictive of the biological endpoint cancer is a question which is only beginning to be addressed experimentally. Cancer is generally viewed as a multistage process initi-

Tab. 1.   Hemoglobin and DNA adducts in female beagle dogs after chronic dosing with 4-aminobiphenyl

Animal	Hb adduct (nmol/g)	DNA adduct per 10^6 nucleotides
1	400	10.7
2	730	12.5
3	480	20.4
4	490	9.0
5	540	8.4
6	240	18.5

ated by DNA adduct formation.   Therefore, to the extent that levels of
protein adducts reflect DNA damage in the target organ they will also represent the process of initiation.   If initiation is the limiting step in the
carcinogenic process then adduct level should be proportional to risk.   If
late-stage processes predominate, adduct levels will not be revealed as a
risk determinant.

There are various experimental approaches which could test the predictive value of protein adducts.   The most revealing, perhaps, is to
identify cancer cases and to measure adduct levels in those cases for
comparison with levels in appropriate controls.   This is a conventional
epidemiological approach which has not yet been attempted.   Its effective-
ness should be greatest if adduct levels can be measured in specimens
collected during the latency period of the cancer rather than when cases
are identified.

A second approach to testing the predictive value of protein adducts
is to measure levels in representative subgroups of larger populations for
comparison with incidence rates in those populations.   A shortcoming of
this approach is that it fails to test association of cases with elevated
adduct levels.   The principle advantage is that it can readily be undertaken.   The finding of an association between adduct levels and incidence
rates provides strong evidence for an etiological role for the chemical under study.   While that may seem trivial in the case of a compound with
known carcinogenicity, it should be emphasized that few human cancers
can be associated with exposure to any one chemical or class of chemicals.
Thus, this may be a powerful tool with which to test the role of suspected
carcinogens.

A study of the second type involving 4-ABP has been undertaken
(3).   Adduct levels of 4-ABP were measured in subjects drawn from a
population previously characterized with respect to bladder cancer incidence (19).   Bladder cancer had been found to be strongly associated
with cigarette smoking, and could be further associated with the type
(flue cured or air cured) of tobacco smoked.   In this population the relative risk (RR) for smokers of flue cured tobacco was 1.7 for those under

50 years of age and 2.8 for those aged 50-59. In the same age groups, the RR for smokers of air cured tobacco was 7.1 and 8.2, respectively.

Hemoglobin adducts of 4-ABP were measured in subgroups from this population distinct from those used to determine the relative risks. The subjects were age 55 or less. Among flue cured tobacco smokers the adduct level averaged 3.5-fold higher than in nonsmokers drawn from this population. Among air cured tobacco smokers the adduct levels were 5.6-fold higher. Clearly, although not strictly proportional to the relative risks, 4-ABP adduct levels rank in the same order as risk in the three different risk groups.

These results provide the first experimental support for the concept of using protein adducts as biomarkers of effect. A case-control study of bladder cancer is currently being developed to test this concept more rigorously.

Other Applications of Protein-Carcinogen Adducts

Our laboratory has applied or plans to apply methods similar to those described above to a variety of other important environmental carcinogens. A summary of the current status of each of these biomarkers is given below.

Aflatoxin B_1

This potent hepatocarcinogen binds to albumin in the rat in vivo but not to hemoglobin. On this basis we developed an analytical approach using monoclonal antibodies and fluorescence detection which has been applied to human serum albumin in China (6,7,15). An extremely good dose-response relationship was found between aflatoxin B_1 intake and albumin adducts and also between albumin adducts and urinary metabolites. The measurement of these albumin adducts has the potential to resolve the issue of relative risk for aflatoxin versus Hepatitis B virus for liver cancer in China and parts of Africa.

IQ (2-amino-3-methylimidazo[4,5-f]quinoline) and Related Compounds

This class of compounds (aminoimidazoquinolines and quinoxalines) is metabolized in a manner similar to the aromatic amines, i.e., DNA binding proceeds via oxidation of the amino group to the hydroxylamine. In the rat metabolites of IQ have been found to bind to both albumin and hemoglobin. The albumin adduct of IQ has been fully characterized as cysteine sulfinamide which can release IQ upon mild basic hydrolysis (18). Thus this class of compounds may be amenable to an approach similar to that which we have used for 4-aminobiphenyl. Since exposure data are completely lacking for these compounds a biomarker will be particularly useful for epidemiological studies.

Polycyclic Aromatic Hydrocarbons

Recently our most intense efforts have been directed to the development of a suitable biomarker for polycyclic aromatic hydrocarbon (PAH) exposure and metabolism. Using benzo(a)pyrene as a model compound we have determined that the major mechanism of binding of PAH diolepoxides to human hemoglobin is via ester formation on a carboxylic acid side-chain. These esters are hydrolytically stable in the intact protein but release tetrols following protein hydrolysis (16). All PAH epoxides we have examined behave in this manner, so it appears we have discovered a generic biomarker for PAH exposure.

Analysis of released diols and tetrols has been accomplished by an approach which uses monoclonal antibodies for isolation and concentration, gas chromatography and mass spectrometry, and fluorescence for final characterization. Using this approach we have been able to identify oxidation products of benzo(a)pyrene, benzo(e)pyrene, and chrysene in human hemoglobin from cigarette smokers. This work is still prelimininary and has not yet been published.

The PAH problem is exceedingly complex because of the wide variety of structures encompassed by this group of compounds and because there are different sources of exposure, each of which has a unique PAH array. It is most probable that hemoglobin adducts will yield information on systemic dose for a number of PAH components. How valuable this information will be for estimation of the biologically effective dose to different tissues remains to be seen.

ACKNOWLEDGEMENTS

This work was supported by PHS Grant Nos. ES00597, ES02109, ES01640, and ES04675, from the National Institutes of Health.

REFERENCES

1. Bailey, E., A.G.F. Brooks, C.T. Dollery, P.B. Farmer, B.J. Passingham, M.A. Sleightholm, and D.W. Yates (1988) Hydroxyethylvaline adduct formation in haemoglobin as a biological monitor of cigarette smoke intake. Arch. Tox. 62:247-253.
2. Bryant, M.S., P.L. Skipper, S.R. Tannenbaum, and M. Maclure (1987) Hemoglobin adducts of 4-aminobiphenyl in smokers and nonsmokers. Cancer Res. 47:602-608.
3. Bryant, M.S., P. Vineis, P.L. Skipper, and S.R. Tannenbaum (1988) Hemoglobin adducts of aromatic amines: Associations with smoking status and type of tobacco. Proc. Natl. Acad. Sci., USA 85:9788-9791.
4. Conney, A.H. (1982) Induction of microsomal enzymes by foreign chemicals and carcinogenesis by polycyclic aromatic hydrocarbons: G.H.A. Clowes memorial lecture. Cancer Res. 42:4875-4917.
5. Ehrenberg, L., K.D. Hiesche, S. Osterman-Golkar, and I. Wennberg (1974) Evaluation of genetic risks of alkylating agents: Tissue doses in the mouse from air contaminated with ethylene oxide. Mutat. Res. 24:83-103.
6. Gan, L.-S., M.S. Otteson, M.M. Doxtader, P.L. Skipper, R.R. Dasari, and S.R. Tannenbaum (1989) Quantitation of carcinogen bound protein adducts by fluorescence measurements. Spectrochimica ACTA 45A:81-86.
7. Gan, L.-S., P.L. Skipper, X. Peng, J.D. Groopman, J.-s. Chen, G.N. Wogan, and S.R. Tannenbaum (1988) Serum albumin adducts in the molecular epidemiology of aflatoxin carcinogenesis: Correlation with aflatoxin B_1 intake and urinary excretion of aflatoxin M_1. Carcinogenesis 9:1323-1325.

8. Harris, C.C., A. Weston, J.C. Willey, G.E. Trivers, and D.L. Mann
    (1987) Biochemical and molecular epidemiology of human cancer: In-
    dicators of carcinogen exposure, DNA damage, and genetic
    predisposition. Env. Health Persp. 75:109-119.
9. Hill, P., N.J. Haley, and E.L. Wynder (1983) Cigarette smoking:
    Carboxy-hemoglobin, plasma nicotine, cotinine and thiocyanate vs
    self-reported smoking data and cardiovascular disease. J. Chron.
    Dis. 36:439-449.
10. Kadlubar, F.F., K.L. Dooley, G. Talaska, D.R. Roberts, F.A.
    Beland, C.H. Teitel, J.F. Young, P.L. Skipper, S.R. Tannenbaum,
    G.D. Stoner, and N. Shivapurkar (1988) Determinants of susceptibil-
    ity to adduct formation after multiple dosing with 4-aminobiphenyl.
    Proc. Am. Assn. Cancer Res. 29:354.
11. Maclure, M., R.B.A. Katz, M.S. Bryant, P.L. Skipper, and S.R.
    Tannenbaum (1989) Elevated blood levels of carcinogens in passive
    smokers. Am. J. Pub. Health (in press).
12. Maclure, M., M.S. Bryant, P.L. Skipper, and S.R. Tannenbaum
    (1990) Decline of the hemoglobin adduct of 4-aminobiphenyl during
    withdrawal from smoking. Cancer Res. (in press).
13. National Research Council Committee on Biological Markers (1987)
    Biological markers in environmental health research. Env. Health
    Persp. 74:3-9.
14. Patrianakos, C., and D. Hoffman (1979) Chemical studies on tobacco
    smoke LXIV. On the analysis of aromatic amines in cigarette smoke.
    J. Analyt. Tox. 3:149-154.
15. Sabbioni, G., P.L. Skipper, and S.R. Tannenbaum (1987) Isolation
    and characterization of the major serum albumin adduct formed by
    aflatoxin B_1 in vivo in rats. Carcinogenesis 8:819-824.
16. Skipper, P.L., S.N. Naylor, L.-S. Gan, B.W. Day, R. Pastorelli,
    and S.R. Tannenbaum (1989) Origin of the tetrahydrotetrols derived
    from human hemoglobin adducts of benzo(a)pyrene. Chem. Res. in
    Tox. 2:280-281.
17. Tannenbaum, S.R., M.S. Bryant, P.L. Skipper, and M. Maclure
    (1986) Hemoglobin adducts of tobacco-related aromatic amines: Appli-
    cation to molecular epidemiology. In Banbury Report 23: Mecha-
    nisms in Tobacco Carcinogenesis, D. Hoffmann and C.C. Harris,
    eds., Cold Spring Harbor Laboratory, New York, pp. 63-75.
18. Turesky, R.J., P.L. Skipper, and S.R. Tannenbaum (1987) Binding
    of 2-amino-3-methylimidazo[4,5-f]quinoline to hemoglobin and albumin
    in vivo in the rat. Identification of an adduct suitable for
    dosimetry. Carcinogenesis 8:1537-1542.
19. Vineis, P., J. Esteve, and B. Terracini (1984) Bladder cancer and
    smoking in males: Types of cigarettes, age at start, effect of
    stopping and interaction with occupation. Int. J. Cancer
    34:165-170.

EVALUATION OF DNA BINDING IN VIVO FOR LOW-DOSE

EXTRAPOLATION IN CHEMICAL CARCINOGENESIS

Werner K. Lutz, P. Buss, A. Baertsch,
and M. Caviezel

Institute of Toxicology
Swiss Federal Institute of Technology
and University of Zurich
Schwerzenbach, Switzerland

INTRODUCTION

The exposure of the human population to unavoidable carcinogens is in a dose range which would not give rise to a significant increase of the tumor incidence in a standard bioassay for carcinogenicity with laboratory animals. This lack of sensitivity of the bioassay is largely due to the limitations in the number of animals used and the variability of the tumor incidence both in control and treated groups (5). Therefore, in order to estimate the "virtually safe" dose which should not lead to more than one additional tumor in one million people exposed, four to five orders of magnitude must be spanned with extrapolations.

A large number of mathematical models have been proposed, based on either tolerance distributions (e.g., probit) or on stochastic models based in part on biological considerations (20). Parameters of the functions are derived from the tumor data of the bioassay and are used for the dose range of extrapolation. This type of extrapolation therefore assumes unrealistically that the reactions of cells, tissues, and organisms to low dose levels of a test compound follow the same rules established from the high, often toxic dose levels. Enormous variations between the various models can be the consequence, and the wide use today of a very simple linear extrapolation of the tumor incidence to the spontaneous level is a sign of the inadequacy of a mathematical approach. A mechanistic approach promises better results. This consists of a search for biological reactions assumed to be related to the process of carcinogenesis and an evaluation of the respective dose-response relationship down to the limit of detection.

A large number of well-known carcinogens are metabolized via chemically reactive intermediates which can bind covalently to DNA. Because carcinogenesis involves an accumulation of a number of genetic changes, it is a general belief that the resulting DNA adducts represent an early

Genetic Toxicology of Complex Mixtures
Edited by M. D. Waters *et al.*
Plenum Press, New York, 1990

lesion which could, under appropriate conditions, be fixed as a mutation in a critical gene.

A semiquantitative correlation exists between the carcinogenic potency and the DNA-binding potency; for instance, expressed in the dose-normalized units of the Covalent Binding Index, CBI = µmol chemical bound per mol DNA nucleotide/mmol chemical administered per kg b.w. (12,13). With the analytical techniques available today (radiolabelled carcinogen, antibodies against DNA-carcinogen adducts, ^{32}P-postlabeling of adducted nucleotides), a determination of DNA adducts is possible after the administration of low doses of carcinogen which would not give rise to an observable increase in tumor incidence. Therefore, the formation of DNA adducts has been investigated for a number of chemicals down to the limits of detection.

DOSE RESPONSE FOR DNA ADDUCT FORMATION

Two reviews of the low-dose response for DNA adduct formation by various carcinogens in vivo are available. Lutz (14) has compiled the data for carcinogens where the dose range investigated covered at least one order of magnitude below the daily dose resulting in 50% tumor incidence (TD_{50}). The main finding was proportionality of the formation of DNA adducts after single exposure at lowest dose levels. At high dose levels, saturation effects were often noted. Swenberg et al. (18) collected evidence of nonlinearities in the dose-response curves both for DNA adduct formation and for cell proliferation, an important modulator of the DNA damage. Mechanisms which generate a sublinear or superlinear shape are compiled in Fig. 1 (top). In the following paragraphs, a number of examples are given to explain mechanistically the shape of the dose-response curve found.

Single-dose Experiments

Proportionality at low dose. In the majority of cases investigated, linear relationships between low dose and DNA adduct formation has been reported (14). This is in line with the understanding that the rates of the processes leading to covalent DNA binding of chemical carcinogens; i.e., diffusion, enzymatic modifications, and chemical reactions, are expected to be proportional to concentration levels far below the K_M values of the enzymatic reactions.

Sublinearity. A number of notable exceptions has been found, however, in an intermediate dose range: A sigmoidal shape of the dose-response curve is seen with formaldehyde-induced DNA-protein crosslinks (8) and with the level of 0^6-methylguanine determined after administration of a methylating agent (16). The sublinear part of the first example is probably due to a saturation of metabolic inactivation, while the second example shows that the instantaneous repair by methyl transfer to an acceptor protein is exhausted above a certain level of DNA methylation.

The extreme case of a sublinearity, a threshhold, cannot be produced under these circumstances. The kinetic model described by Cornfield (6) is often erroneously interpreted as a pharmacokinetic model and used to state that enzymatic activation and inactivation pathways can generate a true no-effect situation. This is not the case. The model

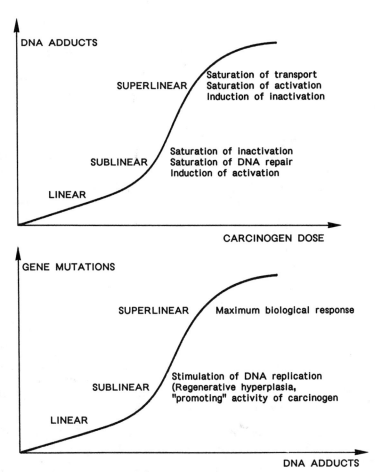

Fig. 1.   Mechanistic possibilities for the generation of nonlinear shapes
in the dose-response relationship for chemical carcinogenesis.
Top, level of DNA adducts as a function of dose; bottom, number of gene mutations as a function of the level of DNA
adducts.

Tab. 1. DNA adduct formation in female F-344 rats after exposure to
[^{14}C]1,2-dichloroethane by two different regimens of inhalation

Exposure profile	Steady low conc.		Peak conc.	
	Rat 1	Rat 2	Rat 3	Rat 4
Chemical dose				
[mg/kg]	33.5	33.7	117	115
Radioactivity dose				
[dpm/kg]	$6.03 \cdot 10^9$	$6.07 \cdot 10^9$	$2.65 \cdot 10^9$	$2.46 \cdot 10^9$
Liver DNA (3rd Purification)				
Spec.act.[dpm/mg]	43.0	42.9	656	723
Percent RA due to				
adduct formation	78	82	94	100
Covalent DNA Binding				
[CBI units]	1.7	1.8	72	91
Lung DNA (2nd Purification)				
Spec.act.[dpm/mg]	28.6	36.0	322	330
Percent RA due to				
adduct formation	56 (pool)		94	94
Covalent DNA Binding				
[CBI units]	0.9		35	39

CBI = µmol adduct per mol DNA nucleotide / mmol DCE per kg bw

produces its well-known hockey stick-like thresholded dose-response
curve under the assumption of an irreversible protective reaction with in-
definitely high rate. This implies that the rate of activation is zero as
long as a toxin-deactivator is available. This situation is unrealistic with
competitive enzymatic reactions.

Superlinearity. Saturation of the metabolic activation of high doses
of carcinogen often leads to a superlinear shape (aflatoxin B_1; the aromatic amine dimethylaminostilbene; the polynuclear aromatic hydrocarbon
benzo[a]pyrene). Since this high dose range is often used in a bioassay
on carcinogenicity (note the superlinear shape for the tumor induction by
vinyl chloride), this aspect is important in a large number of cases (14).
A superlinear shape was also reported at lower dose levels with the
tobacco-specific nitrosamine 4-(N-methyl-N-nitrosamino)-1-(3-pyridyl)-1-
butanone (NNK) and was explained on the basis of a low K_M pathway for
the metabolic activation (2).

Summary. It can be deduced that the relationship between single
exposure dose and level of DNA adduct formation is expected to be proportional in the low-dose range. However, a number of examples can be
put forward to produce a nonlinearity. The dose level at which the nonlinearity is produced depends upon the carcinogen and the cell type and
has to be determined in vivo in the target tissue.

DNA Adducts as a Function of the Dose Rate

1,2-Dichloroethane (DCE) was reported to be carcinogenic in rats in a long-term bioassay using gavage in corn oil (24 and 48 mg/kg/da) but not by inhalation (up to 150-250 ppm, 7 hr/da, 5 da/wk). The daily dose absorbed was similar in the two experiments. We have investigated whether the level of DNA adducts formed in vivo is dependent on the concentration-time course in inhalation exposures. Female F344 rats were exposed to [1,2-^{14}C]DCE in a closed inhalation chamber to either a high initial peak concentration (up to 180 µmol/l for a few minutes, resulting in a metabolized dose of 116 mg/kg after 12 hr) or to a low constant concentration (3 µmol/l ≅ 80 ppm for 4 hr, dose 34 mg/kg after a total of 12 hr). DNA was isolated from liver and lung and purified to constant specific activity. DNA was enzymatically hydrolyzed to the 3'-nucleotides which were separated by reverse phase high performance liquid chromatography (HPLC) to determine the level of nucleotide-DCE adducts. The level of DNA adducts was expressed in the dose-normalized units of the CBI. In liver DNA, the different exposure regimens resulted in highly different CBI values of 82 and two for peak and constantly low DCE exposure levels, respectively. In the lung, the respective values were 37 and 1 (Tab 1). It is concluded that the DNA damage by DCE is largely dependent on the concentration-time profile. The carcinogenic potency determined in the gavage study should not, therefore, be used for other exposure regimes.

These data show that peak blood levels reached with bolus-like administrations can have very different consequences with respect to the formation of reactive metabolites and DNA adducts. With DCE, peak concentrations resulted in a much more dangerous metabolism than low concentrations. In terms of the dose-response relationship, a steep fall of the risk is expected with the reduction of the dose.

Multiple-dose Experiments

The data referred to above are based essentially on single-dose experiments. For a correlation of DNA adduct formation with tumor induction, however, it is the level of dangerous adducts arising from chronic exposure which is the important variable. The few published results deal primarily with the high and intermediate dose range to compare the data with the shape of the dose-response curve in the carcinogenicity study.

With 2-acetylaminofluorene (2-AAF) given for two weeks to mice, the level of DNA adducts in the liver was proportional to the dose [0.5-500 ppm in the diet (9)]. Beland et al. (1) investigated the same compound for DNA adduct formation in liver and bladder. After one month of feeding, the DNA damage was proportional to the dose in both organs. In contrast, the corresponding tumor incidence increased linearly in the liver, but not in the bladder (sublinear shape).

With N-nitrosodimethylamine in the drinking water of mice and rats (10 to 100 ppm for 16 da), the multiple nonlinearities seen with the level of guanine methylations were explained on the basis of induction and depletion of the respective repair capacities (11). With another methylating agent, the tobacco-specific nitrosamine NNK, 12 daily doses of 0.1 to 100 mg/kg per da to male F344 rats generated a superlinear dose-response for

DNA methylation in the lung in the low-dose range, and a sublinear shape at the high-dose levels (2).

Boucheron et al. (4) investigated the ethylation of O^4-thymine in rats by N-nitrosodiethylamine in the drinking water for up to 70 da. This DNA adduct was proportional to the dose up to the equivalent of 1 mg/kg per da. Above this dose, saturation was noted.

In this laboratory, male F344 rats were treated by oral gavage with radiolabelled aflatoxin B_1 for 1 or 10 da at dose levels of 1 ng/kg to 100 µg/kg per day. The lowest dose corresponded to a possible human exposure in high-risk areas. DNA was isolated from the livers at 24 hr after the last dose. Proportionality of the level of DNA adducts was measured over the first four orders of magnitude. A saturation was seen between 10 and 100 µg/kg (Caviezel, PhD thesis no. 7564. Swiss Federal Institute of Technology Zurich, 1984).

"Steady-state" level of DNA adducts. The studies cited above did not show whether a steady state of the DNA adducts had been reached, or whether continued exposure would result in even higher levels of adducts. The periods of treatment had often been chosen on the basis of DNA repair half lives after single administration. This knowledge allows us, in principle, to estimate the time required for a build-up of the DNA adducts. However, since the repair of adducts is not necessarily proportional to the adduct level, we have investigated DNA adduct formation and build-up for various periods of time and as a function of dose.

Groups of three young adult male rats were given radiolabeled aflatoxin B_1 (AFB$_1$; F344 rats) and 2-acetylaminofluorene (2-AAF; Sprague-Dawley rats) in the drinking water at three exposure levels (AFB$_1$: 0.02, 0.6, 20 µg/l; 2-AAF: 0.034, 1.1, 34.3 mg/l) for 4, 6, and 8 wk. With [^{3}H]aflatoxin B_1, the consumption of drinking water resulted in dose levels of 2.2, 73, 2,110 ng/kg per day. With [9-^{14}C]acetylaminofluorene, dose levels of 2.4, 89, 2,170 µg/kg per day resulted. With 2-AAF, at all dose levels, the drinking water contained 5% ethanol to help bring the carcinogen into solution. With both carcinogens, the highest dose would have resulted in a significant tumor induction in a two-year study, being near the TD$_{50}$ value (determined without co-administration of ethanol!). DNA was isolated from the livers after the periods of time indicated and analyzed for DNA-carcinogen adducts.

With AFB$_1$, some increase of the level of DNA adducts, although statistically not significant at the low- and intermediate-dose levels, was seen when going from 4 to 6 to 8 wk (the low value determined after 4 wk of the high dose cannot be explained). Steady state was therefore almost reached at 8 wk (Fig. 2). With 2-AAF, the three lines clearly overlapped within the variability of the replicates. Therefore, steady state had probably been reached by 4 wk already (Fig. 3). Regardless of the time course, the dose-response relationship was linear down to the lowest dose of both carcinogens. The slope in the double-log plots was 1.0, indicating proportionality in a linear scale. In addition, extrapolation to dose zero passed the origin (which cannot be shown in a log scale). We conclude that the level of DNA adducts near the steady state was proportional to the chronic low dose of AFB$_1$ and 2-AAF. It remains to be shown, however, whether the proportionality to dose also holds for the formation of adducts in critical genes.

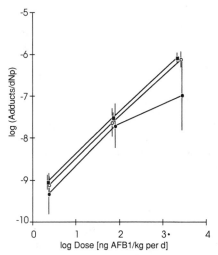

Fig. 2. Formation of DNA adducts in liver DNA of male F344 rats as a function of dose. Administration of [^{3}H]aflatoxin B_1 in the drinking water for 4 (·), 6 (o), and 8 (■) wk. The dose was determined on the basis of the water consumption. Means and one standard deviation of three rats per group.

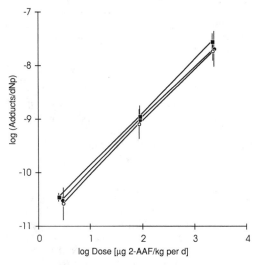

Fig. 3. Formation of DNA adducts in liver DNA of male Sprague-Dawley rats as a function of dose. Administration of [9-^{14}C]2-acetyl-aminofluorene in the drinking water for 4 (·), 6 (o), and 8 (■) wk. The dose was determined on the basis of the water consumption. Means and one standard deviation of three rats per group.

MODULATIONS OF THE DOSE RESPONSE

DNA adducts alone are not sufficient to generate a heritable, initiated phenotype of a cell because adducts cannot be copied onto the progeny. Only upon DNA replication can mutations arise in the new strand opposite an adduct. If this occurs in a critical gene responsible for some step in growth control, the daughter cell could form the origin of a transformed cell clone. Since DNA adducts can be repaired, the probability of a fixation of the DNA adduct in the form of a mutation is dependent on the relative rates of DNA repair and cell division. All acceleration of cell division will therefore have a synergistic effect on the formation of mutations and all effects of a carcinogen on this process will have an effect on the dose-response relationship with respect to tumor incidence (15).

Rate of Cell Division

Regenerative hyperplasia. DNA adduct-forming carcinogens do not bind exclusively to DNA, but always to RNA and protein also. Above a certain level of macromolecular binding, the cell dies. This can induce regenerative hyperplasia in the surrounding tissue. Genotoxic agents at high dose, therefore, can stimulate cell division and accelerate the process of mutagenesis and carcinogenesis in the surviving cells. A more rapid rate of cell division and DNA replication will increase the probability of mutations due to the adducts because the time available for DNA repair is reduced.

If we assume that the level of adducts increases proportionately to the dose but that DNA replication is faster at higher adduct levels, the level of mutations will increase overproportionately in the dose range of cytotoxicity. This means that high-dose exposure to a genotoxic carcinogen (which is the usual situation in a bioassay) will result in a sublinear shape of the dose-response curve for mutations (and tumors) in division-competent cells (Fig. 1, bottom). As an example, the strongly sublinear dose-response relation for nasal tumor induction by formaldehyde between 6 and 14 ppm is probably due to this aspect (17).

It is important to note here that if low-dose levels of DNA adduct forming carcinogen do not induce hyperplasia, the linear relationship seen with the DNA adducts is expected to dominate the dose-response relationship for gene mutation (and tumor incidence; Fig. 1, bottom, linear low-dose part).

Hormonal activity of genotoxic agents. One important aspect of tumor promotion involves clonal expansion of initiated cells. This can be achieved by either a higher rate of cell division as compared with the normal tissue or by giving the initiated cells a survival advantage. The latter can be based on a more efficient detoxication of toxic compounds or on postponing programmed cell death. These cellular reactions are different from the regenerative hyperplasia discussed above, more like reactions to hormones. DNA adduct formation and "promoting activity" of this sort are not mutually exclusive. For instance, both 2-acetylaminofluorene (2-AAF) and trans-4-acetylaminostilbene form DNA adducts in rat liver, but only 2-AAF induces liver tumors. With elegant combination studies by Kuchlbauer et al. (10) it was shown that the activity of 2-AAF must be based upon additional (tumor promoting), receptor-mediated cellular effects. For dose-response considerations, it must be deduced that a linear relationship for DNA adduct formation can be modulated by a hormone-like relationship of a putative promoting activity of the genotoxic carcinogen.

The Multistage Process

Malignant cells have accumulated a number of genetic changes (about four to six independent steps, as deduced from the exponential increase of the age-specific tumor incidence with human age). Gene mutations induced by DNA adducts appear to be effective primarily in the earlier, "initiating" steps. Later, chromosome mutations are seen to be accumulated with tumor development. These mutations are often visible as deletions in cell type-specific genetic loci. Carcinogenesis in the colonic epithelium in humans clearly illustrates and supports this idea (7,19).

The presence of a multi-stage situation alone opens a number of possibilities for nonlinear dose-response relationshps. If we assume that a carcinogen accelerates two steps with an efficacy proportionate to the dose, a second power of the dose will govern the dose-response relationship for tumor induction. If a chemical carcinogen accelerates all four to six steps to the same extent, the dose-response relationship would show the respective exponential power to the dose.

DOSE-RESPONSE WITH MIXTURES

The above analysis indicates that the dose-response relationship for the tumor induction by a single carcinogen is a superposition of a number of curves for the various steps. With mixtures, this complexity is compounded, because a second chemical can influence any of the multiple events governing the carcinogenic activity of the first. The most simple situation imaginable, linear additivity, is possible if the compounds operate independently from each other. Such a situation can be imagined at lowest dose levels, but not in the high-dose range. Indeed, a large low-dose combination experiment with the three genotoxic hepatocarcinogenic N-nitrosamines, N-nitrosodiethylamine, N-nitrosopyrrolidine, and N-nitrosodiethanolamine in 1,800 SD rats revealed linear additivity for tumor induction in the liver but not in secondary target organs (3). At higher dose levels of combination exposures, the nonlinearities observed with single compounds must generate dose-response relationships which will not be amenable to extrapolation. Therefore, we consider it most important to first define the mechanism of action of the individual carcinogen and postulate a dose-response relationship for the single compound. Then, the individual processes, such as metabolic activation and detoxication, DNA adduct formation and repair, stimulation of cell division, and induction of chromosome damage could be investigated in short-term situations in the presence of the second chemical. This approach could guard against erroneous extrapolations derived from the high-dose situation.

CONCLUSIONS

Carcinogenesis is a multistage process with a spontaneous rate in the absence of a defined carcinogen exposure. It involves an accumulation of genetic changes. The dose-response relationship for an increased tumor induction due to a chemical carcinogen is given by a superposition of the individual dose-response curves for the acceleration of the different steps. At higher dose levels used in the bioassay on carcinogenicity, genotoxic carcinogens are active on more than one step. This results in an exponential dose-response relationship with an exponent >1 to the dose and a sublinear part of the dose-response curve. In the very low dose range, where the formation of DNA adducts is expected to be proportional to the dose, it is possible that the gene mutations resulting from miscoding DNA adducts accelerate only one (initial) step in the process of carcinogenesis. The dose-response relationship for the induction of tumors above the spontaneous level could then be linear in this range.

For extrapolating the high-dose tumor data to low levels, the most important feature is to establish whether a genotoxic carcinogen is active beyond the formation of DNA adducts, and in which dose range additional activities come into effect. Mechanistic analysis with respect to the formation of chromosome mutations, cytotoxicity and regenerative hyperplasia, as well as to hormonal effects are required in order to predict the presence of a nonlinear range. The presence of a true threshold is unlikely to exist with DNA adduct-forming carcinogens. Nevertheless, the presence of steep nonlinearities can result in appreciable drops of the risk by only a small decrease in dose.

REFERENCES

1. Beland, F.A., N.F. Fullerton, T. Kinouchi, and M.C. Poirier, (1988) DNA adduct formation during continuous feeding of 2-acetyl-aminofluorene at multiple concentrations. IARC Sci. Publ. 89:175-180.

2. Belinsky, S.A., C.M. White, T.R. Devereux, J.A. Swenberg, and M.W. Anderson (1987) Cell selective alkylation of DNA in rat lung following low dose exposure to the tobacco specific carcinogen 4-(N-methyl-N-nitrosamino)-1-(3-pyridyl)-1-butanone. Cancer Res. 47:1143-1148.

3. Berger, M.R., D. Schmaehl, and H. Zerban (1987) Combination experiments with very low doses of three genotoxic N-nitrosamines with similar organotropic carcinogenicity in rats. Carcinogenesis 8:1635-1643.

4. Boucheron, J.A., F.C. Richardson, P.H. Morgan, and J.A. Swenberg (1987) Molecular dosimetry of 04-ethyldeoxythymidine in rats continuously exposed to diethylnitrosamine. Cancer Res. 47:1577-1581.

5. Cairns, T. (1979) The ED01 study: Introduction, objectives, and experimental design. In Innovations in Cancer Risk Assessment (ED01 Study). Pathotox Publishers, Park Forest South, IL 60466 USA, pp. 1-8.

6. Cornfield, J. (1977) Carcinogenic risk assessment. Science 198:693-699.

7. Fearon, E.R., S.R. Hamilton, and B. Vogelstein (1987) Clonal analysis of human colorectal tumors. Science 238:193-197.

8. Heck, H.A., and M. Casanova (1987) Isotope effects and their implications for the covalent binding of inhaled (3H) and (14C) formaldehyde in the rat nasal mucosa. Tox. Appl. Pharm. 89:122-134.

9. Jackson, C.D., C. Weis, and T.E. Shellenberger (1980) Tissue binding of 2-acetylaminofluorene in BALB/c and C57Bl/6 mice during chronic oral administration. Chem. Biol. Interact. 32:63-81.

10. Kuchlbauer, J., W. Romen, and H.G. Neumann (1985) Syncarcinogenic effects on the initiation of rat liver tumors by trans-4-acetylaminostilbene and 2-acetylaminofluorene. Carcinogenesis 6:1337-1342.

11. Lindamood III, C., M.A. Bedell, K.C. Billings, M.C. Dyroff, and J.A. Swenberg (1984) Dose response for DNA alkylation, (3H) thymidine uptake into DNA, and 06-methylguanine-DNA methyltransferase activity in hepatocytes of rats and mice continuously exposed to dimethylnitrosamine. Cancer Res. 44:196-200.

12. Lutz, W.K. (1979) In vivo covalent binding of organic chemicals to DNA as a quantitative indicator in the process of chemical carcinogenesis. Mutat. Res. 65:289-356.

13. Lutz, W.K. (1986) Quantitative evaluation of DNA binding data for risk estimation and for classification of direct and indirect carcinogens. J. Cancer Res. Clin. Oncol. 112:85-91.

14. Lutz, W.K. (1987) Quantitative evaluation of DNA-binding data in vivo for low-dose extrapolations. Arch. Tox. (Suppl) 11:66-74.

15. Lutz, W.K., and P. Maier (1988) Genotoxic and epigenetic chemical carcinogenesis: One Process, different mechanisms. Trends Pharm. Sci. 9:322-326.

16. Stumpf, R., G.P. Margison, R. Montesano, and A.E. Pegg (1979) Formation and loss of alkylated purines from DNA of hamster liver after administration of dimethylnitrosamine. Cancer Res. 39:50-54.

17. Swenberg, J.A., C.S. Barrow, C.J. Boreiko, H.A. Heck, R.J. Levine, K.T. Morgan, and T.B. Starr (1983) Non-linear biological responses to formaldehyde and their implications for carcinogenic risk assessment. Carcinogenesis 4:945-952.

18. Swenberg, J.A., F.C. Richardson, J.A. Boucheron, F.H. Deal, S.A. Belinsky, M. Charbonneau, and B.G. Short (1987) High to low dose extrapolation: Critical determinants involved in the dose response of carcinogenic substances. Env. Health Persp. 76:57-63.

19. Vogelstein, B., E.R. Fearon, S.R. Hamilton, S.E. Kern, J.L. Bos, M. Leppert, Y. Nakamura, and R. White (1989) Genetic alterations during colorectal tumorigenesis. Proc. Amer. Assoc. Cancer. Res. 30:634-635.

20. Zeise, L., R.Wilson, and E.A.C. Crouch (1987) Dose-response relationships for carcinogens: A review. Env. Health Persp. 73:259-308.

CANCER RISKS DUE TO OCCUPATIONAL EXPOSURE TO POLYCYCLIC AROMATIC HYDROCARBONS: A PRELIMINARY REPORT

D. Krewski,[1,2] J. Siemiatycki,[3,4] L. Nadon,[3]
R. Dewar,[3] and M. Gérin[5]

[1]Health Protection Branch
Health & Welfare Canada
Ottawa, Ontario, Canada K1A 0L2

[2]Department of Mathematics & Statistics
Carleton University
Ottawa, Ontario, Canada K1S 5B6

[3]Epidemiology of Preventive Medicine Research Centre
Institute Armand Frappier
Montreal, Quebec, Canada H7V 1B7

[4]Department of Epidemiology & Biostatistics
McGill University
Montreal, Quebec, Canada H3A 2K6

[5]Department of Environmental Hygiene
University of Montreal
Montreal, Quebec, Canada H3C 3S7

ABSTRACT

A case-control study was undertaken in Montreal to investigate the possible associations between occupational exposures and cancers of the following sites: esophagus, stomach, colorectum, liver, pancreas, lung, prostate, bladder, kidney, melanoma, and lymphoid tissue.   In total, 3,726 cancer patients were interviewed to obtain detailed lifetime job histories, which were translated into a history of occupational exposures. This article provides a preliminary report on the risk of 19 different types of cancer in relation to occupational exposure to polycyclic aromatic hydrocarbons (PAHs). Consideration is given to PAHs derived from coal, petroleum, wood, and other sources as well as all sources combined. Benzo(a)pyrene, the single most potent PAH, was used to form a sixth exposure variable. Separate statistical analyses were carried out for each category of PAH and for each type of cancer using control series drawn from among the other cancer sites in the study. Over three-quarters of

Genetic Toxicology of Complex Mixtures
Edited by M. D. Waters *et al.*
Plenum Press, New York, 1990

Tab. 1.   Number of cases and controls for each case series

Cancer Case Series (ICD code)	Cancer Sites Excluded from Control Series[a]	No. of cases	No. of controls
Esophagus (150)	Lung, stomach	107	2,514
Stomach (151)	Lung, esophagus	250	2,514
Colorectum (153,154)	Lung	787	1,081
-Colon (153, except 153.3)	Lung other colorectal	364	2,081
-Rectosigmoid (153.3, 154.0)	Lung other colorectal	233	2,081
-Rectum (154, except 154.0)	Lung other colorectal	190	1,315
Liver (155)	Lung	50	2,806
Pancreas (157)	Lung	117	2,741
Lung (162)		857	1,523
-Oat cell	Other lung	159	1,523
-Squamous cell	Other lung	359	1,523
-Adenocarcinoma	Other lung	162	1,523
-Other cell types[b]	Other lung	177	1,523
Prostate (185)	Lung	452	1,733
Bladder (188)	Lung, kidney	486	2,196
Kidney (189)	Lung, bladder	181	2,196
Melanoma of the skin (172)	Lung	121	2,737
Hodgkin's lymphoma (201)	Lung, other lymphoma	53	2,599
Non-Hodgkin's lymphoma (200,202)	Lung, Hodgkin's	206	2,599

[a]   For each case series, all cancer patients interviewed served as referents with the exceptions listed in this column. Furthermore, for rectum, lung and prostate, only those subjects interviewed during the same ascertainment periods as the three respective site series were used as referents.

[b]   This is a heterogeneous grouping which includes large cell, spindle cell, adenosquamous and "carcinoma, not otherwise specified".

all subjects had some occupational exposure to PAHs. At the levels of exposure experienced, the preliminary analysis reported here revealed no clear evidence of an increased risk of any type of cancer among exposed men. Further analyses are underway to allow for occupational exposure to other industrial chemicals which may act as confounders, and to evaluate any effects of intensity and duration of exposure.

INTRODUCTION

Polycyclic aromatic hydrocarbons (PAHs) are naturally present in fossil fuels and can be formed by thermal decomposition of organic materials such as wood and plastic. This class of chemicals includes a wide variety of compounds composed of three or more benzene rings configured in different ways (6). People may be exposed to complex mixtures of PAHs or to single compounds such as benzo(a)pyrene (10,11).

Although PAHs are ubiquitous in the environment (3), the highest exposures occur in industrial settings to mixtures of airborne PAHs (1). Epidemiological evidence of the carcinogenic effects of PAHs dates back to observations by Sir Percival Pott in 1775 on the high rate of occurrence of scrotal cancer in chimney sweeps. More recently, elevated lung cancer risks have been noted among workers involved in aluminum refining, coal gasification, coke production, and iron and steel founding (7). Occupational exposure to coal tars, shale oils, and soots has also been shown to increase cancer risk (8). Toxicological studies have also demonstrated increased cancer risks in animals following exposure to a number of specific PAHs, including benzo(a)pyrene (BaP) (4).

METHODS

Study Design

This study is based on analysis of data on occupational exposure derived from a large-scale population-based case-control study conducted by the Institut Armand-Frappier in the Montreal metropolitan area between September, 1979, and December, 1985, the details of which have been previously reported (5,12). This database contains information on occupational exposures to over 300 chemical substances, and affords an opportunity to study agents of particular interest such as PAHs.

Briefly, completed interviews were obtained with 3,726 male cancer cases. Case series were classified using the three-digit International Classification of Diseases coding, with the following exceptions. Colorectal cancer was divided into three topographic subcategories--colon (excluding sigmoid), sigmoid colon plus rectosigmoid junction, and rectum. The following four histologic subcategories of lung cancer were also analyzed: oat cell carcinoma, squamous cell carcinoma, adenocarcinoma, and other histologic types (including those of unspecified morphology). The 78 cases which were diagnosed with primary tumors at two different sites have been included in both case series. This led to the 19 lesions in the 12 tissues listed in Tab. 1.

Controls for each case series were selected from among other cancer cases, with three exceptions. First, lung cancer cases were excluded from all control series. Second, anatomically contiguous sites were excluded from the control series for each case series. Third, for the three sites not ascertained in certain years of the study--lung, rectum, and prostate--controls recorded during those years were excluded. A smaller group of population controls was also obtained, although these will not be considered in this preliminary report. The number of cases and controls considered for each of the 19 types of cancer studied is given in Tab. 1.

Questionnaire

Case interviews were based on a structured questionnaire to elicit information on important nonoccupational confounders and on a semi-structured probing interview to obtain detailed information on the individual's lifetime job history. In this report, only the following four non-occupational confounders will be considered: age (35-54 and 55-70 years), cigarette consumption (nonsmoker, 0-600 pack-years, and over 600 pack-years), socioeconomic status (two categories defined by median income), and ethnicity (French, other).

Exposure Assessment

The information obtained during the interviews on job histories provided the basis for evaluating exposure to PAHs. A team of chemists and industrial hygienists examined each questionnaire and determined the likely exposure of each individual to some 300 industrial agents. Although categories of exposure were established with respect to frequency and level, we will restrict our attention to comparisons of ever-exposed versus never-exposed individuals in this report. Of the 300 occupational exposures, consideration will also be given only to exposures involving PAHs.

Ascertaining exposure to a diverse class of substances such as PAHs is more difficult than with a single chemical entity. In the past, either the benzene-soluble or cyclohexane-soluble fraction of total particulate matter or an indicator compound such as benzo(a)pyrene has been used as a proxy for the mixture.

Since BaP was one of the 300 substances on our checklist, our first attempt at assessing exposure to PAHs used BaP as an indicator of PAH exposure. While BaP may be a good surrogate for PAH exposure in some industrial settings, the concentrations of BaP in complex mixtures of PAHs can vary widely (11). Nonetheless, an analysis of BaP alone, representing one of the most potent PAHs (13), is of some interest.

Due to a lack of hygiene data, it was not possible to evaluate exposure to other specific PAH compounds. However, because the composition of PAH mixtures depends to some degree on the sources of PAH exposure, we considered indicators of PAH exposure defined in terms of the source material. These included PAHs derived from coal products, petroleum products, wood combustion, and pyrolysis of other organic matter such as plastic and rubber. An additional indicator of exposure was based on exposure to PAHs from all sources combined.

Statistical Analysis

For each of the 19 types of cancer, the relative risk due to exposure to each of the six PAH exposure variables was approximated by the odds ratio estimated using the Mantel-Haenzel procedure, adjusting for age, tobacco consumption, socio-economic status, and ethnicity (2). This involved a relatively large number of separate analyses, leading to inflationary pressure on the study-wise false-positive rate. Rather than attempt to strictly control overall specificity, we adopted a hypothesis-generating philosophy in which possible associations would be liberally sought as candidates for further evaluation in subsequent analyses of the

Tab. 2.   Number of cancer cases exposed to PAHs

Category of PAH	Number of Cases Exposed	Percentage of Cases Exposed[a]
Benzo(a)pyrene	1,013	27.2
Coal-Derived PAH's	383	10.3
Petroleum-Derived PAH's	2,720	73.0
Wood-Derived PAH's	191	5.1
PAH's Derived from Other Sources	886	2.4
All PAH's	2,915	78.2

[a] Percentage of 3,726 men exposed to PAH's.

present data, or in independent epidemiologic investigations. Thus, 90% confidence limits were computed for each of the 19 x 6 = 114 computed odds ratios.

RESULTS

The percentage of the 3,726 subjects exposed to each of the six categories of PAHs is given in Tab. 2. While over three-quarters (78.2%) of the subjects were exposed to any PAH, only one-quarter (27.2%) were exposed to BaP. Petroleum products were by far the most common source

Tab. 3.   Occupations most often associated with exposure to PAHs

Category of PAH	Occupation[a]
Benzo(a)pyrene	Mechanics; machinists; farmers; stationary engineers
Coal-Derived PAH's	Stationary engineers; railway trackmen; other railway workers; roofers; ferrous foundry workers
Petroleum-derived PAH's	Truck drivers; commercial travellers/driver salesmen; mechanics and repairmen; machinists
Wood-Derived PAH's	Farmers; cooks; firefighters
PAH's Derived from Other Sources	Machinists; mechanics; cooks; plumbers; other metal workers; ferrous foundry workers
All PAH's	Truck drivers; commercial travellers/driver salesmen; mechanics

[a] Includes all occupations in which the 5% or more of those exposed to PAH's had worked.

Tab. 4.   Odds ratios for exposure to benzo(a)pyrene

Cancer Site/Lesion	Number of Exposed Cases	Odds Ratio	90% Confidence Interval
Esophagus	30	1.3	0.9 - 1.9
Stomach	59	1.1	0.8 - 1.4
Colorectal	166	0.9	0.8 - 1.1
-Colon	83	1.0	0.8 - 1.3
-Rectosigmoid	46	0.8	0.6 - 1.1
-Rectum	41	0.8	0.6 - 1.1
Liver	10	0.8	0.4 - 1.4
Pancreas	33	1.4	1.0 - 1.9
Lung	235	1.0	0.9 - 1.2
-Oat Cell	42	1.0	0.7 - 1.4
-Squamous Cell	109	1.1	0.9 - 1.4
-Adenocarcinoma	38	0.8	0.5 - 1.1
-Other	46	1.0	0.7 - 1.3
Prostate	127	1.3	1.0 - 1.6
Bladder	111	1.0	0.8 - 1.2
Kidney	31	0.7	0.5 - 1.0
Melanoma	17	0.7	0.5 - 1.2
Lymphoma	39	0.8	0.6 - 1.1
Hodgkins	13	1.2	0.7 - 2.1

of exposure to PAHs. Exposure to PAHs derived from combustion of coal and wood was less prevalent, but nonetheless provided appreciable numbers of exposed cases for analysis.

The occupations most often associated with exposure to PAHs are listed in Tab. 3, which include all occupations in which at least 5% of workers exposed to one of our PAH categories had worked. While truck drivers represented the occupation most often exposed to any PAH, mechanics appear in four of the six PAH categories considered.

Tab. 5.   Odds ratios for exposure to coal-derived PAHs

Cancer Site/Lesion	Number of Exposed Cases	Odds Ratio	90% Confidence Interval
Esophagus	12	1.4	0.8 - 2.3
Stomach	26	1.3	0.9 - 1.8
Colorectal	50	0.7	0.6 - 1.0
-Colon	22	0.7	0.5 - 1.0
-Rectosigmoid	17	0.8	0.5 - 1.2
-Rectum	11	0.6	0.3 - 1.0
Liver	3	0.6	0.2 - 1.7
Pancreas	12	1.3	0.7 - 2.1
Lung	91	1.1	0.9 - 1.5
-Oat Cell	20	1.5	1.0 - 2.3
-Squamous Cell	39	1.1	0.8 - 1.6
-Adenocarcinoma	10	0.6	0.3 - 1.1
-Other	22	1.4	0.9 - 2.2
Prostate	62	1.5	1.1 - 2.0
Bladder	36	0.8	0.6 - 1.0
Kidney	15	1.0	0.6 - 1.7
Melanoma	3	0.4	0.2 - 1.1
Lymphoma	16	0.9	0.6 - 1.4
Hodgkins	1	0.2	0.0 - 1.2

Tab. 6.   Odds ratios for exposure to petroleum-derived PAHs

Cancer Site/Lesion	Number of Exposed Cases	Odds Ratio	90% Confidence Interval
Esophagus	69	1.1	0.7 - 1.5
Stomach	165	1.2	0.9 - 1.5
Colorectal	490	1.1	1.0 - 1.3
-Colon	224	1.0	0.8 - 1.3
-Rectosigmoid	147	1.1	0.9 - 1.4
-Rectum	126	1.2	0.9 - 1.5
Liver	20	0.4	0.3 - 0.7
Pancreas	74	1.0	0.7 - 1.4
Lung	589	1.1	0.9 - 1.3
-Oat Cell	105	1.0	0.7 - 1.3
-Squamous Cell	264	1.4	1.1 - 1.7
-Adenocarcinoma	114	1.1	0.8 - 1.4
-Other	106	0.8	0.6 - 1.0
Prostate	292	1.1	0.9 - 1.4
Bladder	294	0.9	0.8 - 1.1
Kidney	111	1.0	0.8 - 1.3
Melanoma	67	0.9	0.7 - 1.3
Lymphoma	128	1.0	0.8 - 1.3
Hodgkins	29	0.7	0.5 - 1.1

The odds ratios for exposure to each of our six PAH categories are given in Tab. 4-9, respectively. Of the 114 odds ratios computed, none suggest strong evidence of an increased cancer risk as a result of exposure to PAHs. Nearly all of the odds ratios are close to one, with 90% confidence intervals including unity. (In only six of the 114 cases did the lower confidence limit exceed 1.0.) These preliminary results are consistent with the global null hypothesis of no increased cancer risk as a result of exposure to PAHs for the 19 lesions and six categories of PAH considered here.

Tab. 7.   Odds ratios for exposure to wood-derived PAHs

Cancer Site/Lesion	Number of Exposed Cases	Odds Ratio	90% Confidence Interval
Esophagus	8	2.0	1.0 - 3.8
Stomach	17	1.7	1.1 - 2.7
Colorectal	38	1.1	0.8 - 1.6
-Colon	17	1.1	0.7 - 1.8
-Rectosigmoid	14	1.3	0.8 - 2.2
-Rectum	7	0.7	0.4 - 1.4
Liver	3	1.3	0.5 - 3.7
Pancreas	8	1.7	0.9 - 3.1
Lung	43	1.1	0.8 - 1.6
-Oat Cell	8	1.1	0.6 - 2.1
-Squamous Cell	20	1.1	0.7 - 1.8
-Adenocarcinoma	9	1.2	0.6 - 2.2
-Other	6	0.8	0.4 - 1.6
Prostate	16	0.9	0.5 - 1.4
Bladder	20	0.9	0.6 - ·1.3
Kidney	3	0.3	0.1 - 0.9
Melanoma	2	0.4	0.1 - 1.3
Lymphoma	7	0.8	0.4 - 1.4
Hodgkins	2	0.8	0.2 - 2.9

Tab. 8.   Odds ratios for exposure to PAHs derived from other sources

Cancer Site/Lesion	Number of Exposed Cases	Odds Ratio	90% Confidence Interval
Esophagus	17	0.7	0.5 - 1.1
Stomach	53	1.1	0.8 - 1.4
Colorectal	149	1.0	0.8 - 1.1
-Colon	75	1.1	0.8 - 1.4
-Rectosigmoid	40	0.8	0.6 - 1.1
-Rectum	36	0.9	0.6 - 1.1
Liver	8	0.8	0.4 - 1.4
Pancreas	20	0.8	0.5 - 1.2
Lung	194	1.1	0.9 - 1.3
-Oat Cell	41	1.3	0.9 - 1.8
-Squamous Cell	87	1.2	0.9 - 1.5
-Adenocarcinoma	41	1.2	0.9 - 1.6
-Other	25	0.6	0.4 - 0.9
Prostate	110	1.4	1.1 - 1.7
Bladder	95	1.0	0.8 - 1.3
Kidney	34	1.0	0.7 - 1.4
Melanoma	19	0.8	0.5 - 1.2
Lymphoma	42	1.1	0.8 - 1.4
Hodgkins	9	0.8	0.5 - 1.6

Tab. 9.   Odds ratios for exposure to all PAHs

Cancer Site/Lesion	Number of Exposed Cases	Odds Ratio	90% Confidence Interval
Esophagus	73	1.0	0.7 - 1.4
Stomach	177	1.2	0.9 - 1.5
Colorectal	525	1.1	0.9 - 1.3
-Colon	238	1.0	0.8 - 1.2
-Rectosigmoid	159	1.1	0.9 - 1.4
-Rectum	136	1.2	0.9 - 1.6
Liver	24	0.5	0.3 - 0.7
Pancreas	82	1.1	0.8 - 1.6
Lung	628	1.1	0.9 - 1.3
-Oat Cell	118	1.2	0.9 - 1.7
-Squamous Cell	278	1.4	1.1 - 1.7
-Adenocarcinoma	120	1.1	0.8 - 1.4
-Other	112	0.7	0.5 - 1.0
Prostate	313	1.1	0.9 - 1.4
Bladder	318	0.9	0.7 - 1.1
Kidney	119	1.0	0.8 - 1.3
Melanoma	70	0.9	0.6 - 1.2
Lymphoma	138	1.0	0.8 - 1.3
Hodgkins	30	0.7	0.4 - 1.1

DISCUSSION

The carcinogenicity of PAHs has been studied extensively in the past, with several sources of PAHs considered by the International Agency for Research on Cancer as recognized human carcinogens (9). Nonetheless, not all industrial cohorts exposed to PAHs demonstrate elevated cancer risks, nor have all PAHs been shown to be animal carcinogens. Since both occupational and environmental exposure to PAHs is widespread, any additional information on cancer risk is of interest, particularly with a complex mixture of PAHs which can vary considerably depending on exposure conditions.

The complexity of mixtures of PAHs presents problems in exposure ascertainment. In this study, BaP is the only single PAH for which adequate exposure data are available. In an attempt to evaluate exposures to mixtures of PAHs, we also considered exposure to PAHs produced by combustion of coal, petroleum, wood, and other sources, as well as PAHs derived from any of these sources. This provided six exposure variables for evaluation as potential risk factors for the 19 types of cancer included in the case-control study.

The results of this preliminary analysis are reassuring in that there was no strong evidence of elevated cancer risk. Although a number of odds ratios were significantly greater than one, some positive findings would be expected purely by chance due to the relatively large number of statistical tests performed. Nonetheless, these findings may warrant further consideration when viewed in the context of positive findings reported in other epidemiological investigations of cohorts of workers exposed to PAHs in specific industries.

Further statistical analyses to evaluate the effects of intensity and duration of exposure are currently underway. This analysis will also allow for consideration of the confidence placed on individual exposure data by our team of chemists. In addition to a number of nonoccupational factors such as diet or health status, there are over 300 occupational variables reflecting exposure to chemicals other than PAHs which will be screened as potential confounders. The results of this analysis, to be reported elsewhere, will provide the basis for a more detailed assessment of the potential carcinogenic risks associated with occupational exposure to PAHs among men in this study.

ACKNOWLEDGEMENTS

We are grateful to Terry Chernis for her editorial review and bibliographic assistance in preparing this article. The participation of Lesley Richardson, Ramzan Lakhani, and Jean Pellerin is also gratefully acknowledged.

REFERENCES

1. Bjorseth, A., and G. Becher (1986) PAH in Work Atmospheres: Occurrence and Determination. CRC Press, Boca Raton, Florida.
2. Breslow, N.E., and N.E. Day, eds., (1980) Statistical Methods in Cancer Research Vol. 1. The Analysis of Case-Control Studies.

IARC Scientific Publications No. 32. International Agency for Research on Cancer, Lyon, France.

3. Environment Canada (1989) Polycyclic Aromatic Hydrocarbons in the Ambient Air of Toronto, Ontario and Montreal, Quebec. Environment Canada, Pollution Measurement Division (PMD 89-14), Ottawa, Canada.

4. Environmental Protection Agency (1984) Health Effects Assessment for Polynuclear Aromatic Hydrocarbons. EPA 540/1-86-013, Environmental Protection Agency, Cincinnati, Ohio.

5. Gérin, M., J. Siemiatycki, H. Kemper, and D. Bégin (1985) Obtaining occupational exposure histories in epidemiologic case-control studies. J. Occ. Med. 27:420-426.

6. International Agency for Research on Cancer (1983) Polynuclear Aromatic Compounds. Part I. Chemical, Environmental and Experimental Data. IARC Monographs on the Evaluation of the Carcinogenic Risk of Chemicals to Humans, Vol. 32. IARC, Lyon, France.

7. International Agency for Research on Cancer (1984) Polynuclear Aromatic Compounds. Part 3. Industrial Exposures in Aluminum Production, Coal Gasification, Coke Production, and Iron and Steel Founding. IARC Monographs on the Evaluation of the Carcinogenic Risk of Chemicals to Humans, Vol. 34. IARC, Lyon, France.

8. International Agency for Research on Cancer (1985) Polynuclear Aromatic Compounds. Part 4. Bitumens. Coal-Tars and Derived Products, Shale-oils and Soots. IARC Monographs on the Evaluation of the Carcinogenic Risk of Chemicals to Humans, Vol. 35. IARC, Lyon, France.

9. International Agency for Research on Cancer (1987) Overall Evaluations of Carcinogenicity: An Updating of IARC Monographs, Volumes 1 to 42. IARC Monographs on the Evaluation of Carcinogenic Risks to Humans, Supplement 7. IARC, Lyon, France.

10. Krewski, D., T. Thorslund, and J. Withey (1989) Carcinogenic risk assessment of complex mixtures. Toxicol. Ind. Health 5:851-867).

11. National Research Council (1988) Complex Mixtures, National Academy Press, Washington, D.C.

12. Siemiatycki, J., S. Wacholder, L. Richardson, R. Dewar, and M. Gérin (1987) Discovering carcinogens in the occupational environment: Methods of data collection and analysis of a large case-referent monitoring system. Scandinavian J. of Work, Environment & Health 13:486-492.

13. Thorslund, T., and G. Charnley (1988) Comparative Potency Approach for Estimating the Cancer Risk Associated with Exposure to Mixtures of Polycyclic Aromatic Hydrocarbons. ICF-Clement Associates, Washington, D.C.

FUTURE DIRECTIONS IN RESEARCH ON THE

GENETIC TOXICOLOGY OF COMPLEX MIXTURES

Joellen Lewtas

Genetic Toxicology Division
U.S. Environmental Protection Agency
Research Triangle Park, North Carolina   27711

INTRODUCTION

The recognition that most human exposure to environmental chemicals occurs as exposures to complex mixtures has stimulated research on the toxicology of complex mixtures (36).   The future assessment of complex mixtures of environmental pollutants will increasingly rely on new interdisciplinary strategies including state-of-the-art genetic and molecular methodologies.   Integrated multidisciplinary studies are now beginning to assess human exposure, dosimetry, and cancer risk relative to complex mixtures of pollutants, and to apportion the human exposure and risk to various sources.   Strategies which hold promise for incorporation of genetic bioassay data in the risk assessment of complex mixtures include: (i) biomonitoring environmental levels, fate, transformation, and human exposure; (ii) characterization of the genotoxic components of complex mixtures using advanced chemical and bioassay methods; (iii) molecular dosimetry of complex mixtures; and (iv) mechanistic studies of the effects of complex mixtures induced by both genetic and nongenetic mechanisms. Important advances in understanding the genetic and carcinogenic effects of complex environmental mixtures will increasingly rely on the successful implementation of multidisciplinary integration of environmental, laboratory, and human studies using state-of-the-art biological, chemical, and molecular methods.

PREDICTABILITY OF SHORT-TERM BIOASSAYS

One important issue for genetic toxicologists in applying these strategies is the issue of predictability.   Evidence has accumulated since the 1960s to support the theory that electrophilic chemicals will react covalently with nucleophilic centers in DNA and will subsequently induce mutations.   Extensive studies by the Millers (34) showed that these electrophilic chemical mutagens were also animal carcinogens.   The basis for using short-term genetic bioassays to detect carcinogens is the theory that electrophilic chemicals covalently binding or reacting with DNA will

Genetic Toxicology of Complex Mixtures
Edited by M. D. Waters *et al.*
Plenum Press, New York, 1990

result in the induction of mutations and subsequently serve as the initiating event in the induction of cancer. Studies of hundreds of chemicals in the 1970s demonstrated a high correlation between mutagenicity in bacteria and carcinogenicity in animals (4,31,41). Despite a preponderance of evidence for the utility of genetic bioassays in the detection of potential carcinogens, a recent report on 73 chemicals tested in both animal cancer bioassays and short-term tests for genetic toxicity by Tennant et al. (42) has called into question the predictablity of genetic bioassays. This study, and the many papers published in response to these questions (3,7), have been evaluated by DeMarini et al. (12) in the context of the utility of genetic bioassays for the evaluation of complex mixtures. Issues that pertain to the utility of short-term genetic bioassays include: (i) concordance between carcinogenicity results in rats and mice, as well as the concordance between these rodents and short-term genetic bioassays (7); (ii) influence of chemical class on the sensitivity and specificity of genetic bioassays (10); and (iii) genotoxic and nongenotoxic carcinogens (2). We cannot expect bioassays that detect, by definition, genotoxic agents to detect agents that may induce cancer by a mechanism which does not involve interaction with DNA, mutation induction, or other genetic effects. This issue of the predictability of genetic bioassays needs to be kept in mind when considering the nature of the complex mixture under study and the specific questions being addressed in the research.

BIOMONITORING HUMAN EXPOSURE TO MUTAGENS

The development of short-term genetic bioassays in the mid-1970s rapidly led to the use of these assays in environmental monitoring. The essential elements of these bioassays that facilitate their application to environmental monitoring is their simplicity, rapidity, low cost, and small sample size requirements. The genetic bioassay most widely used has been the Salmonella typhimurium plate incorporation assay which Dr. Bruce Ames et al. described in 1975 (1) and initially validated as a bioassay to screen for carcinogens (31). This bioassay was rapidly adopted across the world and applied to environmental studies of complex mixtures. In air and water pollution monitoring studies, this bioassay was used to detect and quantitate the mutagenic activity associated with environmental pollutants (26).

The relatively small sample mass requirements (milligram quantities) was one of the important attributes that facilitated application of bacterial mutation assays to environmental monitoring. Recent advances in personal exposure monitoring have resulted in the need for assay methods that could be applied to even smaller, microgram quantities of environmental samples (25,29). Microsuspension bacterial mutation assays have been developed to meet these needs through simple modifications of the S. typhimurium preincubation assay in which the assay is conducted in a 10- to 100-fold smaller volume (21,29). These microsuspension assays have been applied to indoor air pollution monitoring (28), human exposure monitoring (30), and in monitoring trends in air pollution over relatively short time periods (26).

Biomonitoring of ambient air using bacterial mutation assays is being used for the first time in source apportionment studies to determine which emission sources are the major contributors to the mutagenic activity of ambient air (10). Source apportionment is a combination of mathematical and analytical procedures which are used to determine the contribution of specific emission sources to the whole mixture and has been used most extensively in air pollution studies. Two different methods have been used to apportion the contributions of source emissions to ambient air quality: source dispersion modeling which relies on monitoring source emissions and modeling ambient dispersion and receptor modeling which relies on ambient monitoring and source tracer species (17). The receptor-modeling approach to source apportionment has been combined with biomonitoring and is being employed in the U.S. Environmental Protection Agency's Integrated Air Cancer Project (23,25). This approach has been shown to be effective in apportioning the mutagenicity in a simple airshed with mobile sources and woodstove emissions. In this case, the number of sources are small and well characterized with respect to tracer chemicals. In addition, emissions from mobile sources and wood burning contain unique elemental tracers which improve the accuracy of the source apportionment calculations (23). The unique feature in these recent receptor modeling studies is that mutagenicity data from biomonitoring is being used directly in the model together with elemental tracers (e.g., K and Pb). This has made possible the direct apportionment of mutagenic activity in the woodstove impacted airsheds. Future field studies are in progress to extend this methodology to more complex airsheds.

IDENTIFICATION OF GENOTOXIC CHEMICALS IN COMPLEX MIXTURES USING BIOASSAY-DIRECTED FRACTIONATION AND CHARACTERIZATION

Although there are a number of approaches that have been taken to identify potential carcinogens in complex mixtures (9), bioassay-directed fractionation has been the most successful method to identify potent mutagens in complex mixtures (40). In this approach, the complex mixtures are fractionated and each fraction is bioassayed. Mutagenic fractions are further fractionated, bioassayed, and characterized until the major class or specific compounds responsible for the mutagenicity are identified. The S. typhimurium plate incorporation assay (1) has been extensively used in this approach to identify mutagens in air, water, food, and specific emission sources (40). This approach, combined with the use of bacterial tester strains selectively sensitive to certain classes of chemicals (e.g., nitroarenes), led to the identification of nitrated polycyclic aromatic hydrocarbons (nitro-PAH) as potent mutagens in diesel exhaust particles (24,40) and kerosene heater particle emissions (22). The identification of chlorinated furanones as potent genotoxic by-products of water chlorination resulted from the application of bioassay-directed fractionation (32,33).

There are several methodological advances which could have a significant impact on research in this area in the future. First is the development of micromutagenesis bioassays (24,25) which can be coupled directly to analytical fractionation procedures to identify the mutagens in complex mixtures (35,40). Second is the development of new chemical-analytical techniques which can facilitate the identification of trace species (e.g., dinitropyrenes), polar organic species (e.g., hydrolated and nitrated aromatic hydrocarbons), and labile compounds (e.g., organic peroxides) (40). The power of these new methods has recently been demonstrated in the identification of hydroxylated nitroaromatic compounds in urban air particulate matter (35).

The classical way to identify the carcinogenic compounds in complex mixtures, e.g., a hazardous waste site, is to make a list of known carcinogenic compounds and analyze the mixture for the presence of these compounds. This approach rarely leads to the identification of new compounds, thereby potentially underestimating the number and types of carcinogens present. This approach, however, is readily coupled to quantitative risk assessment (43) if the compounds quantitated in the mixture have human cancer potency values that have been previously published (e.g., cancer unit risk numbers). This is a practical and rapid assessment method which makes it possible to conduct quantitative risk assessments of complex mixtures. The additivity assumption is often invoked when using this approach (43). The disadvantage of this method is that it will miss the identification and assessment of previously unidentified and potentially important carcinogens.

As a research tool, bioassay-directed fractionation has the power to identify biologically reactive components in complex mixtures. If bacterial mutagenesis assays are used to direct the fractionation and analysis, then potent mutagens can be identified. It is possible that these mutagens will not be carcinogens due to their mechanism of action, metabolic detoxification, etc. Several food mutagens (e.g., quercetin) do not appear to be animal carcinogens due to in vivo detoxification. On the other hand, this approach has identified potent mutagens that have been shown to be potent carcinogens such as the dinitropyrenes (20). Often the compounds identified are ones which have never been synthesized or studied in biological systems. To confirm the identification and quantitate the concentrations of these compounds, it is then necessary to synthesize standards. These new compounds can then be studied in mammalian bioassays to determine the mechanism of genotoxicity and whether the identified compounds are carcinogens in animals and to demonstrate the relationships between exposure, dose, and response.

MOLECULAR DOSIMETRY OF COMPLEX MIXTURES

Highly sensitive methods are now being used to measure protein and DNA adducts which may result from exposures to environmental pollutants. Several combustion emission constituents, including polycyclic aromatic hydrocarbons [e.g., benzo(a)pyrenes (BaP)] and nitrogen-containing aromatic compounds (e.g., nitrosamines and 4-aminobiphenyl), can be detected either as protein or DNA adducts (18). The postlabeling assay for DNA adducts is particularly applicable to complex mixtures, since the assay does not depend upon prior identification of the specific chemical in the mixture which may form adducts (38).

DNA isolated from humans, animals, or cells exposed to complex mixtures shows evidence of multiple DNA adducts (15,27,37). This is consistent with bioassay-directed fractionation studies which gave evidence for many genotoxic compounds and isomers in several chemical fractions (40). A challenge for the future will be the separation and identification of specific adducts after exposures to the multiple chemicals present in complex mixtures. The level of DNA adducts detectible by postlabeling, one-adduct in 10^9 to 10^{10} multitudes, is far below the current detection limits for direct chemical identification of the adducts.

A number of methods are currently being explored to identify specific types of DNA adducts by selectively using enzymatic digestion methods to identify enzyme-sensitive adducts. We have used this approach to identify arylamine adducts which can form from either nitroaromatic compounds or aromatic amines (14). Gallagher et al. (16), using this approach, have shown that the formation of specific nitroarene DNA adducts can be enriched through incubation of diesel particle extracts and xanthine oxidase with calf thymus DNA. These adducts have also been shown to be nuclease P1 sensitive (16). A second approach is to use DNA adduct standards in co-elution experiments to identify specific adduct regions. Adducts resulting from treatment with complex mixtures and appearing in the same chromatographic region where a major BaP adduct appears were quantitated in DNA isolated from skin and lung tissue of mice treated in a tumor-initiation protocol (15). The complex mixtures were combustion-related emissions containing a known concentration of BaP. Future research will be aimed at resolving and identifying the nature and source of these adducts.

DNA adduct dosimetry studies using this postlabeling assay are being conducted simultaneously in comparative microbial mutagenesis and tumorigenesis assays of combustion emission and airborne particle extracts (27). Complex mixtures which are more potent mutagens and tumor initiators form higher concentrations of adducts per nucleotide (Gallagher, unpubl. data). These new dosimetry methods for complex mixtures may provide a critical data link between exposure and dose in human, animal, and in vitro studies, further reducing the large uncertainties in current risk assessment methodologies.

MECHANISTIC STUDIES OF THE EFFECTS OF COMPLEX MIXTURES INDUCED BY BOTH GENETIC AND NONGENETIC MECHANISMS

The predictability of short-term genetic bioassays has become an issue (12) that is critical to the application of these bioassays in the assessment of complex mixtures. It is becoming clear that bioassays which detect genetic effects (e.g., mutations, DNA damage, chromosomal effects) will not be able to predict the potential carcinogenicity of chemicals or mixtures which induce cancer through a nongenetic mechanism (e.g., induction of hormones, viruses, etc.). Certain classes of chemicals, particularly chlorinated hydrocarbons, are often not detected as gene mutagens (12). Recently, DeMarini and co-workers (11,19) have reported that a number of the carcinogenic chlorinated pesticides and chlorophenols which are not mutagenic in Salmonella, do induce prophage lambda in Eschericia coli. The mechanism responsible for this phage induction may involve the formation of free radicals by the quinone metabolites (11). Free radicals are known to cause DNA strand breaks which are thought to result in prophage induction. It is critical to have some knowledge of the nature of the chemicals present in a complex mixture or to evaluate an unknown mixture for toxicity and potential carcinogenicity using a series of bioassays which detect a wide range of genetic and nongenetic events. Often, a battery of bioassays is selected to detect the same genetic events, albeit in different biological systems. Recent studies make it clear that it may be even more important to select a battery of bioassays which detect different genetic and nongenetic effects related to carcinogenesis when using these assays in a predictive role.

Recent advances in molecular methods using the polymerase chain reaction (PCR) (39) have made it possible to rapidly amplify sections of genes to facilitate the direct DNA sequencing of mutants. Several investigators are now sequencing the mutations occurring at the hisD gene in S. typhimurium using this method (5,8,13). Bell et al. (6) have recently determined the mutational spectra for 1-nitropyrene-induced revertants of TA98 and spontaneous mutants. This approach is now being applied to understand the mutational spectra induced by complex mixtures (D.M. DeMarini, pers. comm.).

CONCLUSIONS: RISK ASSESSMENT OF COMPLEX MIXTURES

Risk assessment of environmental mixtures relies on exposure, dosimetry, and effects data for either individual chemicals present in the mixture, the whole mixture, or fractions thereof. In the case of very complex mixtures, it is impossible to quantitate all of the constituent chemicals and their human exposure, dose, and effects for risk assessment. New genetic and molecular methods are being applied to each aspect of research and assessment of cancer risk resulting from exposure to complex mixtures. Short-term genetic bioassay methods utilizing newly engineered bacterial strains are being used to assess total human exposure to mutagens in the environment (30). New DNA adduct dosimetry methods are being applied to assess human exposure to complex mixtures. In addition, new advances in sequencing the genetic mutations induced by environmental mutagens will improve our understanding of the relationships between DNA adducts, DNA damage and repair, mutation induction, and tumor initiation. New research on the molecular mechanisms involved in cell proliferation and the events critical to tumor progression also need to be incorporated to improve the risk assessment procedures for complex mixtures.

Studies of complex mixtures (e.g., soots and tars) in the early 1900s first employed the use of animal cancer bioassays. The development of short-term genetic bioassays has fostered emerging methodologies and new assessment strategies for complex mixtures. These new strategies are characterized by their multidisciplinary nature. Methodologies developed by engineers, chemists, physicists, and biologists have been integrated into new strategies. Strategies being pursued and developed as a part of the integrated studies provide new evidence on the sources and nature of airborne carcinogens. The use of bioassay methods in exposure and dosimetry assessment will provide direct data on human exposure to genotoxic carcinogens. Molecular methods are being developed to understand the mechanisms of cancer and mutation induction by complex mixtures.

REFERENCES

1. Ames, B.N., J. McCann, and E. Yamasaki (1975) Methods for detecting carcinogens and mutagens with the Salmonella mammalian microsome mutagenicity test. Mutat. Res. 31:347-364.
2. Ashby, J. (1988) The separate identities of genotoxic and non-genotoxic carcinogens. Mutagenesis 3:365-366.
3. Ashby, J., and R.W. Tennant (1988) Chemical structure, Salmonella mutagenicity and extent of carcinogenicity as indicators of genotoxic carcinogenesis among 222 chemicals tested in rodents of the U.S. NCI/NTP. Mutat. Res. 204:17-115.
4. Bartsch, H. (1976) Predictive value of mutagenicity tests in chemical carcinogenesis. Mutat. Res. 38:177-190.

5. Bell, D.A., J.E. Lee, and D.M. DeMarini (1989) Use of polymerase chain reaction to amplify a segment of the hisD3052 gene of Salmonella typhimurium for DNA sequence analysis. Env. Molec. Mutagen. 14(Suppl. 15):18.

6. Bell, D.A., J.G. Levine, and D.M. DeMarini (1990) Rapid DNA sequence analysis of revertants of the hisD3052 allele of Salmonella typhimurium TA98 using the polymerase chain reaction and direct sequencing: Application to 1-nitropyrene-induced revertants. Mutat. Res. (in press).

7. Brockman, H.E., and D.M. DeMarini (1988) Utility of short-term tests for genetic toxicity in the aftermath of the NTP's analysis of 73 chemicals. Env. Molec. Mutagen. 11:421-435.

8. Cebula, T.A., and W.H. Koch (1990) Analysis of spontaneous and psoralen-induced Salmonella typhimurium hisG46 revertants by oligodeoxyribonucleotide colony hybridization: Use of psoralens to cross-link probes to target sequences. Mutat. Res. (in press).

9. Claxton, L.D. (1982) Review of fractionation and bioassay characterization techniques for the evaluation of organics associated with ambient air particles. In Genotoxic Effects of Airborne Agents, R.R. Tice, D.L. Costa, and K.M. Schaich, eds. Plenum Press, New York, pp. 19-33.

10. Claxton, L.D., A.G. Stead, and D. Walsh (1988) An analysis by chemical class of Salmonella mutagenicity tests as predictors of animal carcinogenicity. Mutat. Res. 205:197-225.

11. DeMarini, D.M., H.G. Brooks, and D.G. Parkes (1990) Induction of prophage lambda by chlorophenols. Env. Molec. Mutagen. Vol. 15 (in press).

12. DeMarini, D.M., J. Lewtas, and H.E. Brockman (1989) Utility of short-term tests for genetic toxicity. Commentary in Cell Biology and Toxicology 5:189-200.

13. Fuscoe, J.C., R. Wu, N.H. Shen, S.K. Healy, and J.S. Felton (1988) Base-change analysis of revertants of the hisD3052 allele in Salmonella typhimurium. Mutat. Res. 201:241-251.

14. Gallagher, J.E., M.A. Jackson, M.H. George, J. Lewtas, and I. Robertson (1989) Differences in detection of DNA adducts in the ^{32}P-postlabeling assay after either 1-butanol extraction or nuclease P1 treatment. Cancer Lett. 45:7-12.

15. Gallagher, J.E., M.A. Jackson, M.H. George, and J. Lewtas (1990) Dose related differences in DNA adduct levels in rodent tissues following skin application of complex mixtures from air pollution sources. Carcinogenesis 11:63-68.

16. Gallagher, J.E., M.J. Kohan, M.H. George, M.A. Jackson, and J. Lewtas (1990) Validation/application of ^{32}P-postlabeling analysis for the detection of DNA adducts resulting from complex air pollution sources containing PAHs and nitrated PAHs. In Nitroarenes: Occurence, Metabolism and Biological Impact, P. Howard, ed. Plenum Press, New York (in press).

17. Gordon, G. (1988) Receptor models. Env. Sci. Tech. 22:1132-1142.

18. Harris, C.C. (1985) Future directions in the use of DNA adducts as internal dosimeters for monitoring human exposure to environmental mutagens and carcinogens. Env. Health Persp. 62:185-191.

19. Houk, V.S., and D.M. DeMarini (1987) Induction of prophage lambda by chlorinated pesticides. Mutat. Res. 182:193-201.

20. IARC (1989) IARC Monograph on the Evaluation of the Carcinogenic Risk of Chemicals to Humans, Vol. 46, Diesel and Gasoline Engine Exhausts and Some Nitroarenes. Lyon, France, pp. 189-387.

21. Kado, N.Y., D. Langley, and E. Eisenstadt (1983) A simple modification of the Salmonella liquid incubation assay: Increased sensitivity for detecting mutagens in human urine. Mutat. Res. 121:25-32.

22. Kinouchi, T., K. Nishifuji, H. Tsutsui, S. Hoare, and Y. Ohnishi (1988) Mutagenicity and nitropyrene concentration of indoor air particulates exhausted from a kerosene heater. Japan J. Cancer Res. (Gann) 79:32-41.

23. Lewis, C.W., R.E. Baumgardner, L.D. Claxton, J. Lewtas, and R.K. Stevens (1988) The contribution of woodsmoke and motor vehicle emissions to ambient aerosol mutagenicity. Env. Sci. Tech. 22:968-971.

24. Lewtas, J. (1988) Genotoxicity of complex mixtures: Strategies for the identification and comparative assessment of airborne mutagens and carcinogens from combustion sources. Fund. Appl. Tox. 10:571-589.

25. Lewtas, J. (1989) Emerging methodologies for assessment of complex mixtures: Application of bioassays in the Integrated Air Cancer Project. J. of Tox. and Ind. Health 5:839-850.

26. Lewtas, J. (1990) Environmental monitoring using genetic bioassays. In Genetic Toxicology: A Treatise, A.P. Li and R.H. Heflich, eds. The Telford Press, Caldwell, NJ (in press).

27. Lewtas, J., and J. Gallagher (1990) Complex mixtures of urban air pollutants: Identification and comparative assessment of mutagenic and tumorigenic chemicals and emission sources. In Experimental and Epidemiologic Applications to Risk Assessment of Complex Mixtures IARC, WHO, France (in press).

28. Lewtas, J., S. Goto, K. Williams, J.C. Chuang, B.A. Petersen, and N.K. Wilson (1987) The mutagenicity of indoor air particles in a residential pilot field study: Application and evaluation of new methodologies. Atmos. Env. 21:443-449.

29. Lewtas, J., L. Claxton, J. Mumford, and G. Lofroth (1990) Bioassay of complex mixtures of indoor air pollutants. In IARC Monographs on Indoor Air Pollution Methods (in press).

30. Matsushita, H., S. Goto, and Y. Takagi (1990) Human exposure to airborne mutagens indoors and outdoors using mutagenesis and chemical analysis methods (This volume).

31. McCann, J., E. Choi, E. Yamasaki, and B.N. Ames (1975) Detection of carcinogens as mutagens in the Salmonella/microsome test: Assay of 300 chemicals. Proc. Natl. Acad. Sci., USA 72:5135-5139.

32. Meir, J., R. Knohl, W. Coleman, H. Ringhand, J. Munch, W. Kaylor, R. Streicher, and F. Kopfler (1987) Studies on the potent bacterial mutagen: 3-Chloro-4-(dichloromethyl)-5-hydroxy-2(5H)-furanone:aqueous stability, XAD recovery and analytical determination in drinking water and in chlorinated humic acid solutions. Mutat. Res. 189:363-373.

33. Meir, J. (1988) Genotoxic activity of organic chemicals in drinking water. Mutat. Res. 196:211-245.

34. Miller, E.C., and J.A. Miller (1971) The mutagenicity of chemical carcinogens: Correlations, problems and interpretations. In Chemical Mutagens: Principles and Methods for their Detection, Vol. 1, A. Hollaender, ed. Plenum Press, New York, pp. 83-94.

35. Nishioka, M.G., C.C. Howard, and J. Lewtas (1988) Detection of hydroxylated nitro aromatic and hydroxylated nitro polycyclic aromatic compounds in an ambient air particulate extract using bioassay directed fractionation. Env. Sci. and Tech. 22:907-915.

36. NRC (National Research Council) (1988) Complex Mixtures, National Academy Press, Washington, DC, 227 pp.
37. Perera, F.P., R Schulte, R.M. Santella, and D. Brenner (1990) DNA adducts and related biomarkers in assessing the risk of complex mixtures (This volume).
38. Randerath, K., E. Randerath, H.P. Agrawal, R.C. Gupta, E. Schurdak, and M.V. Reddy (1985) Postlabeling methods for carcinogen-DNA adduct analysis. Env. Health Persp. 62:57-65.
39. Saiki, R.K., D.H. Gelfand, S. Stoffel, S.J. Scharf, R. Higuchi, G.T. Horn, K.B. Mullis, and H.A. Erlich (1988) Primer-directed enzymatic amplification of DNA with a thermostable DNA polymerase. Science 239:487-491.
40. Schuetzle, D., and J. Lewtas (1986) Bioassay-directed chemical analysis in environmental research. Analyt. Chem. 58:1060A-1075A.
41. Sugimura, T., S. Sato, M. Nagao, T. Yahagi, T. Matsushima, Y. Seino, M. Takeuchi, and T. Kawachi (1976) Overlapping of carcinogens and mutagens. In Fundamentals of Cancer Prevention, P.N. Magee, S. Takayama, T. Sugimura, and T. Matsushima, eds. University Park Press, Baltimore, MD, pp. 191-213.
42. Tennant, R.W., B.H. Margolin, M.D. Shelby, E. Zeiger, J.K. Haseman, J. Spalding, W. Caspary, M. Resnick, S. Stasiewicz, B. Anderson, and R. Minor (1987) Prediction of chemical carcinogenicity in rodents from in vitro genetic toxicity assays. Science 236:933-941.
43. U.S. EPA (1986) Guidelines for the health risk assessment of chemical mixtures. Fed. Reg. 51(185):34014-34025.

DISCLAIMER

The research described in this paper has been reviewed by the Health Effects Research Laboratory, U.S. Environmental Protection Agency, and approved for publication. Approval does not signify that the contents necessarily reflect the views and policies of the Agency, nor does mention of trade names or commercial products constitute endorsement or recommendation for use.